AF411369

Advanced Techniques for Design and Manufacturing in Marine Engineering

Advanced Techniques for Design and Manufacturing in Marine Engineering

Editors

Antonio Mancuso
Davide Tumino

MDPI • Basel • Beijing • Wuhan • Barcelona • Belgrade • Manchester • Tokyo • Cluj • Tianjin

Editors
Antonio Mancuso
Università degli Studi di Palermo
Italy

Davide Tumino
Università degli Studi di Enna Kore
Italy

Editorial Office
MDPI
St. Alban-Anlage 66
4052 Basel, Switzerland

This is a reprint of articles from the Special Issue published online in the open access journal *Journal of Marine Science and Engineering* (ISSN 2077-1312) (available at: https://www.mdpi.com/journal/jmse/special_issues/technique_design_engineering).

For citation purposes, cite each article independently as indicated on the article page online and as indicated below:

LastName, A.A.; LastName, B.B.; LastName, C.C. Article Title. *Journal Name* **Year**, *Volume Number*, Page Range.

ISBN 978-3-0365-3114-4 (Hbk)
ISBN 978-3-0365-3115-1 (PDF)

Contents

About the Editors

Antonio Mancuso is a Full Professor of Design and Methods in Industrial Engineering in the Department of Engineering at the University of Palermo (Italy). He achieved his PhD at the University of Palermo. His research is mainly focused on design and optimization methodologies, sailing yacht design, and numerical simulations. Some of his recent research has focused on parametric sailing hull designs, the topological optimization of internal hull frames, and the development of foiling dinghies. He is the coordinator of the university sailing team (named Zyz), which is involved in a European framework aiming to design and manufacture an eco-friendly sailing yacht.

Davide Tumino is an Associate Professor of Design and Methods in Industrial Engineering in the Faculty of Engineering and Architecture at the University of Enna Kore (Italy). He achieved his PhD at the University of Palermo. His research is mainly focused on design methodologies and numerical methods applied to structural mechanics. In recent years, he has developed numerical procedures to optimize the structural performance of sailing yachts. His other research interests concern the application of additive manufacturing techniques in the field of lattice core structures.

Journal of
Marine Science and Engineering

Editorial

Advanced Techniques for Design and Manufacturing in Marine Engineering

Antonio Mancuso [1,*] and Davide Tumino [2,*]

1   Dipartimento di Ingegneria, Università degli Studi di Palermo, 90128 Palermo, Italy
2   Facoltà di Ingegneria e Architettura, Università degli Studi di Enna Kore, 94100 Enna, Italy
*   Correspondence: antonio.mancuso@unipa.it (A.M.); davide.tumino@unikore.it (D.T.);
Tel.: +39-91-23897269 (A.M.); +39-935-536491 (D.T.)

Citation: Mancuso, A.; Tumino, D. Advanced Techniques for Design and Manufacturing in Marine Engineering. *J. Mar. Sci. Eng.* **2022**, *10*, 122. https://doi.org/10.3390/jmse10020122

Received: 11 January 2022
Accepted: 13 January 2022
Published: 18 January 2022

Publisher's Note: MDPI stays neutral with regard to jurisdictional claims in published maps and institutional affiliations.

Modern engineering design processes are driven by the extensive use of numerical simulations, and naval architecture as well as ocean engineering are no exception. Structural design or fluid dynamic performance evaluation can only be carried out by means of several dedicated pieces of software. Classical naval design methodology can take advantage of the integration of these pieces of software, giving rise to more robust design in terms of shape, structural and hydrodynamic performances, and manufacturing processes.

This Special Issue (SI) on "Advanced Techniques for Design and Manufacturing in Marine Engineering", published in the *Journal of Marine Science and Engineering*, aimed to invite researchers and engineers from both academia and industry to publish the latest progress in design and manufacturing techniques in marine engineering as well as to debate current issues and future perspectives in this research area.

After a rigorous peer review process we accepted 11 papers [1–11], covering a wide range of topics related to the themes proposed in the Special Issue. In [1], machine-learning-based algorithms are developed in order to enhance the real-time decision process of an AUV sailing yacht. In [2], topology optimization techniques and laser powder bed fusion manufacturing have been synergically adopted to redesign the bulb of sailing yachts, leading to drag reduction and improving overall boat performance. In [3], the topic of sail design is discussed by means of numerical fluid structure interaction methods and a practical tool is proposed to support the analyst during the design process of a yacht sail plan. The sail design process is also investigated in [4] but using different tools, such as combining a velocity prediction program, RANS computations, and analytical approaches. The problem of grid generation in a CFD model has been studied in [5], where the authors propose, for the particular shape of a submarine, an automated procedure based on Cartesian adaptive grids. The applicability of a CFD numerical technique to a complex biphase fluid medium is the key point of [6], where the robustness of the method is tested to simulate the ventilation phenomenon occurring in stepped hull planing motor yachts. In [7], an analytical tool incorporating the main dimensional naval coefficients of a sailing boat is adopted during the early design stage, with the additional aim of quickly predicting the overall resistance of the hull. In [8], different pieces of sensor information have been used by the authors to train an algorithm able to control water sample collection in deep water. Computational methods have been used in [9] to determine the resistance of ship fuel tanks when subjected to an increased internal pressure. In [10], a simulation model has been used to design a platform able to compensate for the wave action on a vessel, with particular attention to the shape optimization of the structure in order to reduce the total weight. Finally, in [11], CFD tools using moving meshes have been adopted to simulate turbulent flows that originate in an oscillating water column device and move a Savonius turbine.

Author Contributions: A.M. and D.T. contributed equally to this work. All authors have read and agreed to the published version of the manuscript.

Funding: This research received no external funding.

Acknowledgments: We want to express our sincere thanks to all the authors and the reviewers. A particular thank also goes to Zara Liu, from the Editorial Office of *JMSE*, who has supported us during the whole editing process of the Special Issue.

Conflicts of Interest: The authors declare no conflict of interest.

References

1. Yuan, J.; Wang, H.; Zhang, H.; Lin, C.; Yu, D.; Li, C. AUV Obstacle Avoidance Planning Based on Deep Reinforcement Learning. *J. Mar. Sci. Eng.* **2021**, *9*, 1166. [CrossRef]
2. Scarpellini, A.; Finazzi, V.; Schito, P.; Bionda, A.; Ratti, A.; Demir, A.G. Laser Powder Bed Fusion of a Topology Optimized and Surface Textured Rudder Bulb with Lightweight and Drag-Reducing Design. *J. Mar. Sci. Eng.* **2021**, *9*, 1032. [CrossRef]
3. Cirello, A.; Ingrassia, T.; Mancuso, A.; Nigrelli, V.; Tumino, D. Improving the Downwind Sail Design Process by Means of a Novel FSI Approach. *J. Mar. Sci. Eng.* **2021**, *9*, 624. [CrossRef]
4. Cella, U.; Salvadore, F.; Ponzini, R.; Biancolini, M.E. VPP Coupling High-Fidelity Analyses and Analytical Formulations for Multihulls Sails and Appendages Optimization. *J. Mar. Sci. Eng.* **2021**, *9*, 607. [CrossRef]
5. Di Angelo, L.; Duronio, F.; De Vita, A.; Di Mascio, A. Cartesian Mesh Generation with Local Refinement for Immersed Boundary Approaches. *J. Mar. Sci. Eng.* **2021**, *9*, 572. [CrossRef]
6. Cucinotta, F.; Mancini, D.; Sfravara, F.; Tamburrino, F. The Effect of Longitudinal Rails on an Air Cavity Stepped Planing Hull. *J. Mar. Sci. Eng.* **2021**, *9*, 470. [CrossRef]
7. Ingrassia, T.; Mancuso, A.; Nigrelli, V.; Saporito, A.; Tumino, D. Parametric Hull Design with Rational Bézier Curves and Estimation of Performances. *J. Mar. Sci. Eng.* **2021**, *9*, 360. [CrossRef]
8. Xiao, J.; Chen, J.; Tian, Z.; Zhu, H.; Wang, C.; Yang, J.; Sheng, Q.; Zhang, D.; Fang, J. Visible Fidelity Collector of a Zooplankton Sample from the Near-Bottom of the Deep Sea. *J. Mar. Sci. Eng.* **2021**, *9*, 332. [CrossRef]
9. Lee, D.-H.; Cha, S.-J.; Kim, J.-D.; Kim, J.-H.; Kim, S.-K.; Lee, J.-M. Practical Prediction of the Boil-Off Rate of Independent-Type Storage Tanks. *J. Mar. Sci. Eng.* **2021**, *9*, 36. [CrossRef]
10. Zhan, Y.; Tian, H.; Xu, J.; Wu, S.; Fu, J. A Novel Three-SPR Parallel Platform for Vessel Wave Compensation. *J. Mar. Sci. Eng.* **2020**, *8*, 1013. [CrossRef]
11. Dos Santos, L.A.; Fragassa, C.; Santos, A.L.G.; Vieira, R.S.; Rocha, L.A.O.; Conde, J.M.P.; Isoldi, L.A.; Dos Santos, E.D. Development of a Computational Model for Investigation of Oscillating Water Column Device with Savonius Turbine. *J. Mar. Sci. Eng.* **2022**, *10*, 79. [CrossRef]

Journal of
Marine Science and Engineering

Article

A Novel Three-SPR Parallel Platform for Vessel Wave Compensation

Yong Zhan *, Huichun Tian, Jianan Xu, Shaofei Wu and Junsheng Fu

College of Mechanical and Electrical Engineering, Harbin Engineering University, Harbin 150001,
Heilongjiang Province, China; tianhuichun1220@163.com (H.T.); xujianan@hrbeu.edu.cn (J.X.);
wsfhrbeu@163.com (S.W.); fujunsheng0517@163.com (J.F.)
* Correspondence: zhanyong@hrbeu.edu.cn

Received: 9 November 2020; Accepted: 8 December 2020; Published: 10 December 2020

Abstract: A wave compensation platform based on 3-SPR parallel platform is designed for marine ships with a dynamic positioning system. It can compensate for the heave, rolling, and pitching movement of a vessel under level 4 sea state. The forward kinematics of the mechanism is used to draw the central point position workspace and the attitude workspace of the moving deck of the compensation platform. The compensation effects of the 3-RPS parallel compensation platform and the 3-SPR parallel compensation platform are compared, and the feasibility and superiority of the compensation scheme using the 3-SPR parallel compensation platform are proved. To lower the working height of the upper deck of the compensation platform and reduce the extension range of the support legs, the structure of the compensation platform is optimized, and a novel 3-SPR parallel platform is designed. Finally, a simulation model was established. Using the inverse kinematic model as a compensation movement solver which can online calculate the length of branch legs based on the measured heaving, rolling, and pitching values of vessels, the compensation effect of the new structure under a certain sea state is simulated. The result demonstrated the efficiency of the ship motion decoupling movement of the newly designed compensation platform and proved the competence of it.

Keywords: wave compensation platform; 3-SPR parallel platform; 3-RPS parallel platform; structure optimization; workspace analysis; level 4 sea state

---

1. Introduction

A considerable amount of energy sources are stored in oceans which occupy 71% of the surface area of the earth. Due to increasing continental energy usage leading to the shortage of traditional energy, the necessity of ocean energy excavation becomes progressively important [1]. However, rough sea circumstances, caused by sea wind and waves, renders plenty of problems that make not only the transportation of cargo or human beings between vessels but the landing or taking off the process of helicopter much more hazardous than that on land [2,3]. For example, cargo on decks may slide from one side to another as a result of ship rolling; a normal rigid gangway maybe tear apart due to the variable distance between two ships, and a landing copter may crash on the deck of a rising ship.

The heave compensation platform is a system designed particularly for decoupling ship motions under diverse ocean circumstances [4]. This system structurally consists of two decks and several mobile cylinders, by using ship motion prediction algorithms [5] and an active control system to module these cylinders' length, the heave compensation platform system can maintain its upper deck motionless [6,7].

Under the affection of sea waves, the motion of a ship is a six-degrees-freedom movement which can be divided into simple motions: rolling, yawing, pitching, surging, swaying, and heaving, presented in Figure 1. In consideration of modern vessels, such as the supply ships of the offshore wind farm and the drilling ships, the movement of yawing, surging, and swaying can be drastically alleviated by the ship dynamic positioning system [8]. This means the main purpose of a heave compensation platform shifts to compensating the remaining motions of a ship [9].

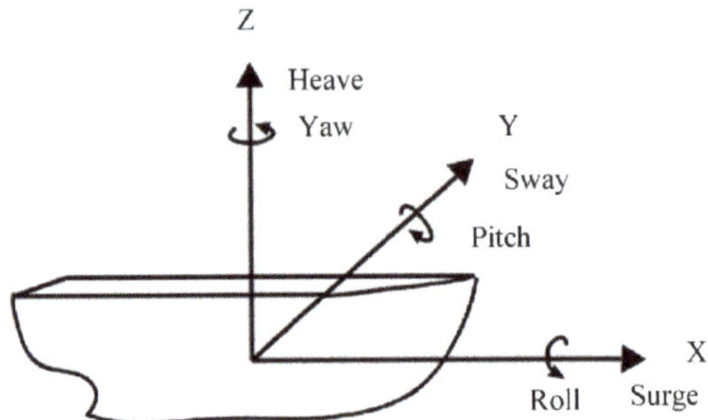

Figure 1. Six-degrees-freedom motions of the vessel.

According to the literature of mechanical structure, a parallel platform has the characteristics of high motion precision, large structural stiffness, and strong spatial motion ability [10], which is suitable for compensating multi-degree-of-freedom ship motion. 3-RPS (R, P, S denote revolute, prismatic, spherical joints, respectively) parallel platform and 3-SPR parallel platform are the most prevail structure which owing a 3-degrees-freedom movement that is similar to the remaining motions of a ship [11,12].

This paper presents a structural design of a novel heave compensation platform which can compensate vessel movement under 4 level sea state. First, the workspace of compensation platforms based on the 3-RPS parallel platform and 3-SPR parallel platform are analyzed. Second, the structure of the compensation platform is optimally designed to get a lower working height and shorter branch leg elongation. At last, the forward and backward kinematic model is established, and a heave compensation simulation using different forms of the platform is done to verify the conclusion.

2. Movement Analysis of Heave Compensation Platforms

To make sure a parallel platform is competent for ship motion compensation, a movement ability analysis is required. Among existing approaches, the working space analysis of a parallel manipulator is a visualized description showing its movement capacity in terms of geometry, which is an important index to measure the performance of a motion platform [13].

2.1. Working Space Analysis of 3-RPS Parallel Compensation Platform

The construction of a 3-RPS parallel compensate platform is presented in Figure 2. In the practical application of this type of compensating platform, the sub-platform fixed on the ship deck is the revolute joint attached platform, and the spherical joint attached platform works as the compensation deck. According to the 3-RPS parallel manufacture's kinematic model [14], the working space of the compensation deck can be acquired.

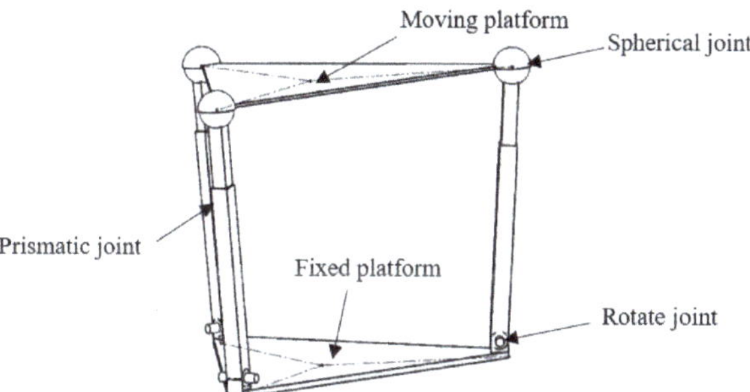

Figure 2. 3-RPS parallel compensate platform.

In terms of decoupling a ship movement under level-4 sea state for which the significant wave height is 2.5 m, the heave compensation platform system needs a vertical moving range of no less than ±1.5 m. For the great capacity of cargo or workers, the area of the compensation deck of this system must be adequate. The prototype was designed with configurations are set as below: both decks of this system are equilateral triangles with a side length of 5 m. The range of the prismatic joint is 3–6 m, and the maximum deflection angle of the spherical joint is 40°.

The forward kinematics of the parallel 3-RPS platform is used to establish the motion model of the compensation platform [15], the constraint range of the movement joints are added, and the workspace of the compensation platform is calculated and drew by MATLAB.

The calculated interval length is 1 cm, and different combinations of prismatic joint lengths are used as inputs to solve the pose of the upper deck of the compensation platform in this case. The images of different attitudes are drawn and shown in Figure 3. In the figure, the lower blue triangle represents the deck fixed platform on the ship deck and the upper part is the moving platform motion space.

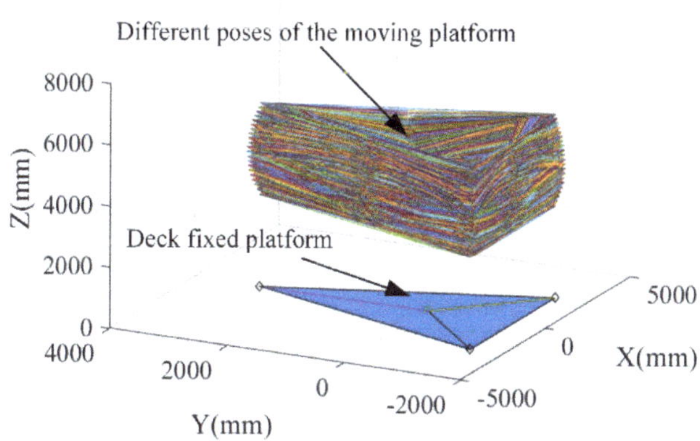

Figure 3. 3-RPS parallel compensate platform motion space.

As can be seen from Figure 3, in the motion space image of the compensation platform directly drawn, the upper and lower parts of the motion space of the moving platform are all flat, the images of the moving platform solved by different inputs overlap with others badly. The posture of the platform inside the motion space cannot be observed clearly.

In this paper, the workspace of the 3-RPS compensation platform comprises the position workspace, which describes the position of the point lying at the center of the moving platform [16], and the attitude workspace is a diagram that describes the deflection of the moving platform.

Figure 4a shows the workspace of the central point of the moving platform of the parallel 3-RPS compensation platform. To facilitate observation, slices of different heights of the position workspace are drawn, as shown in Figure 4b.

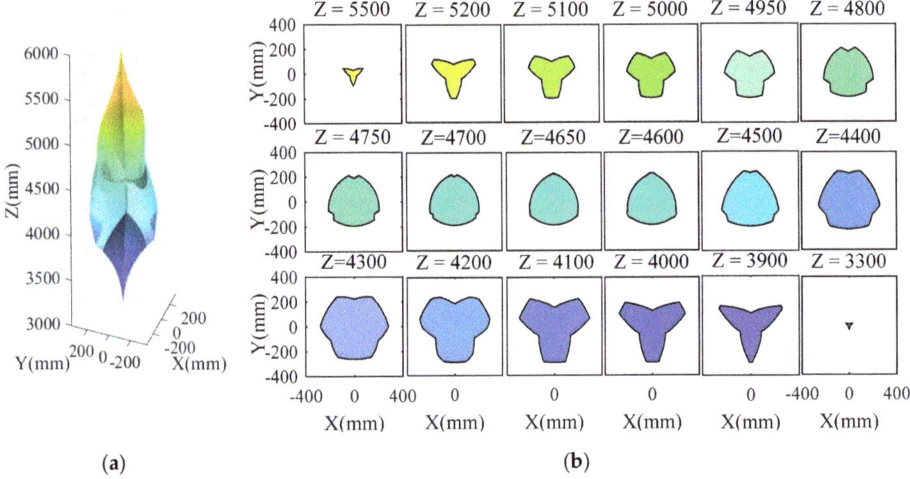

(a)   (b)

Figure 4. Position workspace of 3-RPS compensation platform; (**a**) 3D shape of the position workspace; (**b**) slice figure of different altitudes.

By analyzing Figure 4, the position workspace of the parallel 3-RPS compensation platform has the following characteristics:

1. The highest and lowest point of the position workspace is determined by the range of three prismatic joints; there is only one point at the highest and lowest points of the compensating workspace, where the compensating platform has no deflection ability.
2. The structure of the compensation platform possesses a similar symmetry with its position workspace.
3. The slicing area of the position workspace varies with different heights and increases and then decreases from top to bottom.

Position workspace can only show the variation of platform position, it cannot express the deflection angle of the moving platform, so the attitude workspace is required to show the deflection of the moving platform [17], 3-RPS parallel compensation platform attitude workspace is shown in Figure 5.

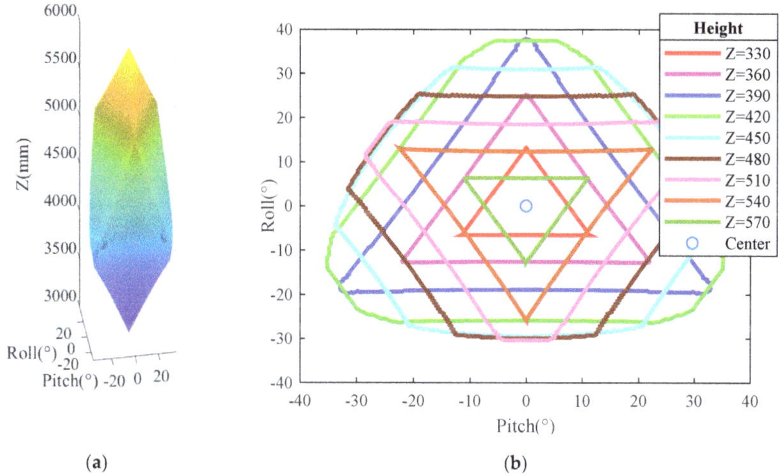

Figure 5. Attitude workspace of 3-RPS compensation platform, (**a**) is the attitude workspace image, (**b**) is a contour map derived from the attitude workspace.

Take a point in the contour map and introduce the vector from the central point to the point. The length of the vector represents the deflection angle degree value and the direction of the vector represents the deflection direction. Such as a point within a contour. It indicates that the angle of this platform pose represented by this point can be realized at this height.

The deflection capability of a parallel 3-RPS compensation platform has the following characteristics:

1. Deflection capability of the platform varies with the height of the moving platform. With the height decreases from high to low, the deflection capability of the platform also increases firstly and then decreases.
2. Deflection capability of platforms at the same height and in different directions is not the same. With the change of height, the direction of the maximum deflection angle also changes, and the deflection capability of each direction near the height median is relatively average.

2.2. Workspace Analysis of 3-SPR Parallel Compensation Platform

The structure demonstration of the parallel 3-SPR compensation platform is shown in Figure 6. The motion model of the parallel 3-SPR compensation platform is established by using the same structure size as the 3-RPS platform above. Comparing with the compensation platform based on the 3-RPS parallel structure, the equation sets of the 3-SPR compensation platform is more complex because the revolute joint constraint of the 3-SPR platform is attached to the moving platform [18]. The kinematic of the 3-RPS compensation platform is relatively simple, and the pose of the moving platform corresponds to the solution of leg length [19] Therefore, we can use a method of "inverse solution" to calculate the workspace of the parallel 3-SPR compensation platform.

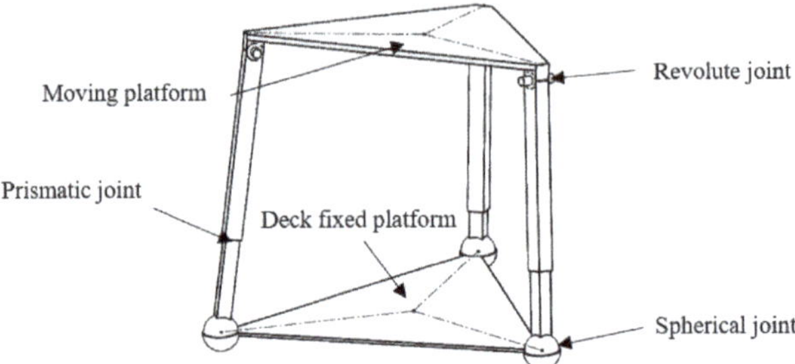

Figure 6. 3-SPR parallel compensation platform.

The inverse solution is illustrated below. First, compute the position of the moving platform of the 3-RPS platform, get the transfer matrix from the mobile platform relative to the deck fixed platform, and then calculate the inverse matrix of this transfer matrix, this process is equivalent to swap the motion platform and fixed platform in the model. This inverse matrix is the transfer matrix represents a movement from the deck fixed platform relative to the moving platform, the last step is using this inverse matrix to perform a point coordinate transformation of the original fixed platform, and the position solution of 3-SPR compensation platform moving platform will show itself. The nonlinear least square method is used to obtain the position workspace and attitude workspace of the 3-SPR compensation platform [20], as shown in Figures 7 and 8.

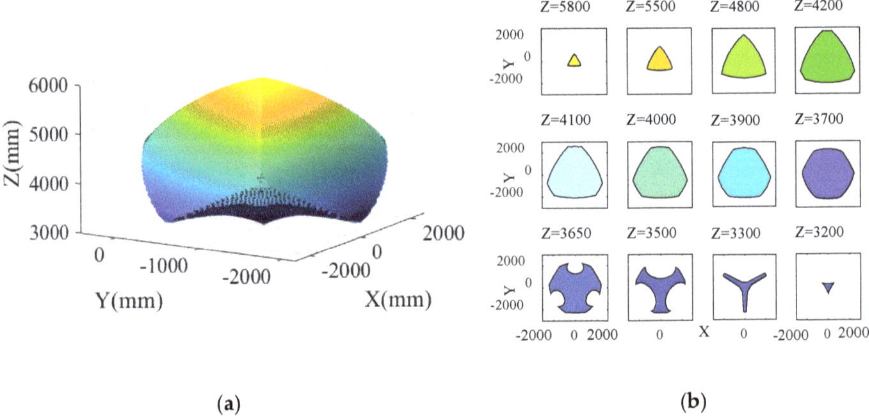

Figure 7. Position workspace of the 3-SPR compensation platform. (**a**) The 3D shape of the position workspace; (**b**) slice figure of different altitudes.

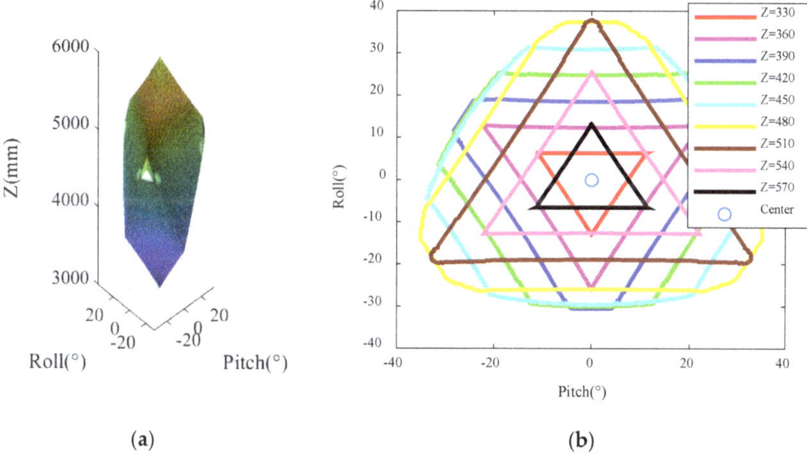

(a)   (b)

Figure 8. Attitude workspace of 3-SPR compensation platform, (**a**) is the attitude workspace image, (**b**) is a contour map derived from the attitude workspace.

According to Figures 7 and 8, the workspace of a parallel 3-SPR compensation platform has the following characteristics:

(1) section shape and section area of the position workspace and attitude workspace varies with the height of the moving platform;

(2) shape of the position workspace changes dramatically in the area with lower height, while the shape of the position workspace is inert in the area with higher height;

(3) overall shape of the position workspace and attitude workspace possesses a trilateral symmetry.

2.3. Compensation Feasibility Comparison Analysis

The parallel 3-SPR compensation platform and parallel 3-RPS compensation platform are similar in structure, both of them have the motion capability of three degrees of freedom. However, the volume and shape of the position workspace and the movement capacity of the moving platform vary. The platform which is more suitable for wave compensation operation is selected through comparative analysis.

Firstly, the comparison diagram of attitude workspace is shown in Figure 9. The blue geometry is a 3-SPR compensation platform attitude workspace. The golden one represents the attitude workspace of 3-RPS.

It can be seen that the geometry shape of the attitude compensation space of the two compensation platforms is the same, and the difference lies in the different directions. The maximum deflection capacity of the two platforms is the same, and the deflection capacity varies inversely to the height.

Secondly, moving platform position workspace is compared, as shown in Figure 10. In the figure, the centrally located golden geometry is the position workspace of the 3-RPS compensation platform. The blue transparent geometry is a parallel 3-SPR compensation platform position workspace.

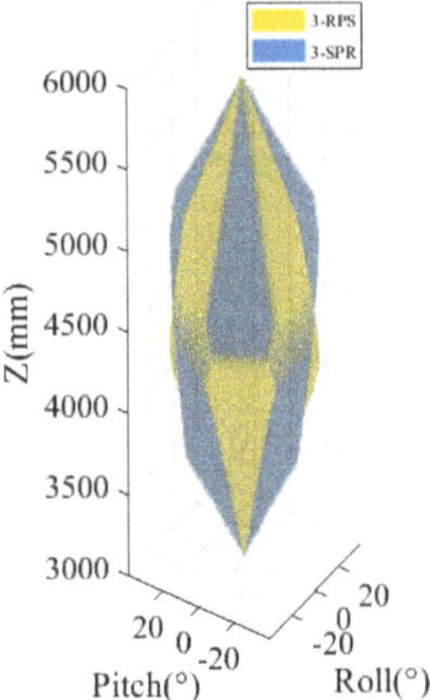

Figure 9. Attitude workspace comparison diagram of two platforms.

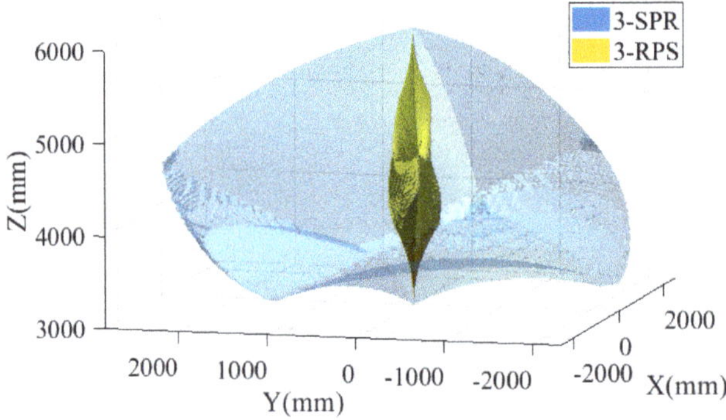

Figure 10. Position workspace comparison of two platforms.

As can be seen from the figure, with the same structural size parameters, the two platforms' workspaces have the same total height. The position workspace volume of the central point of the 3-SPR platform is much larger than that of the 3-RPS platform, indicating that the 3-SPR compensation platform has stronger horizontal moving capability than the 3-RPS compensation platform.

In the motion space diagram of the moving platform, the position information and attitude information can be better combined, and the work of the two compensation platforms can be more directly compared. In the motion space of the moving platform, different platform poses will render image overlap. Enlarging the calculation interval step can decrease the number of platform positions in the image and using the sidelines of the moving platform to contrast two platforms' movement. The motion space contrast diagram is shown in Figure 11.

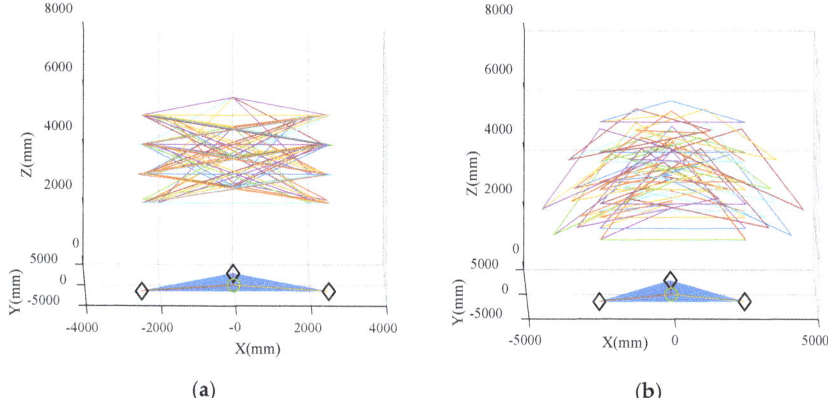

Figure 11. Moving platform motion space comparison: (**a**) is a parallel platform moving platform 3-RPS compensation movement space; (**b**) is the 3-SPR parallel compensation platform's motion space. In both figures, the solid triangular pattern at the bottom indicates the fixed platform, while the solid triangular lines at the top forms the moving platform under different poses.

It can be seen that the parallel 3-SPR compensation platform has a large motion space. When the moving platform of the 3-SPR compensation platform performs a compensation movement, the central point of the moving platform will have a large horizontal displacement relative to the vertical line passing through the central point of the fixed platform. This movement is called platform parasitic movement [18]. The amplitude of platform parasitic movement is proportional to the deflection angle of the platform. The parasitic movement of the moving platform also exists in the deflection of the moving platform of the 3-RPS compensation platform, but the amplitude of the parasitic movement is much smaller.

From this we can conclude that: those two compensation platforms possess the same deflection capability but the orientation of the attitude workspace is inversed; The 3-SPR parallel compensation platform covers a bigger room of position workspace than the 3-RPS parallel compensation platform does; 3-SPR parallel compensation platform owns a stronger horizontal moving ability and its moving platform deflection simultaneously causes obvious platform parasitic movement.

It should be pointed out that, due to the finite area of the ship's deck, a compensation platform cannot be installed on the center of the ship deck, and there is an inevitable distance between the moving platform and fixed platform, so the ship rolling and pitching movement will consequently make moving platform produce an associated sway and surge movement relative to the fixed platform; thus, a compensation system shall have the ability to compensate this movement. By comparing the workspace of the 3-RPS compensation platform and the 3-SPR compensation platform, we can know that the ability of the 3-RPS compensation platform to compensate for horizontal movement is far less than that of the 3-SPR compensation platform.

Under the same ship attitude condition, both the 3-RPS platform and the 3-SPR platform of the same size produce an ideal compensation effect. The comparison of the compensation effect is shown in Figure 12.

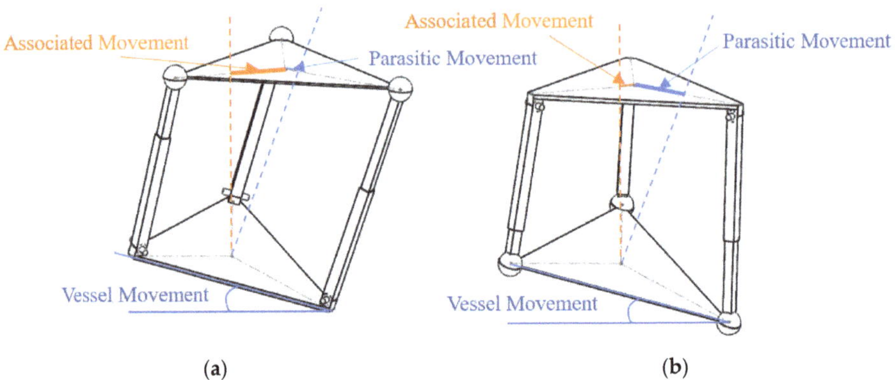

Figure 12. Compensation effect comparison; (**a**) is the compensation effect of the 3-RPS compensation platform; (**b**) is the compensation effect of the 3-SPR compensation platform.

From the ideal comparison effect of the two platforms, it can be indicated that the 3-RPS compensation platform cannot effectively compensate for the associated sway and surge caused by the rolling and pitching of the ship movement. In contrast, the 3-SPR compensation platform can reduce this movement by its horizontal moving ability. Therefore, in the practical wave compensation system installation, the parallel 3-SPR compensation platform is more competent.

3. Structural Optimization Design of Parallel 3-SPR Compensation Platform

For waves with a significant wave height of 2.5 m, the heave compensation platform should have a vertical motion range of 3 m to compensate for the ship movement. The vertical distance between the lowest pose of the platform and the ship deck should be at least 3 m, and the vertical distance should be 4.5 m when the compensation system is in the ready phase, and the vertical height can reach 6 m at the highest pose. In the actual system, such a height is very dangerous. Meanwhile, the cylinder rod is getting more easily to lose stability with the cylinder's elongation extend. Therefore, on the premise of not changing the total height of the position workspace of the platform, the height of the moving platform and the elongation of the cylinder rod should be reduced as much as possible.

Structural optimization method 1: changing the area ratio of upper and lower platforms. By changing the area ratio between the upper platform and the lower platform, the dexterity of the platform can be increased, but the area of the upper platform should not be too small, and the optimal situation is that the upper platform area is half of the area of the lower platform [14]. However, this method has no obvious effect on the reduction of the working height of the heave compensation platform and the shortening of the cylinder rod elongation.

Structural optimization method 2: involving the motion of the revolute joint in heave compensation. The mechanism of the revolute joints participating in heave compensation is shown in Figure 13. The vertical height of the end of the leg can be directly affected by the angle change of the revolute joint when the length of the rod is unchanged. In the previous heave compensation platform motion, the rotation ability of the revolute joint was not fully utilized. Therefore, the structure of the compensation platform could be changed to make the revolute joint fully participate in the compensation motion [21].

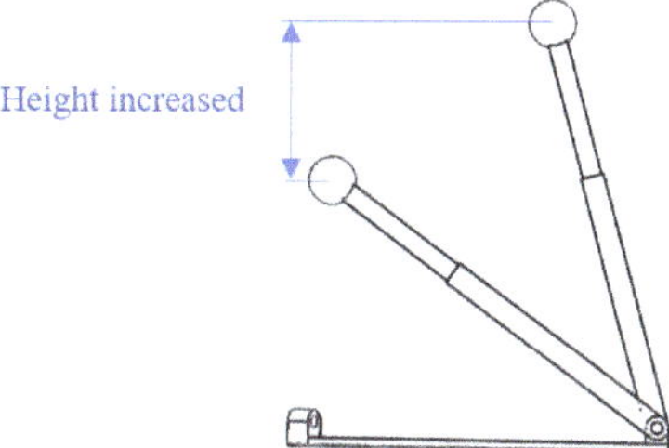

Figure 13. Mechanism of revolute joints participating in compensation to heave motion.

The first optimization structure is changing the corresponding mode of the spherical joints and the rotating joints. Rotate the moving platform by 180° around the axis passing the vertical direction of the central point without changing the corresponding relationship between revolute joints and spherical joint. To avoid the collision of the three branch support legs, each leg is divided into two sections in the middle and interlaced with each other. The specific structure is shown in Figure 14.

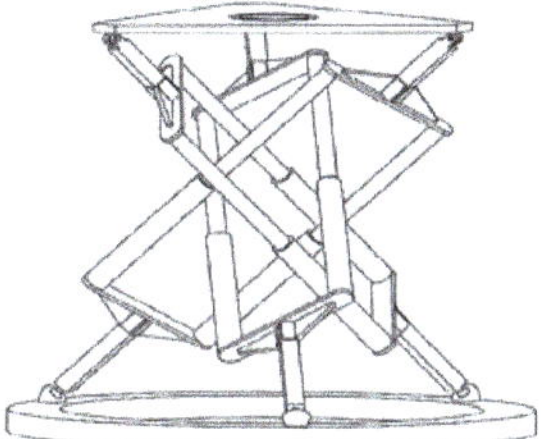

Figure 14. The first type of optimized structure.

This structure has solved the problems in the original platform such as the extent of the high altitude of the upper deck and length of branch legs. To compensate ship movement under level 4 sea state, the platform height of neutral position goes down from 4.5 m to 3.5 m, prismatic joints elongation can be reduced from the original 3 m down to 2.2 m, and the platform still has the extremely good deflection ability. However, the structure of this type of platform is complex, and it is also more difficult to synchronize the two cylinders on the same branch leg.

In the second optimization structure, the arrangement of the revolute joints and the corresponding mode between the rotation joints and the spherical joints are changed [21]. The axes of those three revolute joints rotate 30° counter-clockwise (or clockwise) around their vertical direction at the same time. Meanwhile, the corresponding modes between the rotation joints and the spherical joints are transferred counterclockwise (or clockwise), too. The structure diagram of the improved platform is shown in Figure 15.

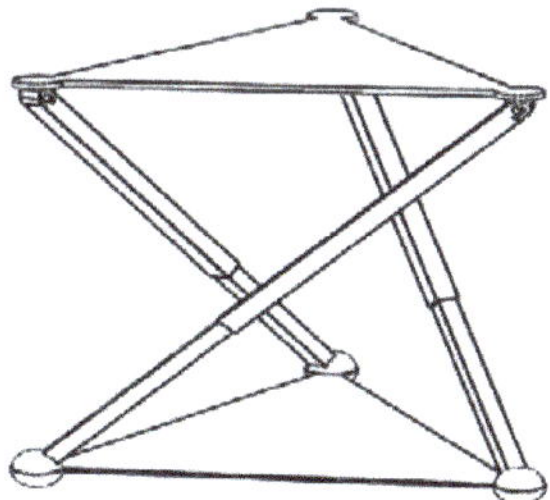

Figure 15. The second optimization structure.

The second form of structure optimization is simple in structural expression and easy to build in practice, and the revolute joint fully participates in the compensation movement of the platform. However, the motion model of the compensation platform has changed greatly. A new calculation of its working space is required. The motion model of this novel compensation platform is analyzed below.

4. Modeling and Analysis of the Optimized 3-SPR Platform

To facilitate explaining, the platform connected with the spherical joint is named as S-deck, and the other platform connected with rotation joints is called R-deck. Three branch legs are represented by $S_1P_1R_1$, $S_2P_2R_2$, $S_3P_3R_3$, with S_i, P_i, R_i respectively indicate spherical joints, prismatic joints and rotation joints. Establishing coordinate system, make the origin located at the center of the R-deck, with Y axis pointing to vertex R_3, Z axis pointing vertically down, and the X axis is defined by the right-hand rule.

As shown in Figure 16, the revolution axis of the R_1 point rotation joint of the new platform is perpendicular to the line connecting R_2, and R_1 point. The revolution axis of the rotation joint of R_2 point is perpendicular to the line connecting R_3 R_2 point, and the revolution axis of R_3 point rotation joint is perpendicular to the line connecting R_1 R_3 point.

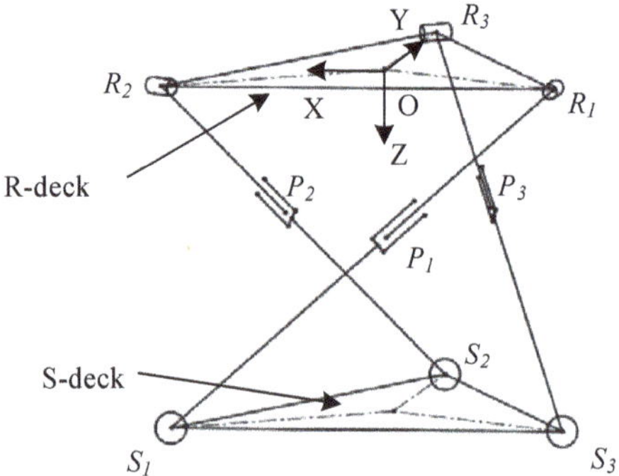

Figure 16. Structure demonstration of optimized compensation platform.

Set R-deck as equilateral triangle, with side length L, and set the coordinates of the three vertices as

$$R_1=\begin{bmatrix} \frac{-L}{2} \\ \frac{-\sqrt{3}L}{6} \\ 0 \end{bmatrix} \quad R_2=\begin{bmatrix} \frac{L}{2} \\ \frac{-\sqrt{3}L}{6} \\ 0 \end{bmatrix} \quad R_3=\begin{bmatrix} 0 \\ \frac{\sqrt{3}L}{3} \\ 0 \end{bmatrix} \tag{1}$$

Set the length of the three branch chains be respectively as g_1, g_2, g_3.

Name the point of spherical joint that connected on the same branch leg with rotation joints R_i as S_i. and its coordinates are assigned as

$$S_1=\begin{bmatrix} x_1 \\ y_1 \\ z_1 \end{bmatrix} \quad S_2=\begin{bmatrix} x_2 \\ y_2 \\ z_2 \end{bmatrix} \quad S_3=\begin{bmatrix} x_3 \\ y_3 \\ z_3 \end{bmatrix} \tag{2}$$

Due to the movement of rotation joints only possess one degree of freedom, the vertex S_1 must be located in a plane which is perpendicular to R-deck and with line R_1R_2 on it. The same as S_2 to line R_2R_3 and S_3 to line R_1R_3. These three planes are named as plan $R_1R_2S_1$, plan $R_2R_3S_2$ and plan $R_3R_1S_3$. Therefore, the x-coordinate value in S_1, S_2, S_3 has the following relationship with the y-coordinate value

$$\begin{cases} y_1=-\frac{\sqrt{3}L}{6} \\ y_2=-\sqrt{3}x_2+\frac{\sqrt{3}L}{3} \\ y_3=\sqrt{3}x_3+\frac{\sqrt{3}L}{3} \end{cases} \tag{3}$$

and the coordinate of three vertices is changed into

$$S_1=\begin{bmatrix} x_1 \\ -\frac{-\sqrt{3}L}{6} \\ z_1 \end{bmatrix} \quad S_2=\begin{bmatrix} x_2 \\ -\sqrt{3}x_2+\frac{\sqrt{3}L}{3} \\ z_2 \end{bmatrix} \quad S_3=\begin{bmatrix} x_3 \\ \sqrt{3}x_3+\frac{\sqrt{3}L}{3} \\ z_3 \end{bmatrix} \tag{4}$$

Since the length of those three branch legs are represented as g_1, g_2, g_3, so the trajectory of vertex S_1 is a special circle that uses vertex R_1 as the center, g_1 as the radius, and be on the plane comprising vertex R_1, R_2, and S_1. The trajectory of vertex S_2 and S_3 can also be obtained by this way. Those three trajectories' functions are shown below.

$$\begin{cases} (x_1+\frac{L}{2})^2+z_1^2=g_1^2 \\ 4(x_2-\frac{L}{2})^2+z_2=g_2^2 \\ (4x_3)^2+z_3^2=g_3^2 \end{cases} \tag{5}$$

There is another constrain that the length of the side of deck-R is L, so the functions that expressing this relationship are shown below.

$$\left| \vec{S_1S_2} \right|=\left| \vec{S_2S_3} \right|=\left| \vec{S_3S_1} \right|=L \tag{6}$$

Calculate the side length using vertices' coordinate and then simplify those equations. We can get the functions

$$\begin{cases} (x_1-x_2)^2-3(x_2-\frac{1}{2}L)^2+(Z_1-Z_2)^2=L^2 \\ (x_2-x_3)^2-3(x_2+x_3)^2+(Z_2-Z_3)^2=L^2 \\ (x_3-x_1)^2-3(x_3+\frac{1}{2}L)^2+(Z_3-Z_1)^2=L^2 \end{cases} \tag{7}$$

Combining Equations (5) and (7), we can get

$$\begin{cases} (x_1+\frac{L}{2})^2+z_1^2=g_1^2 \\ 4(x_2-\frac{L}{2})^2+z_2=g_2^2 \\ (4x_3)^2+z_3^2=g_3^2 \\ (x_1-x_2)^2-3(x_2-\frac{1}{2}L)^2+(Z_1-Z_2)^2=L^2 \\ (x_2-x_3)^2-3(x_2+x_3)^2+(Z_2-Z_3)^2=L^2 \\ (x_3-x_1)^2-3(x_3+\frac{1}{2}L)^2+(Z_3-Z_1)^2=L^2 \end{cases} \tag{8}$$

Equation (8) is a nonlinear system of equations, using the nonlinear least square method to solve it and the coordinate of those three spherical joint centers can be derived. Using the coordinate values to calculate the transform matrix representing the movement from the R-deck to the S-deck. Reverse the matrix and using the reversed matrix can we derive the coordinates of rotate joint centers after the movement of the R-deck to the S-deck.

The position workspace, shown in Figure 17a, and attitude workspace, shown in Figure 17b, of this new rotated 3-SPR compensation platform can be drawn by calculating all the feasible positions of the center point of the R-deck and all feasible angles that the R-deck can reach.

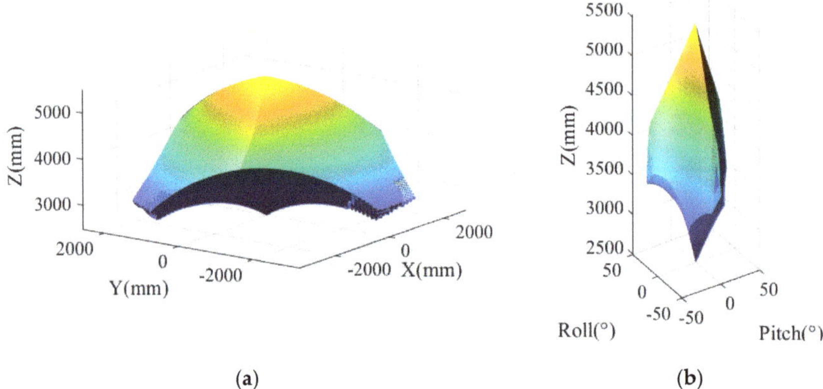

Figure 17. Position workspace and attitude workspace of the novel compensation platform. (**a**) is the position workspace of the new platform. (**b**) is the attitude workspace of the new platform.

It can be seen that the workspace of the 3-SPR compensation platform mentioned above and the workspace of the new rotated 3-SPR compensation platform possesses similar volume and shape, which means the new platform is competent for compensating the sway and surge movement concomitant to the roll and pitch movement of a vessel. It also can be seen that the height of the compensation platform workspace has been reduced.

5. Results and Discussion

5.1. Working Height and Side Length Optimization

The working height and altitude of the new compensation platform is no longer determined by the prismatic joint alone, but influenced by the angle of the rotating joint and prismatic joint together. At the same time, the branch legs do not support the R-deck vertically, so the effective component of support force also changes with the height of the platform. Finding out the relationship between those elements will facilitate the equipment configuration design and establishment of a control system.

First, the correlation of branch leg elongation and the neutral height with different side lengths is discussed. The neutral height is the height of the medium position of a compensation platform, to completely decouple the ship motion under 4-level sea condition, the height of the position workspace of the novel compensation platform is set to 3000 mm. The relation of those three elements is shown in Figure 18.

It can be seen that, with the same side length of the deck of compensation platform, the increase of the neutral height of the platform leads to the increase in the elongation of the branch leg, and with the same neutral height of the compensation platform, the decrease in side length also leads to increase in elongation of branch legs. Therefore, for the purpose of decreasing the elongation of branch legs, the side length of the compensation platform should be set as long as possible and the neutral height should be set as low as possible.

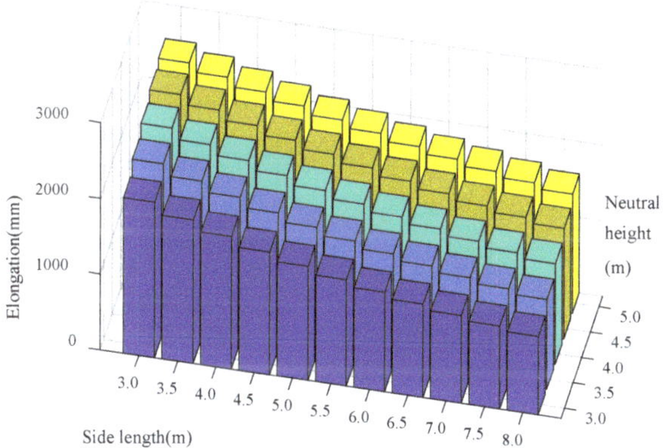

Figure 18. Correlation between branch leg elongation and neutral height under different side length circumstance.

Another aspect to be discussed is the percentage of the efficient support force of the branch legs. Due to the branch legs no more support the moving deck of the compensation platform vertically, the proportion of force that propping up the moving deck is related to the height of the moving deck. The efficient support force can be calculated using Equation (9) below, where F_e represents the efficient force, F_a represents the actuation force, and θ represents the angle of the revolute joint of the branch leg.

$$F_e = F_a \times sin\theta \tag{9}$$

The bar chart in Figure 19 shows the relation between the percentage of efficient support force of a branch leg calculated according to the lowest position of the moving deck of different neutral heights. The side length is also involved in the discussion of different conditions.

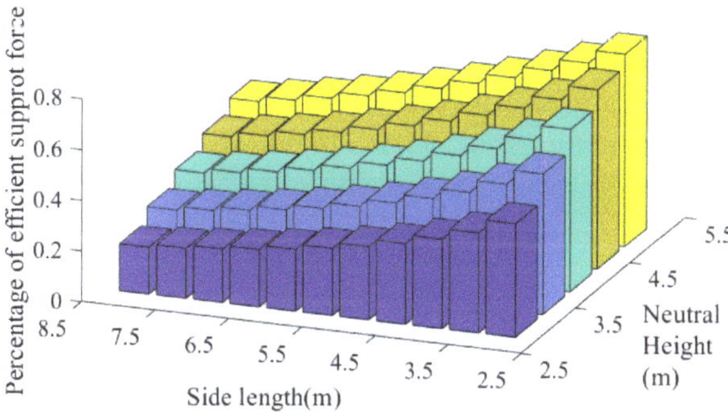

Figure 19. Correlation between the percentage of vertically support force of branch legs and neutral height under different side length circumstances.

As the figure shows, with the increment of the neutral height of the compensation platform under a certain side length, the percentage of efficient support force goes up, and reducing the side length without changing the neutral height can also provide more efficient support force from the branch leg to the moving deck. Thus, we may conclude that: to increase the efficient force, the side length should be reduced and the neutral height should be set at a high value.

Comparing the conclusions discussed above, the effect of side length and neutral height of compensation platform on the elongation of branch legs and percentage of efficient support force is approximately opposite. For the purpose to make the compensation platform possess the same capacity as the original platform set the side length to 5 m. The correlation of platform height, length of leg, and efficient force percentage is shown in Figure 20.

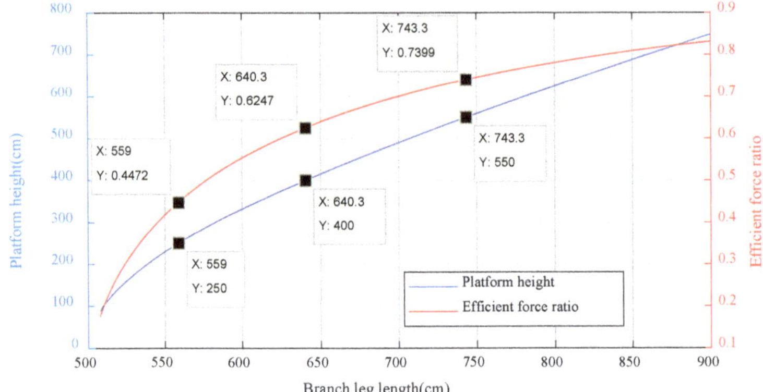

Figure 20. Curve of the relationship between branch leg length and the percentage of efficient support force, and the curve of the relationship between platform height and prismatic joint length.

To make sure the compensation platform has a good active reaction, the proportion of efficient force should not be too low. Therefore, the optimized neutral height of the compensation platform is set to 4 m with the efficient force ratio goes to 62.47%, and due to the height of the workspace is 3 m, the lowest position of the moving deck is 2.5 m with the percentage of efficient support force is 45%.

The new platform also owns a faster movement response than the old one. For instance, when the platform height increases from 250 cm to 400 cm, the branch legs length only needs to increase by 81.3 cm.

The height of workspace of this compensation platform can also be moderated to face different needs. For instance, if the compensation platform is using to decouple the ship movement from a heavy cargo or machine, the neutral height of this platform should be increased to provide more percentage of efficient support force, and if this platform is used to compensation the wave motion for a lightweight one or be used as a part of the heave compensation gangway for transportation, the neutral height can be set lower to make the movement of the moving platform more flexible and get a faster actuation response.

5.2. Compensation Movement Simulation

The kinematic simulation model is built to analysis the effect of the wave compensation and compare the performance of the novel platform with the original one. The simulation model in Simulink is shown in Figure 21.

Solidworks is used to draw the structure model of the new 3-SPR compensation platform and import the assembly model into the Simscape module of Simulink. This simulation includes an inverse kinematic [22] solver which is designed according to the structure model message. This solver can calculate, online, branch legs' length that compensates the vessel movement by inputting the value of heave motion, pitch, and roll of the vessel.

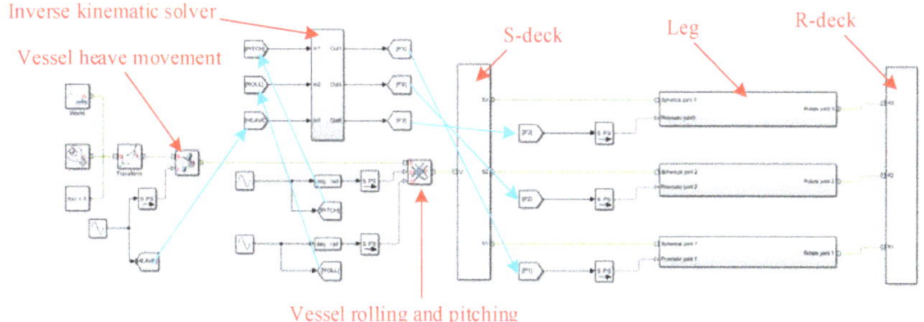

Figure 21. Diagram of simulation in Simulink.

Employing a prismatic joint and a universal joint between the S-deck and the rigid ground coordinate of this Simulink model. Utilize the prismatic joint motion to simulate the heave movement of the vessel and the universal joint revolution to imitate the rolling and pitching. The subsystem of a branch leg in Simulink model is shown in Figure 22.

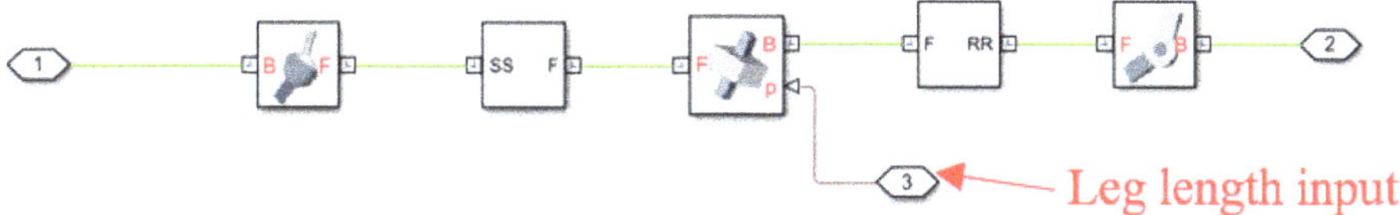

Figure 22. Subsystem diagram under the leg section mask is composed of a spherical joint, a prismatic joint and a revolute joint. Using the prismatic joint as the actuator, and the spherical joint and revolute joint are respectively connected to the S-deck and the R-deck. The direction of different legs is determined by rigid transform module under the R-deck system.

The maximum values of vessel motion are important to this simulation for they directly determine the structural configuration of the compensation platform, and whether the platform is capable for this job. In contrast, the verisimilitude of the vessel motion is less important which means simplifying the component of the wave will not affect the result of simulation distinctly. The significant amplitude of heave motion under 4-level sea states is about 1.25 m, and the maximum rolling angle is proximate 10 degrees, and the pitching angle is smaller. In order to make the results easier to be read and let the figure show more periods, the components of the wave have been simplified and the frequency of all movements has been increased. To make sure the compensation platform can fully compensate the ship motion, all the amplitude of movement has been amplified. The vessel motion under 4-level sea status is simulated using the coefficient written below in Table 1. Figures 23 and 24 show the time history of ship heave, rolling, and pitching motion.

Table 1. Vessel motion simulating configuration.

	Heave	Rolling	Pitching
Amplitude	1500 mm	15 degree	5 degree
Frequency (rad/s)	0.5	1	0.5
Phase (rad)	0	0	0.1

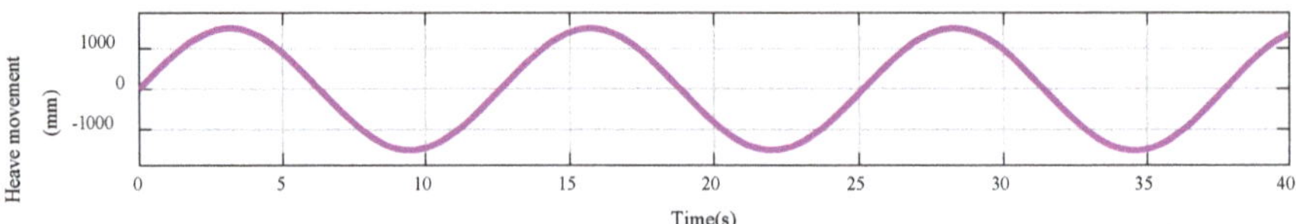

Figure 23. Heave movement of vessel.

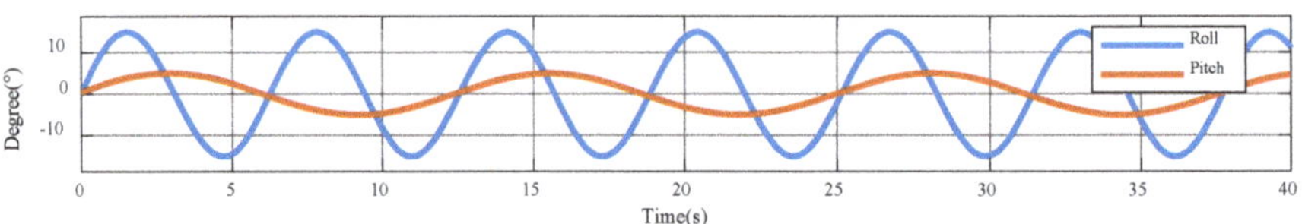

Figure 24. Roll movement and pitch movement of vessel.

Figure 25 shows the leg elongation of the novel structure during the compensation process to the simulative wave above. To compensate a ±1.5 m wave, the support legs of the novel compensation platform need to possess a range of elongation of about 2.3 m.

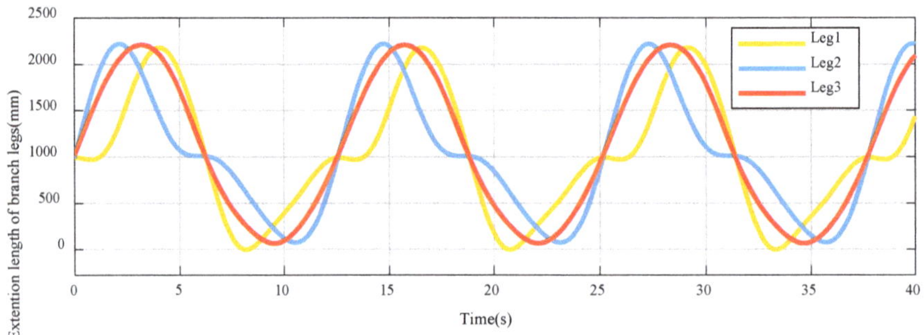

Figure 25. Leg elongation extent of the novel wave compensation platform.

Figure 26 presents the leg elongation of the original platform during the process of compensation to the simulative wave. As shown in the picture, to compensate that wave, the support legs of the old compensation platform need to have a 4 m elongation range at least.

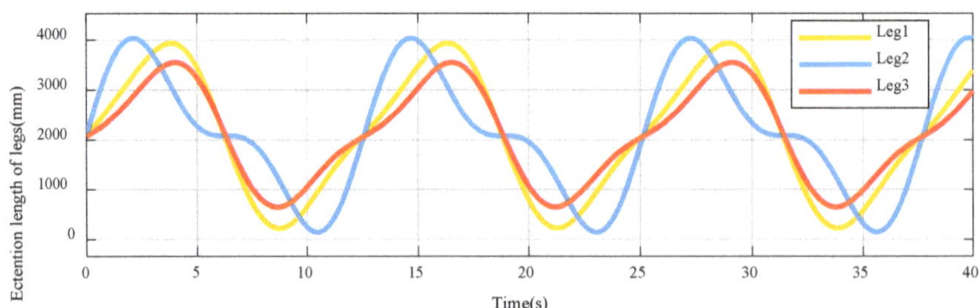

Figure 26. Leg elongation extent of the original wave compensation platform.

Therefore, the simulation of the compensation process of both novel and original platform comes to the conclusion that the novel compensation platform has a smaller range of support legs' elongation than the old one under the same condition of wave height. Meanwhile, a shorter range during the same time indicates a smaller elongation speed. The diagrams contain the velocity of the branch legs' extension movement of both platform structures during the compensation process is below.

By comparing Figures 27 and 28, it can be seen that the maximum velocity of support legs of the novel compensation platform is about 0.9 m/s, but the old one is 1 m/s. In addition, the maximum velocity of the leg 2 of the old platform is up to 1.5 m/s. To compensate the same wave, the support legs of the novel compensation platform need smaller velocity than the old one, which means the movement response of the new platform is faster.

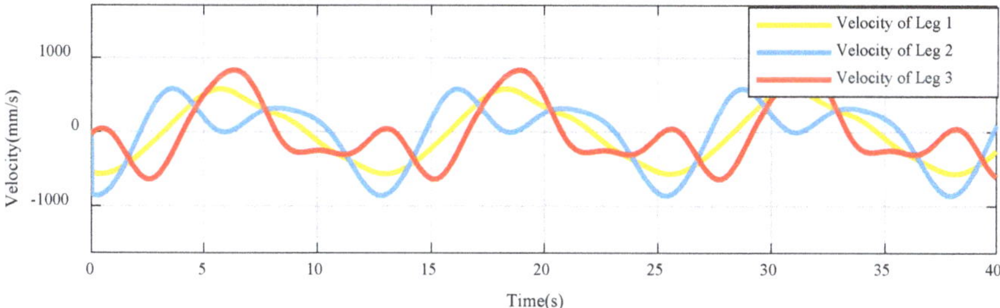

Figure 27. The velocity of support legs of the novel compensation platform during the compensation to the simulative wave.

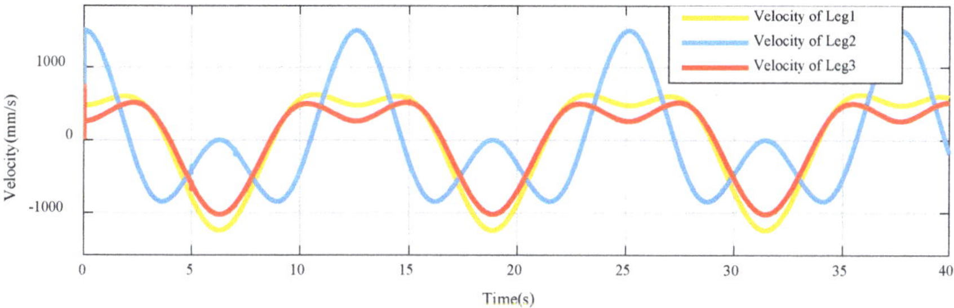

Figure 28. Velocity of branch legs of the original compensation platform during the compensation to the simulative wave.

Due to the structural constraints, the parasitic movement of the parallel platform is unavoidable. A move simulation of the moving platform during the compensation process is done to analyze the compensation effect of the novel compensation platform. The heave roll and pitch movement of the moving platform after compensation is shown in Figures 29 and 30 respectively.

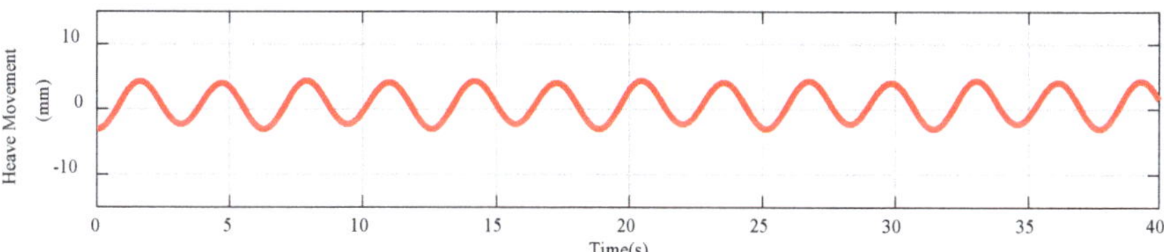

Figure 29. Heave movement of the R-deck of the compensation platform. The final movement range is less than 1 cm.

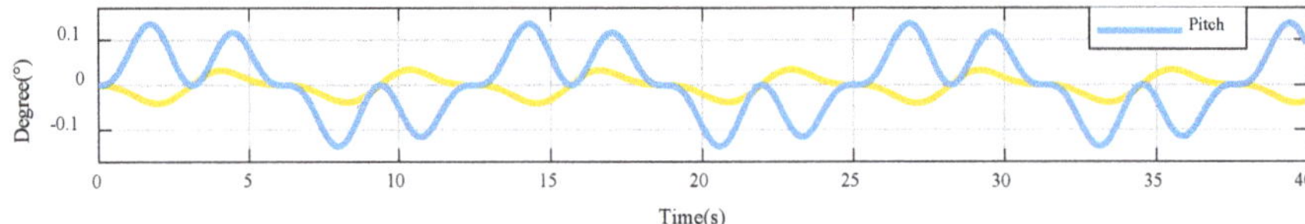

Figure 30. Rolling value and pitching value of the R-deck of the compensation platform. The maximum rolling angle is less than 0.34 degrees and the biggest pitch angle is 0.136 degrees. Both are drastically less than the revolute movement of the vessel.

It can be seen in Figure 31 that the associated movement is about 0.1 mm which is difficult to feel. Generally speaking, the heave movement of vessel has been reduced by 99.96%, rolling movement has been alleviated by 97.74%, and the angle of pitch movement has been reduced by 97.28%.

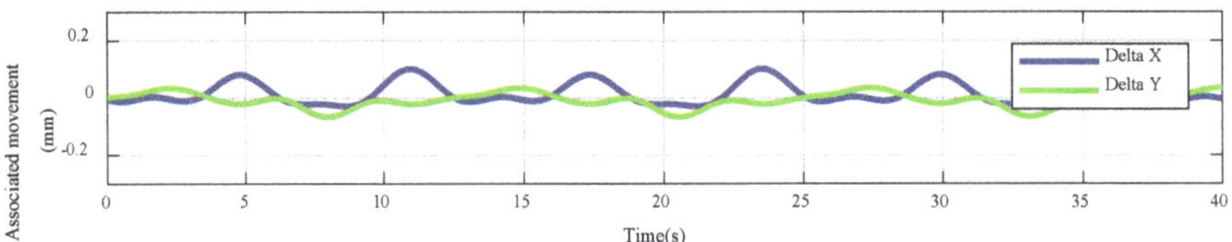

Figure 31. Associated movement of the R-deck along axis x and axis y. The associated movement is caused by the parasitic movement of the parallel platform structure.

6. Conclusions

Considering the range of vessel motion under level-4 sea state, compensation platforms based on 3-RPS parallel structure and 3-SPR parallel structure were presented. Using forward kinematic methodology, we analyzed the compensation ability, and the workspace of both compensation platforms were drawn.

By comparing the position work space and the attitude workspace, associating with the parasitic movement of both platforms, we came to the conclusion that the 3-SPR parallel structure is more competent for vessel movement compensation than the 3-RPS parallel structure.

In order to enhance the safety and actuating response of the compensation platform, a structure optimized 3-SPR parallel platform was designed which has a lower workspace altitude and shorter elongation range of branch legs. Using the forward kinematic method calculated the workspace and verified its feasibility of compensating.

A simulation model of the vessel motion compensation using the novel parallel structure and the original platform were built. Employing the inverse kinematic method as the movement solver, the performance of the compensation processes were compared. The feasibility and efficiency of the novel compensation platform was verified and calculated. After being compensated by the novel compensation platform, the heave movement of vessel was reduced by 99.96%, rolling movement can be alleviated by 97.74%, and the angle of pitch movement was reduced by 97.28%.

Author Contributions: Conceptualization, Y.Z., J.X., H.T., J.F., and S.W.; Methodology, Y.Z., J.X., H.T., J.F., and S.W.; Software, H.T., J.F., and S.W.; Validation, Y.Z. and H.T.; Formal analysis, Y.Z., H.T., and S.W.; Investigation, Y.Z.; Resources, J.X. and Z.Y.; Data curation, J.F.; Writing—original draft preparation, Y.Z. and H.T.; Writing—review and editing, Y.Z., H.T., and J.F.; Visualization, Y.Z.; Supervision, Y.Z. and J.X.; Funding acquisition, Y.Z. and J.X. All authors have read and agreed to the published version of the manuscript.

Funding: The authors gratefully acknowledge the financial support from The Natural Science Foundation of Heilongjiang Province, China (no. E2018022), Information Technology of the People's Republic of China-Floating Security Platform Project (the second stage), and the Opening Fund of Acoustics Science and Technology Laboratory (grant no. SSKF2020009).

Acknowledgments: The authors would like to thank the members of the Marine Electromechanical Systems Research Institute for their continued support and discussion.

Conflicts of Interest: The authors declare that the research was conducted in the absence of any commercial or financial relationships that could be construed as a potential conflict of interest.

References

1. Han, C. Design and Research of Active Heave Compensation System Based on Laser Ranging Sensor. *World J. Eng. Technol.* **2020**, *8*, 13–18. [CrossRef]
2. Li, B.; Zhao, X. Position Analysis of A 3-Spr Parallel Mechanism. *J. Theor. Appl. Inf. Technol.* **2013**, *48*, 8.
3. Wei, L.; Limin, T.; Zhengnan, J. Design and control of cable-drive parallel robot with 6-dof active wave compensation. In Proceedings of the 2017 3rd International Conference on Control, Automation and Robotics (ICCAR), Nagoya, Japan, 24–26 April 2017; pp. 129–133.
4. Gu, Y.-F.; Xie, R.; She, J.-G. Optimization design of heave compensation device platform under six level of sea condition. *Ship Sci. Technol.* **2017**, *39*, 141–145.
5. Markus Richter, E.A.; Klaus, S.; Johannes, K. Oliver Sawodny Model Predictive Trajectory Planning with Fallback-Strategy for an Active Heave Compensation System. In Proceedings of the 2014 American Control Conference (ACC), Portland, OR, USA, 4–6 June 2014.
6. Zhang, Y.; Ma, S.; Duan, W. Linear time-domain strip method for ship motion prediction. *Chin. J. Ship Res.* **2018**, *13*, 1–6, 28.
7. Su, C.; Zheng, W.; Zeng, Y.; Ding, D. An Active Wave Compensation Method for the Gangway of Wind Turbine Maintenance Vessel. *Naval Archit. Ocean Eng.* **2017**, *33*, 22–25.
8. Ngongi, W.; Du, J.; Massami, E.; Chang, W.-J.; Kassembe, E. Dynamic positioning of ships based on robust fuzzy observer. *J. Eng.* **2020**, *2020*, 228–238. [CrossRef]
9. Wang, W. Kinematics Analysis of Wave Motion Compensation Stable Platform. *Shipbuild. Technol. Res.* **2019**, *4*. [CrossRef]
10. Sun, Z.; Zhao, Q.; Wang, N.; Qin, Y. Modeling and Control of 3—RPS Parallel Vibration Isolation Platform. *For. Eng.* **2020**, *36*, 69–76.
11. Nayak, A.; Wenger, P.; Caro, S. Comparison of 3-RPS and 3-SPR parallel manipulators based on their maximum inscribed sigularity-free. *New Trends Mech. Mach. Sci.* **2018**, 121–130.
12. Tang, G.; Hu, C.; Hu, X. Modeling and simulation of three-degree-of-freedom parallel wave compensation platform. *J. Shanghai Marit. Univ.* **2020**, *41*, 20–26.
13. Zhou, B.; Mao, T.; Yang, R. Workspace Analysis of 3- dof RPS Parallel Mechanism. *J. Hunan Univ. Nat. Sci.* **2003**, 58–61.
14. Hongmei, Y. *The Kinematics Analysis of 3-RPS Parallel Machine Tool*; Northeastern University: Liao Ning Province, China, 2013.
15. Lu, Y.; Zhao, Y. Position and workspace analyses of 3-SPR and 3-RPS parallel manipulators. In Proceedings of the International Design Engineering Technical Conferences and Computers and Information in Engineering Conference, Long Beach, CA, USA, 24–28 September 2005; pp. 957–962.
16. Zhang, F.; Jia, Y.; Zhang, C.; Liang, K. Research on Design and Simulation of 3-RPS Parallel Stable Platform. *Mach. Build. Autom.* **2018**, *47*, 145–148. [CrossRef]
17. Zhang, L.-J.; Guo, F.; Liu, S.-Y. Deflection capacity analysis of moving platform of a 3-RPS parallel mechanism. *J. Yanshan Univ.* **2012**, *36*, 196–200.
18. Nayak, A.; Stigger, T.; Husty, M.L.; Wenger, P.; Caro, S. Operation mode analysis of 3-RPS parallel manipulators based on their design parameters. *Comput. Aided Geom. Des.* **2018**, *63*, 122–134. [CrossRef]
19. Shi, Y.; Hao, Q.; Xu, G.; Lu, Y. Analysis about the Relationship Between Platform Size and Workspace of a Plane Symmetry 3-SPR Parallel Manipulator. *Mach. Des. Manuf.* **2012**, 37–39.
20. Rad, C.; Stan, S.; Bălan, R.; Lapusan, C. Forward kinematics and workspace analysis of a 3-RPS medical parallel robot. In Proceedings of the 2010 IEEE International Conference on Automation, Quality and Testing, Robotics (AQTR), Cluj-Napoca, Romania, 28–30 May 2010; pp. 1–6.
21. Shi, Y.L.Y. Analysis about workspace of 3-SPR parallel manipulator influenced by the joints' distribution. *J. Yanshan Univ.* **2008**, 304–310.

22. Hu, B.; Jing, Y.L.; Yu, S.Z. Analyses of inverse kinematic, statics and workspace of a novel3RPS-3SPR Serial-Parallel manipulator. *Open Mech. Eng. J.* **2011**. [CrossRef]

Publisher's Note: MDPI stays neutral with regard to jurisdictional claims in published maps and institutional affiliations.

Journal of *Marine Science and Engineering*

Article

Practical Prediction of the Boil-Off Rate of Independent-Type Storage Tanks

Dong-Ha Lee [1], Seung-Joo Cha [2], Jeong-Dae Kim [1], Jeong-Hyeon Kim [3], Seul-Kee Kim [3] and Jae-Myung Lee [1,3,*]

[1] Department of Naval Architecture and Ocean Engineering, Pusan National University, Busan 46241, Korea; dongha2@pusan.ac.kr (D.-H.L.); jeongdae3416@pusan.ac.kr (J.-D.K.)
[2] Engineering Team, KC LNG Tech, Busan 48058, Korea; sjcha@kclng.co.kr
[3] Hydrogen Ship Technology Center, Pusan National University, Busan 46241, Korea; honeybee@pusan.ac.kr (J.-H.K.); kfreek@pusan.ac.kr (S.-K.K.)
* Correspondence: jaemlee@pusan.ac.kr

Abstract: Because environmentally-friendly fuels such as natural gas and hydrogen are primarily stored in the form of cryogenic liquids to enable efficient transportation, the demand for cryogenic fuel (LNG, LH) ships has been increasing as the primary carriers of environmentally-friendly fuels. In such ships, insulation systems must be used to prevent heat inflow to the tank to suppress the generation of boil-off gas (BOG). The presence of BOG can lead to an increased internal pressure, and thus, its control and prediction are key aspects in the design of fuel tanks. In this regard, although the thermal analysis of the phase change through a finite element analysis requires less computational time than that implemented through computational fluid dynamics, the former is relatively more error-prone. Therefore, in this study, a cryogenic fuel tank to be incorporated in ships was established, and the boil-off rate (BOR), measured considering liquid nitrogen, was compared with that obtained using the finite element method. Insulation material with a cubic structure was applied to the cylindrical tank to increase the insulation performance and space efficiency. To predict the BOR through finite element analysis, the effective thermal conductivity was calculated through an empirical correlation and applied to the designed fuel tank. The calculation was predicted to within 1% of the minimum error, and the internal fluid behavior was evaluated by analyzing the vertical temperature profile according to the filling ratio.

Keywords: cryogenic tank; boil-off gas (BOG); boil-off rate (BOR); finite element analysis (FEA); liquid nitrogen

Citation: Lee, D.-H.; Cha, S.-J.; Kim, J.-D.; Kim, J.-H.; Kim, S.-K.; Lee, J.-M. Practical Prediction of the Boil-Off Rate of Independent-Type Storage Tanks. *J. Mar. Sci. Eng.* **2021**, *9*, 36. https://doi.org/10.3390/jmse9010036

Received: 2 December 2020
Accepted: 29 December 2020
Published: 1 January 2021

Publisher's Note: MDPI stays neutral with regard to jurisdictional clai-ms in published maps and institutio-nal affiliations.

1. Introduction

The International Maritime Organization (IMO) has established emission control areas (ECAs) in the North and Baltic seas to improve the air quality and limit the presence of low-quality residual fuel (or heavy fuel oil, HFO). Marine emission legislations such as the Tier III requirements of the revised MARPOL Annex VI mandate ships to reduce NOx emissions, with an objective of reducing the greenhouse gas emissions by 20% and 50% until 2020 and 2050, respectively [1,2].

Owing to these requirements, alternative fuels such as natural gas and hydrogen gas are being used instead of HFO as ship propellants. In particular, liquefied natural gas (LNG), whose volume can be reduced by cooling natural gas to 110 K, is environmentally friendly as it can reduce the energy efficiency design index by 20% [3–5]. It has been predicted that ships to transport LNG and those that employ LNG as a fuel will undergo accelerated development by 2035 [6,7]. Nevertheless, to enable the efficient storage and transportation of LNG, the storage systems must be insulated to maintain the temperature. Moreover, in LNG storage tanks, boil-off gas (BOG) is generated owing to the vaporization of liquefied gas as the external heat is ingressed. The BOG can deteriorate the structural intensity owing to the pressurization of the storage system. Therefore, this gas is usually

25

purged and subjected to reliquefaction [8–10]. However, because this process results in economic losses, it is desirable to predict and minimize the amount of BOG in the design stage.

LNG tanks can be categorized into membrane and independent types. Membrane-type tanks are mainly implemented in the LNG cargo hold of carriers instead of as fuel tanks [11]. IMO A and B type tanks, which are independent tanks, can resist the sloshing load owing to their stiff structure. Moreover, such tanks have a secondary barrier to prevent the leakage of liquids and equipment to process the BOG [12,13]. Therefore, such tanks are advantageous in terms of inspection and repair capacities. In contrast, type C independent tanks are designed as pressure tanks with a high inner pressure and without the equipment for BOG processing (Table 1) [14]. Such tanks are mainly used as fuel tanks for LNG-propelled ships due to their low space efficiency.

Table 1. IMO independent type tanks for LNG ship [14].

Type	Prismatic Type	MOSS Type	Cylindrical Type
IMO Tank Type	Type A	Type B	Type C
Schematic structure			
Secondary barrier	Complete	Partial	No requirements
Characteristic	Fully refrigerated at atmospheric pressure	Fully refrigerated at atmospheric pressure	Pressurized at ambient or lower temperature
Notes	For small vessels less than approx. 20,000 m^3 capacity	For large vessels	For LNG carriers

In the context of the expansion of the ECAs and more stringent environmental regulations, it is necessary to examine small vessels sailing along the coast as well as large vessels. The heat transfer rate of small pressurized liquid tanks is larger than that of a large liquid tank owing to the larger surface area to volume ratio of the former [15]. However, most of the existing research pertains to large cargo holds [16–18], and the research on fuel tanks is limited. In particular, the research on cryogenic fuel tanks for small vessel applications is inadequate.

Several researchers have employed computational fluid dynamics (CFD) to examine the physical phenomena pertaining to the various types of BOG occurrences [19–23]. However, the boil-off rate (BOR) calculations are challenging owing to the complex heat transfer and fluid flow involved, owing to which, the calculation time to ensure the convergence in the case of small steps is extremely large. In particular, it is difficult to set the initial temperature considering the filling ratio (FR) of a tank because the temperature changes continuously owing to the liquid evaporation, and heat convection occurs in a complex manner [24]. Many researchers performed finite element analyses (FEA) based on the conduction model to address the complex convection behavior and reduce the calculation time compared to that required for CFD computations [25–28].

Considering these aspects, in this study, cryogenic fuel storage tanks for small ships were designed according to the rules of the IMO and ship classification, and the amount of generated BOG was measured experimentally. Based on the measurement data, a numerical analysis was performed to predict the BOR of other small tanks. The BOG was measured considering the change in mass of the cryogenic liquid inside the tank, and the numerical analysis was performed using a commercial FEA code. Variables that are boundary conditions in the process of numerical analysis were derived by empirical correlation. In the verification process of an experiment, it is possible to predict the results or variables used for calculation by the method by recent deep learning. This method can be applied to predict the phase change of cryogenic fluid, and there are cases applied to equipment such as heat exchangers [29,30]. However, in this study, BOR prediction by finite element analysis was performed, and data were verified through comparison with experimental results.

2. Experiment Details

2.1. Design and Manufacturing of the Experiment Structure

IMO type C independent tanks consist of an inner pressure tank and insulation to prevent heat ingress. In the case of the pressure tank, hemispherical parts are attached at both ends of a cylindrical tank in the horizontal orientation to resist the internal pressure. The cylindrical and hemispherical parts are manufactured using stainless steel 304L, which demonstrates excellent characteristics under cryogenic temperatures [31].

For fuel tanks operating at cryogenic temperatures, the thickness of the inner wall should be determined considering the thermal stress and internal pressure. The thickness of internal tanks under pressure is regulated by the international gas carrier (IGC) code and Korea Register (KR) [32]:

$$t_c = \frac{PD}{2fJ - 1.2P} + 1.0$$
$$t_h = \frac{PR}{2fJ - 0.2P} + 1.0 \tag{1}$$

where t_c and t_h denote the minimum thickness of the cylinder plate (mm) and hemisphere plate (mm), respectively; P is the design pressure (MPa), D is the diameter of the cylinder plate (mm); R is the radius of the hemisphere (mm); f is the maximum allowable stress (MPa); and J is the weld efficiency.

The IGC code classifies tanks with an operating pressure of 0.2 MPa or higher as IMO type C independent tanks. However, as the BOR generally decreases with an increase in the internal pressure, most storage containers currently operate in environments greater than 0.5 MPa [33]. Therefore, the design pressure of the tank produced in this study was set to range from 1 MPa to 1.5 MPa, and the thickness of the pressure tank was set as 4 mm.

The insulation system was made of polyurethane foam (PUF) and manufactured as a cuboid box. In general, a saddle support is installed in the case of the conventional cylindrical tank; however, such support is unnecessary when manufacturing the support as a cuboid box, and this aspect can help prevent heat loss. PUF, synthesized with polyol and isocyanate [34,35], was expanded in a plywood box mold. The density of the applied PUF was 96 kg/m^3, to ensure the required insulation performance. The specifications of the tank are listed in Table 2.

Table 2. Specification for IMO Type C tank.

		IMO Type C Tank	
		Length (mm)	1414
Dimension		Breadth (mm)	624
		Height (mm)	612
Pressure		Design (MPa)	1.5
		Operating	-
Insulation thickness		Maximum (mm)	224
		Minimum (mm)	100
	Internal volume (m^3)		0.127
	Test fluid		Liquid nitrogen

2.2. Experimental Apparatus and Procedure

Thermocouples (TCs) and a weighing scale (WS) were sed to measure the internal temperature distributions and BOR of the cuboid insulation type C tank. As shown in Figure 1, the fuel storage tank was filled with liquid nitrogen instead of cryogenic fuel to ensure safe operation. The WS (CAS Corporation, Korea) least count was 0.1 kg, and data were obtained every 10 s considering the period of the experiment. T-type TCs, which are suitable for cryogenic temperatures, were attached to the surface of the inner tank through spot welding. Subsequently, the TCs were connected to the data acquisition (DAQ) system, and the data were transferred to the main computer (Figure 2). The devices were attached at 10% FR intervals to determine the temperature distribution at the FRs at the 10% to 90% height point and further welded at the location of the maximum load (98%) after the 90% location. To avoid the damage of the TCs during the foaming of polyurethane, a coating was applied on the welded area, and an aerogel mat was placed on the inner tank to protect the tank (Figure 3).

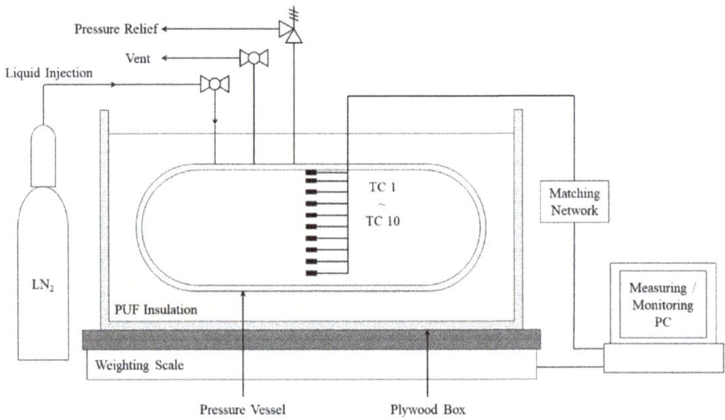

Figure 1. Schematic of experiment and arrangement for TC.

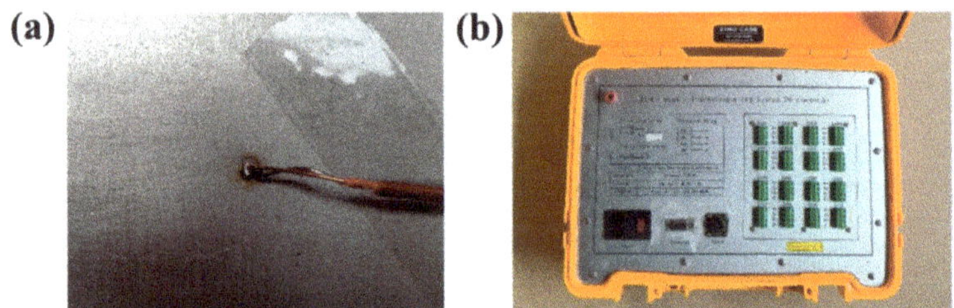

Figure 2. Data acquisition apparatus to measure inner tank surface temperature (**a**) Spot welding of thermal couple (**b**) data acquisition system

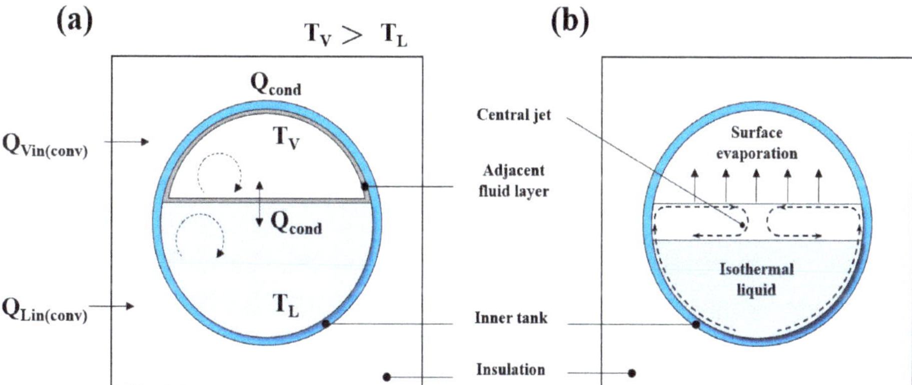

Figure 3. Schematic diagram of (**a**) heat transfer model, (**b**) hydrodynamic model.

In the experiment, the amount of spontaneous vaporization was measured by fully charging the tank fully through the pressure difference of the liquid nitrogen at a certain pressure. Ventilation valves and pressure relief valves were installed in addition to the inlet valves to ensure sufficient discharge at pressures above 1.5 MPa. Each valve operated independently, and the vent valve was installed at the 98% level to prevent 100% filling.

Pre-cooling was performed to prevent the occurrence of structural defects caused by the difference in the thermal expansion coefficient owing to the rapid temperature change in the tank. In general, the cryogenic cargo hold for ships is implemented for one or two days [36]; however, in this study, a period of 6 h was considered, to allow the temperature to converge, considering the size of the tank.

3. Numerical Analysis of the Heat Transfer

3.1. Theoretical Model

The BOR of the IMO independent type C fuel tank was calculated using a thermal conductivity model. In general, the BOR is defined as the amount of liquid evaporated through vaporization during the day relative to the total load in the storage tank. Over time, external heat enters the cryogenic tank (Q), which causes the BOR to increase. The BOR can be calculated as follows:

$$BOR = \frac{Q_{in} \times 24 \times 3600 \times 10}{V \times \rho \times H_{vap}} \times 100 \tag{2}$$

where Q_{in} is the heat inflow from the outside, and ρ and H_{vap} denote the density of the cryogenic fluid and latent heat, respectively.

As shown in Figure 3, the fluid heat transfer in a liquefied gas storage tank in a static state can be interpreted from two perspectives. First, as shown in Figure 3a, assuming that the inner fluid is a conduction model, the BOG generation can be interpreted as conduction to the inner fluid through the heat from the external environment. In the case shown in Figure 3b, the fluid dynamic behavior inside the tank is considered. A stable isothermal distribution exists in the lower part of the liquid state; however, a thermal stratification section appears due to the temperature difference in the upper part, and convection occurs owing to the temperature difference [37–39]. The vapor space in the tank rises upwards, the evaporation continues, and the density increases, in a process known as weathering [4,40]. In the domain of vertical storage tanks, considerable research has been performed on the thermal stratification section owing to the uniform cross-section [41,42]; nevertheless, a horizontal shape has not been applied to ships owing to the geometrical differences.

Considering these aspects, in this study, the analysis was performed using the thermal conductivity model. However, the convection phenomenon of the upper tank due to BOG can cause significant errors when calculating with the thermal conductivity model. Therefore, effective thermal conductivity was applied to increase the precision of the BOR prediction. The effective thermal conductivity was applied to the adjacent layer of gas and inner tank internal, and the heat inflow was calculated by using the empirical correlation for the effective thermal conductivity to reduce the error. The adjacent layer was assumed to be a shell and calculated by considering the geometry and FR of the tank.

3.2. Effective Thermal Conductivity of the Interface of the Liquid and Vapor

The governing equations used in the numerical analysis were derived from the energy conservation equations, and the transfer rate equations were applied. The resulting heat diffusion equation can be expressed as follows:

$$\frac{\partial}{\partial x}\left(k\frac{\partial T}{\partial x}\right) + \frac{\partial}{\partial y}\left(k\frac{\partial T}{\partial y}\right) + \frac{\partial}{\partial z}\left(k\frac{\partial T}{\partial z}\right) = \rho c_p \frac{\partial T}{\partial t} \tag{3}$$

where k is the thermal conductivity, c_p is the specific heat capacity, ρ is the density, and T is the reference temperature.

As the hydrodynamic behavior is not considered in the thermal conductivity model, the temperature of the fluid boundary layer can be considered to be equivalent to that of the adjacent part (no-temperature jump condition). This condition can satisfy the no-slip condition, according to which, the velocity of the fluid in the boundary layer is zero [26]. These conditions can be expressed as in Equation (4):

$$\dot{q}_{conv} = h\left(T_{adj} - T_{BOG}\right) = -k_{conv}\left(\frac{\partial T}{\partial n}\right)$$
$$\dot{q}_{conv} = \dot{q}_{cond} = -k_{fluid}\frac{\partial T}{\partial y}\bigg|_{y=0} \tag{4}$$

where h is the heat transfer coefficient of the BOG, subscript adj represents the adjacent fluid layer of the gas, and n is the normal vector.

The heat from the outside of the tank leads to convection, and the effective thermal conductivity is applied to the adjacent fluid layers to account for this aspect in the conduction model. The effective thermal conductivity can be expressed as in Equation (5):

$$k_{eff} = kNu \tag{5}$$

$$Nu = \frac{hL}{k} \tag{6}$$

Here, Nu is the Nusselt number, defined as the ratio of the convection coefficient to the conduction coefficient.

The convection coefficient depends on the structure; therefore, empirical correlations are applied to obtain the corresponding value. Migliore et al. defined the Nusselt number as in Equation (7) [4]:

$$Nu = 0.116Ra^{0.32} \tag{7}$$

$$Ra = Gr \cdot Pr \tag{8}$$

Ra, that is, the Rayleigh number, is derived from the dimensionless number defined in Equations (9) and (10):

$$Gr = \frac{g\beta(T_L - T_G)L_c{}^3}{v^2} \tag{9}$$

$$Pr = \frac{\mu C_p}{k} \tag{10}$$

where *Gr* and *Pr* represent Grashof Number and Prandtl Number, respectively. Here, *g* is the gravitational acceleration, β is the thermal expansion coefficient, *Lc* is the characteristic length, and *v* and μ represent the kinetic and dynamic viscosity coefficients, respectively.

In general, as the FR decreases, the temperature difference of the gas layer increases, and the Grashof number increases, corresponding to an increase in the effective thermal conductivity. Figure 4 shows the BOR calculation process flow through the numerical analysis. In this study, the temperature obtained experimentally and the effective thermal conductivity for each FR of the tank were used as the input values. The BOR was calculated by solving the heat transfer equation, and a no-temperature jump condition and no-slip condition were implemented considering the heat conduction model.

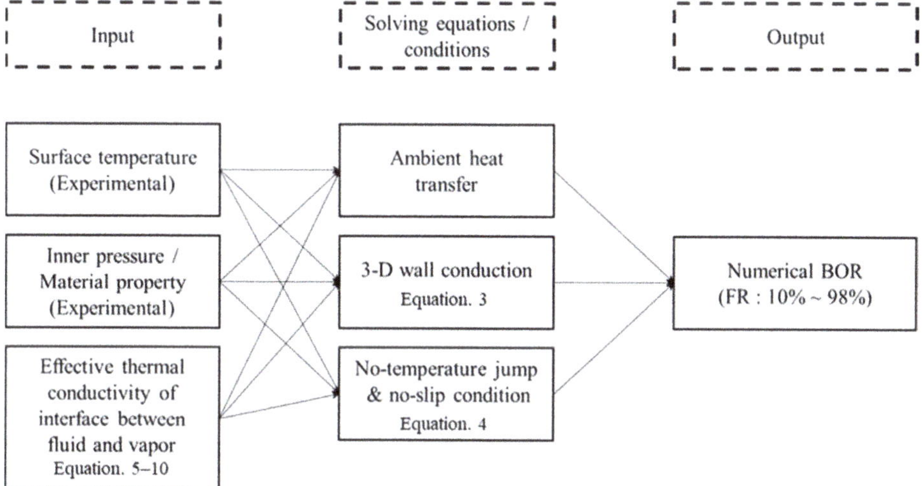

Figure 4. Flow chart depicting solution procedure.

3.3. Numerical Model Description

A commercial finite element code, ABAQUS, was used to predict the BOR in the tank through a computational analysis. As the thermal conductivity model was considered, both the liquid and gas were modeled, and an extremely thin layer was implemented as the boundary layer of the gas. The effective thermal conductivity of the boundary layer, K_{eff} was calculated using Equation (5). Plane symmetry conditions were implemented in the model, to reduce the computation time.

PUF, with a density of 96 kg/m^3, was used as the heat insulation material. The thermal conductivity of PUF according to the temperature is presented in Table 3. The physical properties for the FEA analysis are listed in Table 4. The initial temperature was that obtained experimentally for each FR, and the natural convection condition was assigned to the outside. The BOR was predicted by calculating the aggregate heat flux flowing into the tank. The mesh convergence test was performed to determine the size of the element. The heat flux between the liquid and gas at 50% FR was compared for different mesh sizes. As shown in Figure 5, the heat flux converged for an element size of approximately 10 mm. Therefore, in the heat transfer analysis, the element type was set as DC3D8 in ABAQUS, and the generated mesh is shown in Figure 6.

Table 3. Material property for thermal analysis.

Item	Temperature (K)	Thermal Conductivity (mW/m-K)	Density (kg/m^3)	Specific Heat (J/kg-K)
PUF	100	0.0163	96	1500
	140	0.0212		
	170	0.0246		
	200	0.0266		
	230	0.0248		
Stainless 304L	100	9.75	7860	499
	150	11.55		
	200	12.89		
	250	13.9		
Plywood	-	0.13	880	1260

Table 4. Test results of BOR experiment.

FR (%)	M_m	M_s	BOR(%)
98	93.5		-
90	85.9		28.27
80	76.3		29.74
70	66.8		38.22
60	57.2	95.41	42.61
50	47.7		43.14
40	38.2		43.15
30	28.6		42.68
20	19.1		41.08
10	9.5		36.91

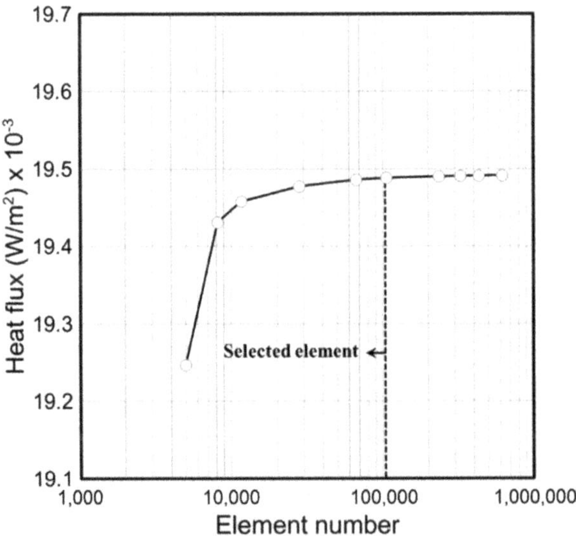

Figure 5. Results of mesh convergence test.

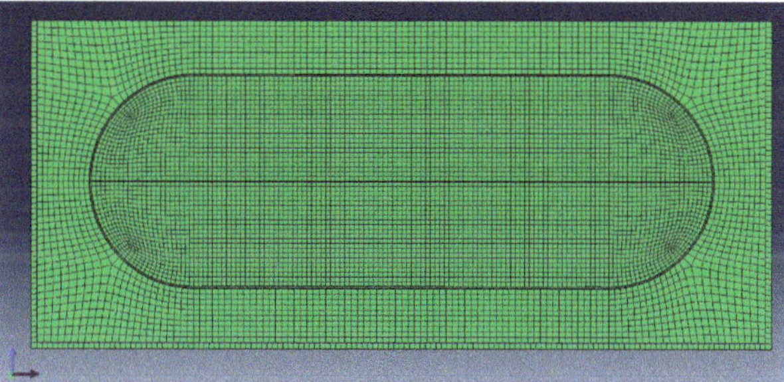

Figure 6. Elements of analysis model.

4. Results and Discussion

4.1. Experiment Results

As shown at Figure 7, liquid nitrogen was present at 77 K at atmospheric pressure; however, the temperature converged at 100 K owing to the increase in the saturation temperature after the pressurization process. During the initial 20 h, the pressure inside the tank increased to a set pressure of 1.5 MPa, and the temperature of the liquid increased accordingly. At this stage, because the convection of the vapor did not occur, isothermal temperature behavior was observed. Vaporized convection occurred near TC 10, corresponding to the 98% level, resulting in a rise in the temperature. Thereafter, due to the circulation caused by the temperature difference in the liquid in the upper and lower parts, the isothermal temperature distribution did not occur. In particular, the temperature variation due to the phase changes was not significant in the bounder layer, owing to the temperature distribution generated by the central jet in the stratified section. After 55 h, the temperature inside the tank rose sharply, and most of the liquid evaporated, thereby forming a pressure tank.

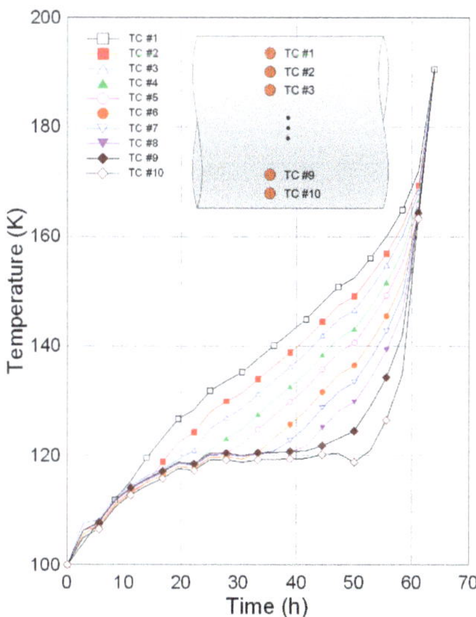

Figure 7. Time dependent temperature distributions of inner tank at each point.

The initial pressurization reduced the weight loss owing to the increase in the temperature of the liquid nitrogen, thereby inhibiting the vaporization, which decreased the BOR. The WS was used to determine the difference in the BOR over time, and the changes in the mass over time were as shown in Figure 8. Except in the stages immediately after the buffering and immediately before the exhaustion, a linear behavior was observed. As time elapsed, the pressure was maintained at 1.5 MPa, and a certain amount of BOG was released. The BOR for the different mass changes was as defined in Equation (11):

$$BOR = \frac{dM_m/dt}{M_s} \times 24 \times 100 (\%) \tag{11}$$

where M_m is the weight of the liquid nitrogen measured in real-time, and M_s is the weight of the total liquid nitrogen.

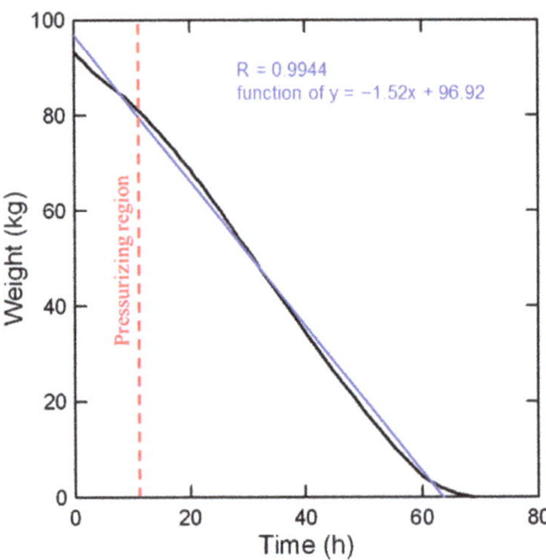

Figure 8. Time dependent liquid nitrogen weight.

In the context of the given period, the least count of the WS was limited; therefore, the time according to the FR for each 10% unit was set as dt, and the *BOR* values according to time are listed in Table 4. The *BOR* increased and later decreased as FR decreased, likely because of the active heat exchange with the liquids due to the augmented convection of the internal gases. As the pressure increased, the temperature of the saturated vapor increased, thereby preventing the evaporation of the liquid. Therefore, when the pressure increased after buffering, the BOR was smaller than that at the other FRs.

Figure 9 shows the temperature profile with time, for the different levels. Immediately after the liquid nitrogen buffering (0 h), the vertical temperature profile was almost identical. After 10 h, despite the formation of internal vaporization gases, the vertical temperature profile pertaining to the increase in the temperature of the liquid did not change significantly during the tank pressurization [43]. Over time, a thermal stratification region occurred at the liquid and gas boundaries, resulting in a certain temperature distribution. When half of the liquid nitrogen was exhausted at 30 h, a temperature difference occurred near the height of 200 mm, and a thermal strain was observed from the tank bottom, even though a certain amount of liquid nitrogen remained after 50 h.

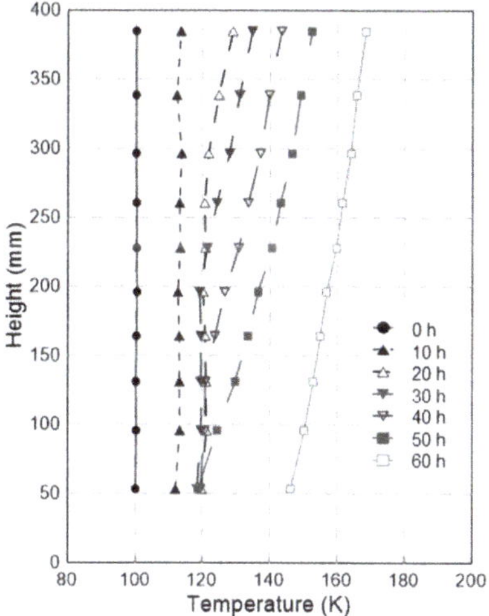

Figure 9. Vertical temperature profile over time.

4.2. Numerical Analysis Results and Prediction of BOR

The thermal analysis of the cubic fuel tank for ships was performed using a numerical method via commercial finite element codes. The temperature of the internal liquid nitrogen was set as the initial boundary condition, as obtained experimentally. Figure 10a–d show the temperature distribution of the tank according to the numerical analysis results for each FR. As shown in Figure 10a, a large temperature gradient occurred in the vapor layer, and a temperature of 140 K was noted at the top of the tank. However, as shown in Figure 10d, the temperature gradient of the gas part was not significant, and a temperature distribution of 110 K to 130 K was observed at the top of the tank. In the insulation temperature distribution, a temperature gradient occurred along the cylindrical part of the tank. At the corners of the insulation system, the temperature was similar to the ambient temperature.

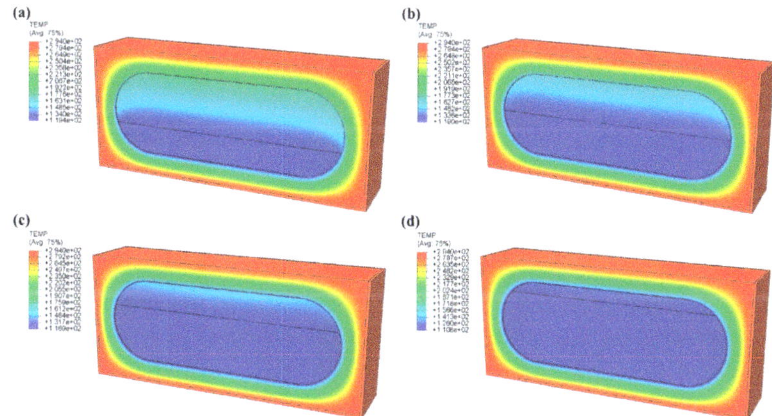

Figure 10. Temperature distributions of analysis model at each FR (**a**) 30% (**b**) 50% (**c**) 70% (**d**) 90%.

Figure 11 shows the comparison of the predicted and experimentally obtained vertical temperature profiles for each FR. The temperature gradient for the gas parts was larger than that for the liquid part, likely owing to the assumptions considered in the thermal conductivity model. Because the momentum of the gas due to the evaporation of the liquid was not considered, the temperature at the top of the tank, as obtained using the thermal conductivity model, was higher than the experimentally obtained value. This tendency was more notable at smaller FRs. Nevertheless, because the effective thermal conductivity was applied to the adjacent layers of the gas boundary to predict the BOR through FEA, a realistic heat ingress value was expected to be obtained.

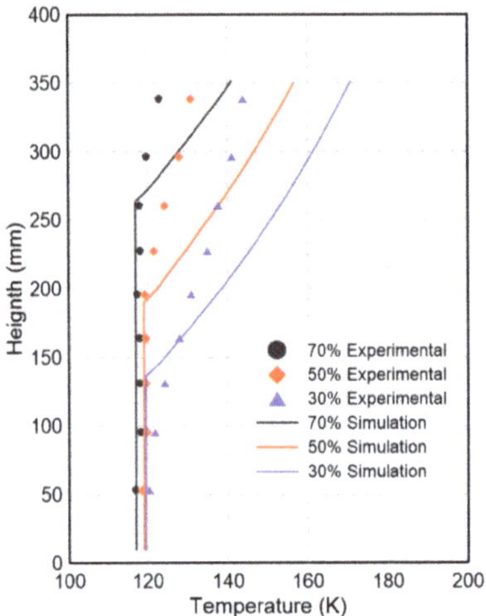

Figure 11. Comparison between experiments and numerical analysis for vertical temperature profile.

The BOR was calculated by combining the heat flux from the contact surface of the tank to the gas. The BOR obtained through the simulation and experiment are shown in Figure 12. In both the settings, as the FR decreased, the BOR increased and later decreased under a certain volume. The experimental BOR was slightly lower than the simulation value for the FR ranging from 70% to 90%. As the initial design pressure was set as 1.5 MPa, the BOG was not generated immediately after the experiment commenced. The maximum error of 19% was observed at an FR of approximately 90%, likely because of the ambiguity of the buffering point in the time measurement during the experiment. As the FR decreased, K_{eff} increased, resulting in a higher BOR. When the FR increased to 50%, the area of contact in the liquid area reduced, and the resulting heat inflow reduced.

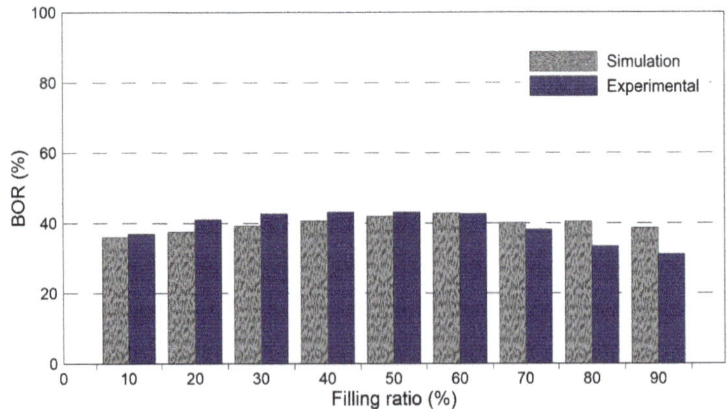

Figure 12. Comparison of BOR with simulation and experimental.

5. Conclusions

Experimental and numerical analyses were conducted to predict the BOR of a cubic IMO type C independent tank for small ships. In the experiment, the BOR was measured using a WS, and TCs were welded on the tank to analyze the surface temperature according to the FR. The BOR of the designed cubic tank was slightly larger than that for the LNG tank for commercial ships by approximately 35; nevertheless, this difference can likely be reduced by changing the thickness and density of the insulation material. The measured BOR using FEA is expected to have an error rate of less than 10%. The following key conclusions were derived:

(1) In the early stage of the BOR experiment, the rise in the pressure inside the tank was dominant due to the gas generation owing to the evaporation. This phenomenon increased the saturation temperature of the liquid nitrogen, and the internal liquid temperature converged to 105 K. Therefore, under high FRs, the amount of BOG generation was the smallest.

(2) When analyzing the finite element of the fuel tank through the thermal conductivity model, the error can be reduced by applying the effective thermal conductivity value to the boundary layer of the gas instead of considering the hydrodynamic behavior. In particular, in this work, the BOR prediction was relatively accurate during the pressure convergence period compared to that in the experiment. In this regard, it is necessary to verify the effective thermal conductivity values depending on various empirical formulas and to consider the parameters for the section in which the pressure changes.

Author Contributions: D.-H.L. and S.-J.C. provided and designed experimental ideas for the wrote article. J.-D.K., J.-H.K. and S.-K.K. contributed in design, analyze and discuss the results and wrote the article. J.-M.L. supervised the entire investigation. All authors have read and agreed to the published version of the manuscript.

Funding: This work was supported by the R&D Platform Establishment of Eco-Friendly Hydrogen Propulsion Ship Program (No. 20006632, 20006644) funded by the Ministry of Trade, Industry & Energy (MOTIE, Korea).

Institutional Review Board Statement: Not applicable.

Informed Consent Statement: Not applicable.

Data Availability Statement: The data presented in this study are available on request from the corresponding author.

Conflicts of Interest: The authors declare no conflict of interest.

References

1. Herdzik, J. Emissions from marine engines versus IMO certification and requirements of tier 3. *J. KONES* **2011**, *18*, 161–167.
2. Azzara, A.A.; Rutherford, D.; Wang, H. Feasibility of IMO Annex VI Tier III implementation using Selective Catalytic Reduction. *Int. Counc. Clean Transp.* **2014**, *9*, 161–167.
3. Lee, C.S.; Cho, J.R.; Kim, W.S.; Noh, B.J.; Kim, M.H.; Lee, J.M. Evaluation of sloshing resistance performance for LNG carrier insulation system based on fluid-structure interaction analysis. *Int. J. Nav. Archit. Ocean Eng.* **2013**, *5*, 1–20. [CrossRef]
4. Migliore, C.; Salehi, A.; Vesovic, V. A non-equilibrium approach to modelling the weathering of stored Liquefied Natural Gas (LNG). *Energy* **2017**, *124*, 684–692. [CrossRef]
5. Yoo, B.Y. The development and comparison of CO_2 BOG re-liquefaction processes for LNG fueled CO_2 carriers. *Energy* **2017**, *127*, 186–197. [CrossRef]
6. Schinas, O.; Butler, M. Feasibility and commercial considerations of LNG-fueled ships. *Ocean Eng.* **2016**, *122*, 84–96. [CrossRef]
7. Lee, H.J.; Yoo, S.H.; Huh, S.Y. Economic benefits of introducing LNG-fuelled ships for imported flour in South Korea. *Transp. Res. Part D Transp. Environ.* **2020**, *78*, 102220. [CrossRef]
8. Kwak, D.H.; Heo, J.H.; Park, S.H.; Seo, S.J.; Kim, J.K. Energy-efficient design and optimization of boil-off gas (BOG) re-liquefaction process for liquefied natural gas (LNG)-fuelled ship. *Energy* **2018**, *148*, 915–929. [CrossRef]
9. Kim, D.; Hwang, C.; Gundersen, T.; Lim, Y. Process design and economic optimization of boil-off-gas re-liquefaction systems for LNG carriers. *Energy* **2019**, *173*, 1119–1129. [CrossRef]
10. Romero Gómez, J.; Romero Gómez, M.; Lopez Bernal, J.; Baaliña Insua, A. Analysis and efficiency enhancement of a boil-off gas reliquefaction system with cascade cycle on board LNG carriers. *Energy Convers. Manag.* **2015**, *94*, 261–274. [CrossRef]

11. Krikkis, R.N. A thermodynamic and heat transfer model for LNG ageing during ship transportation. Towards an efficient boil-off gas management. *Cryogenics* **2018**, *92*, 76–83. [CrossRef]
12. Lee, J.S.; You, W.H.; Yoo, C.H.; Kim, K.S.; Kim, Y. An experimental study on fatigue performance of cryogenic metallic materials for IMO type B tank. *Int. J. Nav. Archit. Ocean Eng.* **2013**, *5*, 580–597. [CrossRef]
13. Kim, D.H.; Jeong, H.B.; Choi, S.H.; Jo, Y.C.; Kim, D.E.; Jeong, T. Structural assessment under sloshing impact for the IMO type b independent LNG tank. *Proc. Int. Offshore Polar. Eng. Conf.* **2014**, *3*, 180–185.
14. Kim, T.W.; Kim, S.K.; Park, S.B.; Lee, J.M. Design of Independent Type-B LNG Fuel Tank: Comparative Study between Finite Element Analysis and International Guidance. *Adv. Mater. Sci. Eng.* **2018**, *2018*. [CrossRef]
15. Adom, E.; Islam, S.Z.; Ji, X. Modelling of Boil-off Gas in LNG tanks: A case study. *Int. J. Eng. Technol.* **2010**, *2*, 292–296.
16. Qu, Y.; Noba, I.; Xu, X.; Privat, R.; Jaubert, J.N. A thermal and thermodynamic code for the computation of Boil-Off Gas–Industrial applications of LNG carrier. *Cryogenics* **2019**, *99*, 105–113. [CrossRef]
17. Kulitsa, M.; Wood, D.A. LNG rollover challenges and their mitigation on Floating Storage and Regasification Units: New perspectives in assessing rollover consequences. *J. Loss Prev. Process Ind.* **2018**, *54*, 352–372. [CrossRef]
18. Lee, J.-S.; Kim, K.-S.; Kim, Y. Development of an insulation performance measurement unit for full-scale LNG cargo containment system using heat flow meter method. *Int. J. Nav. Archit. Ocean Eng.* **2018**, *10*, 458–467. [CrossRef]
19. Gavelli, F.; Bullister, E.; Kytomaa, H. Application of CFD (Fluent) to LNG spills into geometrically complex environments. *J. Hazard. Mater.* **2008**, *159*, 158–168. [CrossRef]
20. Zakaria, M.S.; Osman, K.; Saadun, M.N.A.; Manaf, M.Z.A.; Mohd Hanafi, M.H. Computational simulation of boil-off gas formation inside liquefied natural gas tank using evaporation model in ANSYS fluent. *Appl. Mech. Mater.* **2013**, *393*, 839–844. [CrossRef]
21. Saleem, A.; Farooq, S.; Karimi, I.A.; Banerjee, R. A CFD simulation study of boiling mechanism and BOG generation in a full-scale LNG storage tank. *Comput. Chem. Eng.* **2018**, *115*, 112–120. [CrossRef]
22. Lee, J.H.; Kim, Y.J.; Hwang, S. Computational study of LNG evaporation and heat diffusion through a LNG cargo tank membrane. *Ocean Eng.* **2015**, *106*, 77–86. [CrossRef]
23. Ovidi, F.; Pagni, E.; Landucci, G.; Galletti, C. Numerical study of pressure build-up in vertical tanks for cryogenic flammables storage. *Appl. Therm. Eng.* **2019**, *161*, 114079. [CrossRef]
24. Scurlock, R.G. *Stratification, Rollover and Handling of LNG, LPG and Other Cryogenic Liquid Mixtures*; Springer: Berlin/Heidelberg, Germany, 2015.
25. Kang, M.; Kim, J.; You, H.; Chang, D. Experimental investigation of thermal stratification in cryogenic tanks. *Exp. Therm. Fluid Sci.* **2018**, *96*, 371–382. [CrossRef]
26. Lin, Y.; Ye, C.; Yu, Y.-y.; Bi, S.-w. An approach to estimating the boil-off rate of LNG in type C independent tank for floating storage and regasification unit under different filling ratio. *Appl. Therm. Eng.* **2018**, *135*, 463–471. [CrossRef]
27. Tsili, M.A.; Amoiralis, E.I.; Kladas, A.G.; Souflaris, A.T. Power transformer thermal analysis by using an advanced coupled 3D heat transfer and fluid flow FEM model. *Int. J. Therm. Sci.* **2012**, *53*, 188–201. [CrossRef]
28. Shen, Y.; Zhang, B.; Xin, D.; Yang, D.; Peng, X. 3-D finite element simulation of the cylinder temperature distribution in boil-off gas (BOG) compressors. *Int. J. Heat Mass Transf.* **2012**, *55*, 7278–7286. [CrossRef]
29. Mohanraj, M.; Jayaraj, S.; Muraleedharan, C. Applications of artificial neural networks for thermal analysis of heat exchangers–a review. *Int. J. Therm. Sci.* **2015**, *90*, 150–172. [CrossRef]
30. Giannett, N.; Redo, M.A.; Jeong, J.; Yamaguchi, S.; Saito, K.; Kim, H. Prediction of two-phase flow distribution in microchannel heat exchangers using artificial neural network. *Int. J. Refri.* **2020**, *111*, 53–62. [CrossRef]
31. Kim, J.H.; Park, W.S.; Chun, M.S.; Kim, J.J.; Bae, J.H.; Kim, M.H.; Lee, J.M. Effect of pre-straining on low-temperature mechanical behavior of AISI 304L. *Mater. Sci. Eng. A* **2012**, *543*, 50–57. [CrossRef]
32. Convention, I. *1983 Amendments to the International Convention for the Safety of Life International Code for the Construction and Equipment of Ships Carrying Liquefied Gases in Bulk*; (IGC Code) International Code for the Construction and Equipment of Ships Carrying Liq: London, UK, 1998; p. 15.
33. Lee, J.; Choi, Y.; Jo, C.; Chang, D. Design of a prismatic pressure vessel: An engineering solution for non-stiffened-type vessels. *Ocean Eng.* **2017**, *142*, 639–649. [CrossRef]
34. Park, S.B.; Lee, C.S.; Choi, S.W.; Kim, J.H.; Bang, C.S.; Lee, J.M. Polymeric foams for cryogenic temperature application: Temperature range for non-recovery and brittle-fracture of microstructure. *Compos. Struct.* **2016**, *136*, 258–269. [CrossRef]
35. Kim, J.H.; Choi, S.W.; Park, D.H.; Park, S.B.; Kim, S.K.; Park, K.J.; Lee, J.M. Effects of cryogenic temperature on the mechanical and failure characteristics of melamine-urea-formaldehyde adhesive plywood. *Cryogenics* **2018**, *91*, 36–46. [CrossRef]
36. Ma, H.; Cai, W.; Zheng, W.; Chen, J.; Yao, Y.; Jiang, Y. Stress characteristics of plate-fin structures in the cool-down process of LNG heat exchanger. *J. Nat. Gas Sci. Eng.* **2014**, *21*, 1113–1126. [CrossRef]
37. Li, X.; Xie, G.; Wang, R. Experimental and numerical investigations of fluid flow and heat transfer in a cryogenic tank at loss of vacuum. *Heat Mass Transf. Und. Stoffuebertragung* **2010**, *46*, 395–404. [CrossRef]
38. Woodfield, P.L.; Monde, M.; Mitsutake, Y. Measurement of Averaged Heat Transfer Coefficients in High-Pressure Vessel during Charging with Hydrogen, Nitrogen or Argon Gas. *J. Therm. Sci. Technol.* **2007**, *2*, 180–191. [CrossRef]
39. Joseph, J.; Agrawal, G.; Agarwal, D.K.; Pisharady, J.C.; Sunil Kumar, S. Effect of insulation thickness on pressure evolution and thermal stratification in a cryogenic tank. *Appl. Therm. Eng.* **2017**, *111*, 1629–1639. [CrossRef]

40. Bashiri, A.; Fatehnejad, L. Modeling and simulation of rollover in LNG storage tanks. *Int. Gas Union World Gas Conf. Pap.* **2006**, *5*, 2522–2528.
41. Jazayeri, S.A.; Khoei, E.M.H. Numerical comparison of thermal stratification due natural convection in densified LOX and LN2 tanks. *Am. J. Appl. Sci.* **2008**, *5*, 1773–1779. [CrossRef]
42. Ludwig, C.; Dreyer, M.E.; Hopfinger, E.J. Pressure variations in a cryogenic liquid storage tank subjected to periodic excitations. *Int. J. Heat Mass Transf.* **2013**, *66*, 223–234. [CrossRef]
43. Seo, M.; Jeong, S. Analysis of self-pressurization phenomenon of cryogenic fluid storage tank with thermal diffusion model. *Cryogenics* **2010**, *50*, 549–555. [CrossRef]

Journal of
*Marine Science
and Engineering*

Article

Visible Fidelity Collector of a Zooplankton Sample from the Near-Bottom of the Deep Sea

Jing Xiao [1], Jiawang Chen [1,2,*], Zhenwei Tian [1], Hai Zhu [1], Chunsheng Wang [3], Junyi Yang [4], Qinghua Sheng [4], Dahai Zhang [1] and Jiasong Fang [5]

[1]   Ocean College, Zhejiang University, Zhoushan 310027, China; 21834122@zju.edu.cn (J.X.);
      tzw10@zju.edu.cn (Z.T.); zjuzh@icloud.com (H.Z.); zhangdahai@zju.edu.cn (D.Z.)
[2]   Center for Evolution and Conservation Biology, Southern Marine Science and Engineering Guangdong
      Laboratory (Guangzhou), Guangzhou 511458, China
[3]   Marine Ecology and Environment Laboratory, Second Institute of Oceanography, MNR,
      Hangzhou 310012, China; wangsio@sio.org.cn
[4]   College of Electronics and Information, Hangzhou Dianzi University, Hangzhou 310018, China;
      junyiyang@hdu.edu.cn (J.Y.); sheng7@hdu.edu.cn (Q.S.)
[5]   Shanghai Engineering Research Center of Hadal Science and Technology College of Marine Sciences,
      Shanghai Ocean University, Shanghai 201306, China; jsfang@shou.edu.cn
*   Correspondence: arwang@zju.edu.cn; Tel.: +86-0580-2092214

Abstract: The multi-net visible fidelity zooplankton collector is designed to obtain near-bottom fidelity zooplankton. The collector is sent to the designated sampling location based on the information provided by the camera and altimeter. The host computer sends instructions to control the opening of the net port for sample collection and closing of the sampling cylinder cover after sampling. The collector contains three trawls so that three samples can be collected for each test, and environmental parameters can be collected simultaneously. After sampling, The sample maintains its fidelity, that is, maintaining the temperature and pressure of the seabed sample after sampling. Two experiments were carried out in the Western Pacific, and six bottles of zooplankton samples were successfully obtained. The development of a multi-net visible zooplankton collector is of great significance for the collection of near-bottom zooplankton.

Keywords: near-bottom zooplankton; multi-net; visible sampling; fidelity; deep sea

Citation: Xiao, J.; Chen, J.; Tian, Z.; Zhu, H.; Wang, C.; Yang, J.; Sheng, Q.; Zhang, D.; Fang, J. Visible Fidelity Collector of a Zooplankton Sample from the Near-Bottom of the Deep Sea. *J. Mar. Sci. Eng.* **2021**, *9*, 332. https://doi.org/10.3390/jmse9030332

Academic Editor: Antonio Mancuso and Davide Tumino

Received: 13 January 2021
Accepted: 16 March 2021
Published: 17 March 2021

Publisher's Note: MDPI stays neutral with regard to jurisdictional claims in published maps and institutional affiliations.

1. Introduction

Knowledge of the biology and ecology of deep-sea organisms is still very limited compared with all other marine ecosystems [1,2]. Among all deep-sea habitats and domains, knowledge of the planktonic component is far more limited than for the benthic counterpart [3]. Zooplankton biodiversity decreases with increasing water depth, but the equitability increases [4]. The high cost of shipping times and technologies to operate in deep-sea environments makes it difficult to conduct oceanographic sampling [5]. This is particularly evident for investigations on deep-sea zooplankton [6]. Meso- and macro-zooplankton organisms play a key role in biological processes in all marine ecosystems, being a "linkage" between phytoplankton/micro-zooplankton and the higher trophic levels. In addition, zooplankton organisms are able, through vertical migration, to contribute to the functional linkage between the photic zone and the dark deep ocean [7–9].

Sophisticated sampling systems are now available to quantify the abundance of planktonic organisms [10]. The development of electronically controlled multiple net units designed to allow sampling in discrete depth strata has revolutionized our ability to determine the vertical structure and depth-integrated abundance of zooplankton. Less sophisticated net-based sampling devices, however, remain in widespread use, both because of their low cost and their ease of deployment. The well-known ones were developed by the U.S. GLOBEC program (U.S. GLOBEC is a multi-disciplinary research program designed by

41

oceanographers, fishery scientists, and marine ecologists) with the Bongo and the Multiple Opening and Closing Net Environmental Sensing System (MOCNESS) [11,12] and Bedford Institute of Oceanography Net and Environmental Sampling System (BIONESS) [13]. One sampling type adopts a louvered structure, uses the weight of the net pole to realize the opening and closing of the net port, and uses the diagonal drag method for sampling. The tripping of the net pole is controlled by a device composed of a stepping motor, and the net port is opened when the net pole moves to the bottom of the frame. However, if the sea conditions are not good, the net pole cannot maintain its balance during the fall process and it will get stuck. Another sampling type is Multinet [14], which uses an "inverted L" type opening and closing method.

However, these collectors do not consider the preservation of samples with fidelity, that is, they do not store the collected samples with in-situ insulation and pressure preservation. A sample collected in this way is very different from an in-situ sample. Therefore, research on a visible sampling technique for near-bottom fidelity zooplankton in the deep sea will have important theoretical and practical significance for obtaining live hydrothermal zooplankton.

2. Methods

2.1. Multi-Net Visible Collector

The specific indicators of the developed multi-net visible zooplankton collector are as follows:

- Maximum design working depth: 4000 m;
- Number of trawl nets: 3;
- Network port area: 0.5 m^2
- The effective volume of the sample barrel: $\geq$0.25 L;
- The pressure in the sample storage bin shall not be lower than 80% of the original pressure at the sampling point within 6 h after boarding;
- The temperature rise in the sample storage bin after boarding does not exceed the original temperature of 8 °C;
- Hybrid transmission of the underwater power supply and data images are realized with transmission power $\geq$1.5 kw and transmission distance $\geq$6 km;
- The environmental parameters transmitted in real time include temperature, salinity, depth, turbidity, and dissolved oxygen;
- The total weight of the trailer body: 1.2 tons;
- Outer frame size: 1.5 m × 1 m × 2 m.

2.1.1. Sampling Principle

The multi-net visible zooplankton collector consists of two parts: the main body and the sample collection. The overall structure diagram and main body diagram are shown in Figures 1 and 2. After the research vessel reaches the working position, the collector is lowered into the water and sent to the designated sampling location based on the information provided by the camera and altimeter. The motor is sent a command to open a net port through the spring trigger opening and closing mechanism (Figure 3). After the net port is opened, the opening and closing mechanism sends feedback information, the motor stops rotating, and the sampling operation starts. After sampling, the fuse is energized and blown by the command sent, and the sampling cylinder cover is closed under the action of the torsion spring. After the cylinder cover is closed, the opening and closing net mechanism closes the net port under the drive of the motor, and the first sampling ends. After the first sampling is over, the collector can be dragged to the next sampling location, and the above steps are repeated for the second and third sampling in sequence, or the collector is taken out of the water directly.

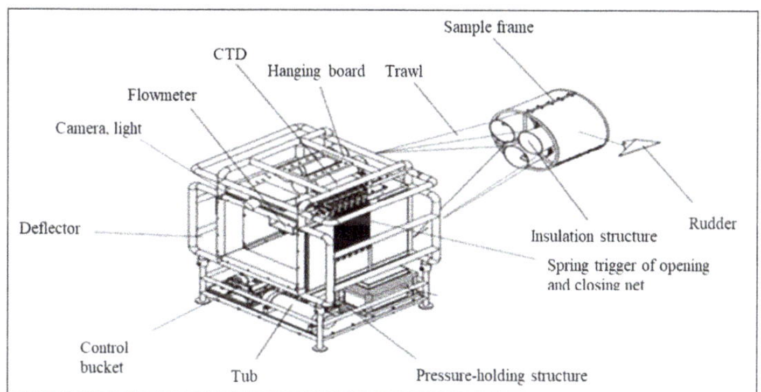

Figure 1. The overall structure of the multi-link visual control large-caliber trawl system.

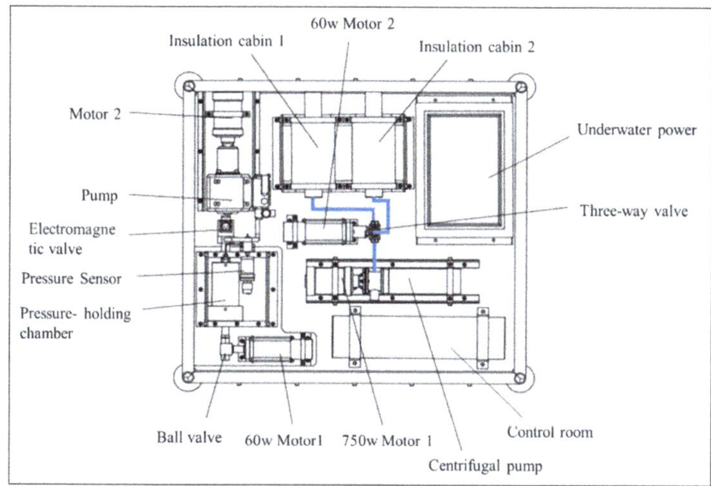

Figure 2. The main body of the multi-link visual control large-caliber trawl system.

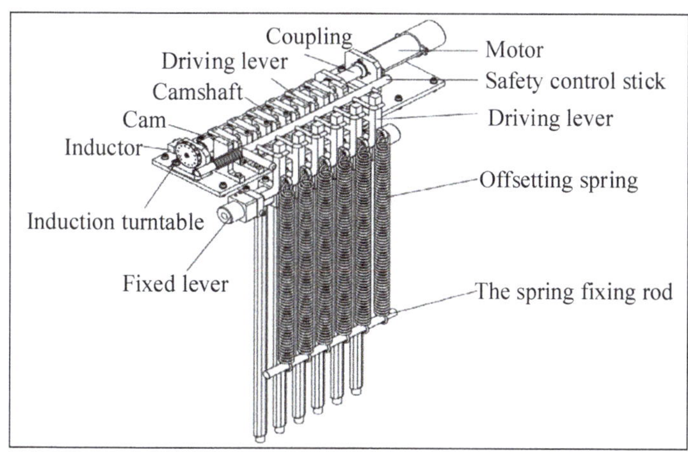

Figure 3. Net opening and closing mechanism.

2.1.2. The Spring-Triggered Switch

The spring-triggered opening and closing structure (Figure 3) includes a frame, a 60 W brushless DC motor, a camshaft with six cams, three nets, six net port levers, an induction turntable with 14 magnets (magnetic N and S poles are installed crosswise), a sensor, and a locking mechanism with six paddles.

The three net ports are four-sided when they are opened. Two sides are fixed on the trawl support rods, and the other two sides are fixed on two net port levers (one is the opening lever and the other is the closing lever). In the driving mechanism of the opening and closing net (Figure 4), the motor is connected to the camshaft through a coupling, and six cams and an empty station are evenly distributed on the camshaft. The other end of the camshaft is connected to the sensing structure (Figure 5), and the sensor is installed on the sensing plate. The locking mechanism (Figure 6) is fixed on the frame, one end is in contact with the cam, and the other end is locked by the hook. The six levers are fixed on the frame through the middle hole to form a lever mechanism. The other side is connected to the frame by a spring, and the spring is in a stretched state in a non-working state.

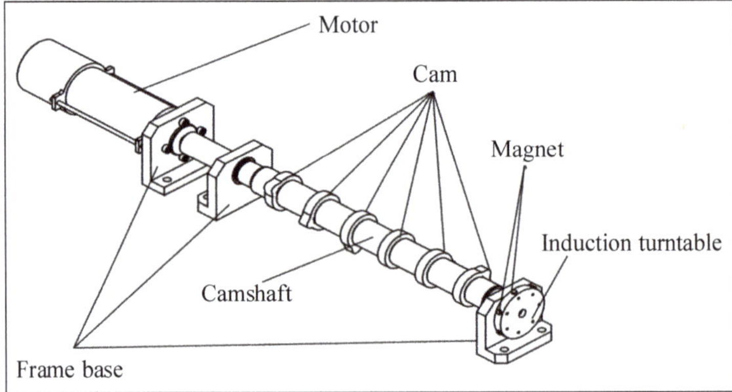

Figure 4. Driving schematic diagram of the net opening and closing mechanism.

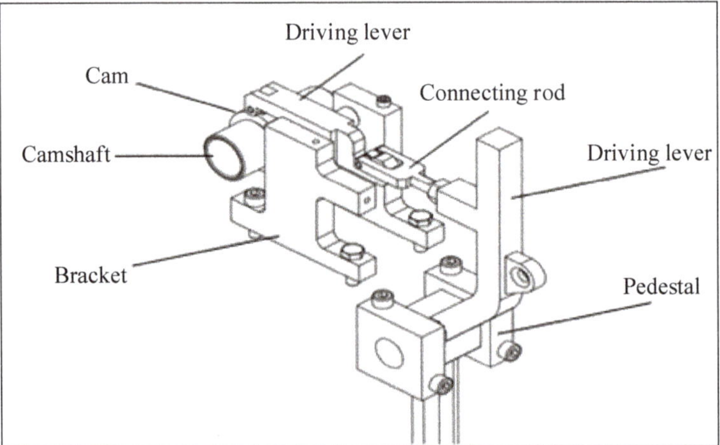

Figure 5. Schematic diagram of locking device of opening and closing net mechanism.

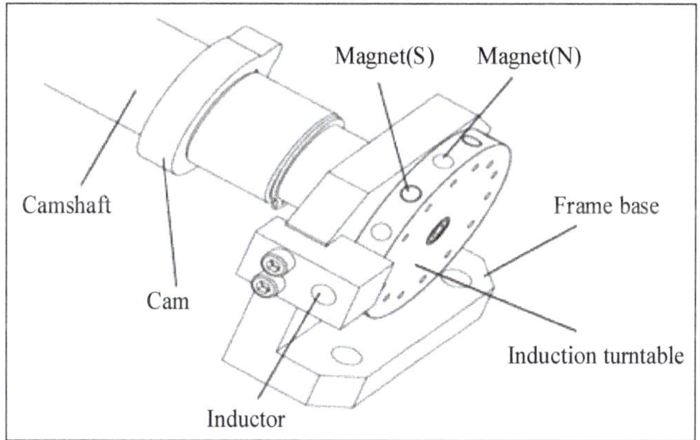

Figure 6. Schematic diagram of the induction device of the net opening and closing mechanism.

2.2. Temperature Retention System

In order to maintain the original temperature of the zooplankton sample, a sample collection cylinder that can maintain temperature is designed as shown in Figure 7 (the left picture is the front and the right is the back). This part includes an outer frame, three insulation barrels, and a steel wire fixing seat. The sample collector is designed based on the principle of double-layer water bath insulation, and the base material adopts engineering plastics, which has a good insulation effect. The tail sample collection device uses plastic with poor thermal conductivity, uses a solenoid valve to trigger the collection of biological samples, and uses a water bath for sample insulation.

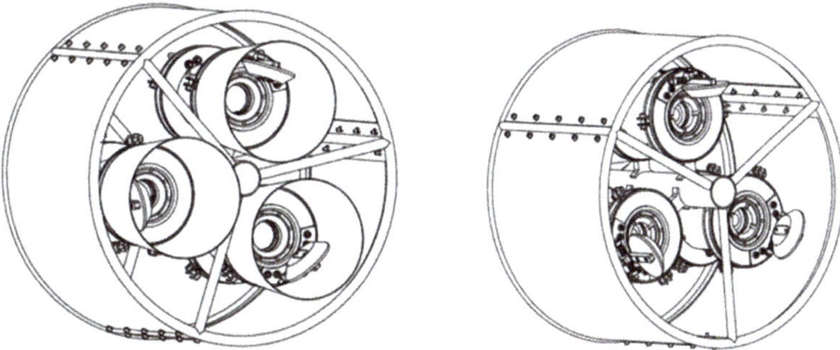

Figure 7. Schematic diagram of the sample collection cylinder structure.

The structural cross-sectional view of the temperature-retaining cylinder is shown in Figure 8. It is divided into an outer tube and an inner tube with a filter screen. A cylinder cover is arranged on each end cover, and a torsion spring is installed between each cylinder cover and the end cover. When the cylinder cover is opened, the two covers are connected by steel wire and a fuse, and the torsion spring is in a compressed state. When the fuse is energized and blown, the two cylinder covers are closed under the action of their respective torsion springs, and the sea water and zooplankton samples are sealed in the cylinder.

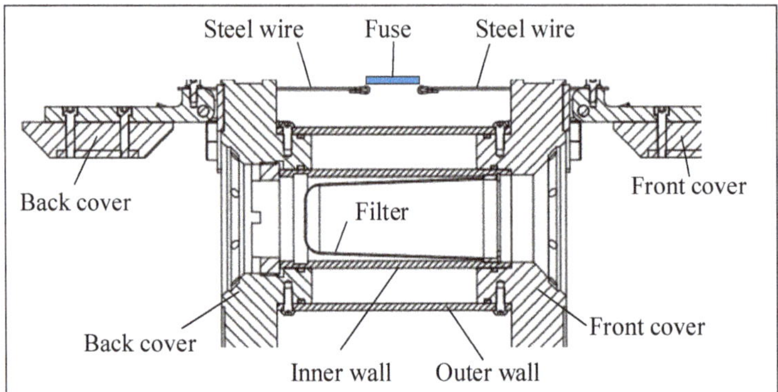

Figure 8. Cross-sectional view of the insulation tube.

2.3. Control System

The monitoring system of the visible collector mainly includes a deck monitoring unit, an optical cable communication machine, an underwater optical fiber communication module, an embedded control system main board, and driving cabins. The system structure diagram is shown in Figure 9.

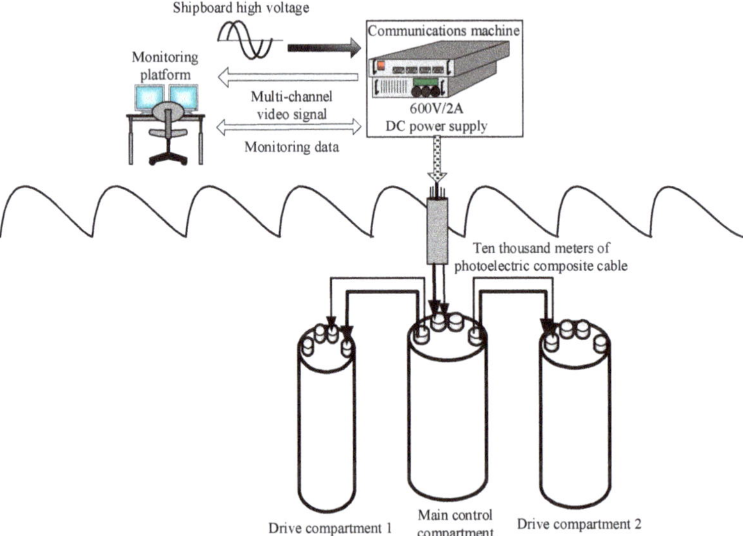

Figure 9. Sampling monitoring system structure diagram.

The power supply system converts the shipborne AC high-voltage power into a 600 V/2 A DC stabilized power supply through a DC stabilized power supply and transmits the power to the subsea equipment through an armored photoelectric composite cable. The video image signal of the seabed, the sensor data of the seabed uplink and the feedback data of the action execution result, and the control command data of the upper computer's downlink are transmitted to the seabed through the optical fiber. The monitoring platform realizes the real-time monitoring of the direct-view sampling of near-bottom zooplankton and is responsible for controlling the operation status of the seabed trawl based on seabed monitoring data and video images, as well as drawing the real-time change curve of CTD temperature, conductivity, and pressure (depth) data.

The principle block diagram of the main control cabin is shown in Figure 10, which mainly shows the transmission of direct current, the detection of water leakage, inclination, and flow conditions, the switch control of underwater lights, underwater cameras and bottom altimeters, and the upload of underwater video signals and underwater data.

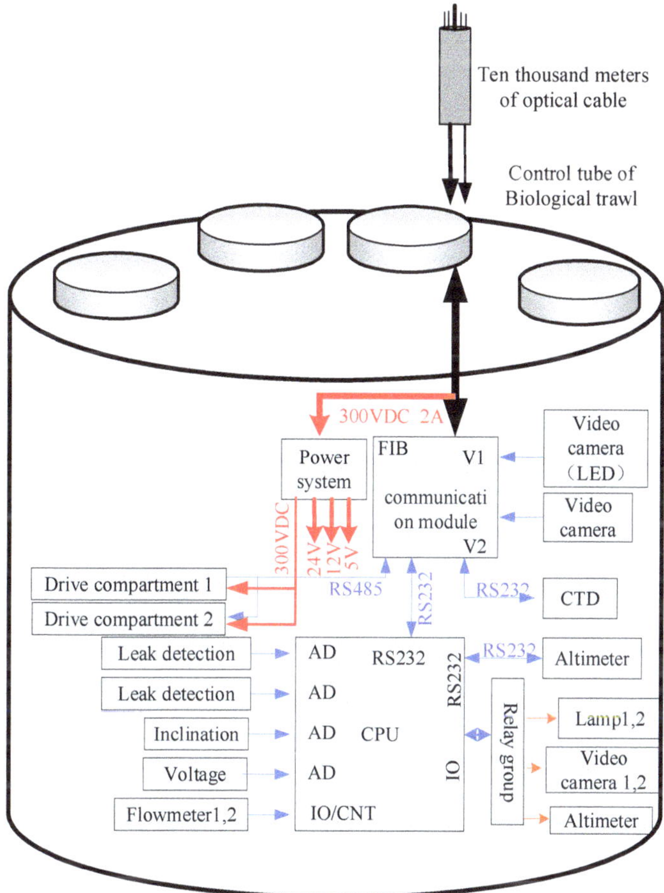

Figure 10. Block diagram of the main control cabin.

3. Results

3.1. The Field Assessment

The multi-net visible zooplankton collector was set up in two sampling stations with water depths exceeding 3000 m in the Northwest Pacific. The sampling device was tested on deck before launching. Through the photoelectric composite cable, the following were carried out: communication test, network port opening and closing test, ball valve opening and closing test, and the altimeter, camera lighting system, flowmeter detection and other functional testing tests. After the research vessel arrived at the predetermined station, its course was adjusted to the top wind and top current, maintaining a speed of 1.5–3 knots with straight sailing. The cable was unwound at a speed of about 30 m/min to a depth of 3000 m, and then the cable was slowly unwound to a distance of 5–10 m from the seabed, after which the cable laying was terminated. The following steps were conducted: Turn on the CTD and other sensors to record and synchronize environmental data during the unwinding process. Keep the speed and heading stable during the near-bottom operation. Turn on the camera and underwater lights, as well as the video and data recording system.

When the multi-net visible zooplankton collector is tested near the seabed, open the first layer of nets, start trawling for 15 min, close the first layer of nets and open the second layer of nets. Continue to trawl for 15 min, close the second layer of nets and open the third layer of nets. After 15 min of trawling, close the third layer of nets. When the sampling is over, start to recover the collector, raise the cable at a speed of about 30 m/min, and slow down when the collector comes out of the water. After the collector is recovered on deck, check the pressure and temperature in the sample cylinder and determine whether the sample is successfully collected to verify the feasibility of the collector.

MACTD16-1-BPTV station carried out the first operation in the seamount area of the Northwest Pacific, with a maximum working depth of 3487 m and an accumulated underwater working time of about 4 h. MACTD16-BPTVA station carried out the second operation in the seamount area of the Northwest Pacific, with a maximum working depth of 3764 m; the cumulative underwater working time was about 4 h. During the two operations, a total of six samples of near-bottom zooplankton were collected, and synchronized video and environmental data were obtained. The collector was kept in good condition during the operation. The test site is shown in Figure 11.

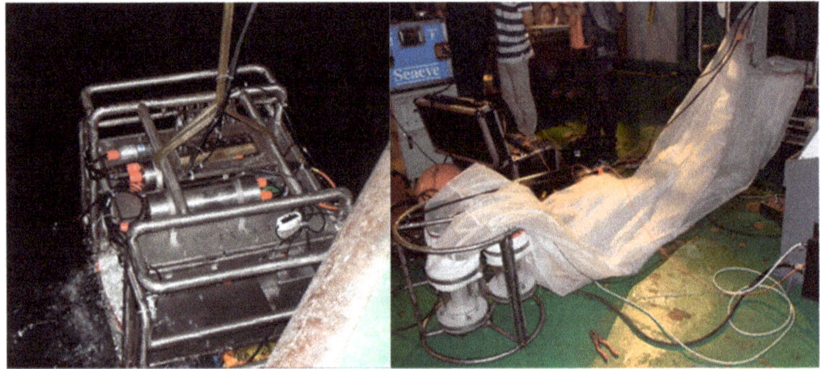

Figure 11. The sea trial site of the collector.

3.2. Experimental Results

The multi-net visible zooplankton collector was used to conduct two sets of experiments at two stations. After the sampling was completed, the data obtained through CTD included the water depth, pressure, and temperature of each station at the time of sampling. The records are shown in Table 1.

Table 1. Data record table.

Sample No.	Sampling Time	Working Depth	Pressure (MPa)	Temperature (°C)	Pressure (MPa) (6 h Later)	Temperature (°C) (6 h Later)	Sampling
1	15 min	2685	26.00	3.369	24.2	10.10	Yes
2	15 min	3487	34.00	3.020	31.0	7.60	Yes
3	15 min	3016	0.35	17.200	/	/	Yes
1	15 min	3205	32.00	3.112	29.8	8.10	Yes
2	15 min	3764	37.00	3.031	34.2	7.78	Yes
3	15 min	3341	33.00	3.117	29.5	7.82	Yes

It can be seen from the data in Table 1 that in the first sampling experiment, the internal pressure was not maintained due to the closure failure of the No. 3 network port, and the temperature was also the sea surface temperature at that time. After inspection and adjustment, the collector was lowered again. In the second sampling, the network ports of the three sample cylinders closed normally. The maximum working depth of the collector was about 3487 m. The sensors in the chambers showed that the pressure-retaining effect was good, and the pressure and temperature data were also obtained successfully. Six hours after the samples were taken, the pressure and temperature data of the sample cylinders were used to verify the performance of the device and the sample acquisition capabilities.

According to the pressure gauge, the pressure of the sample cylinders after 6 h was higher than 80% of the initial pressure. The water temperature on the seafloor was about 3 °C, and the maximum temperature in the sample cylinder after the collector was recovered on deck was 10.1 °C, an increase of 7.1 °C. The pressure and temperature data showed that the samples were maintained at their original pressure and temperature.

The fidelity samples were sent to a biological laboratory for research; the collected samples could not be viewed directly because opening the lid will cause the sample to lose its original temperature and pressure. Through the samples remaining in the trawl net and in the sample cylinder No. 3 in the first test (Figure 12), it was determined that the sample was obtained successfully.

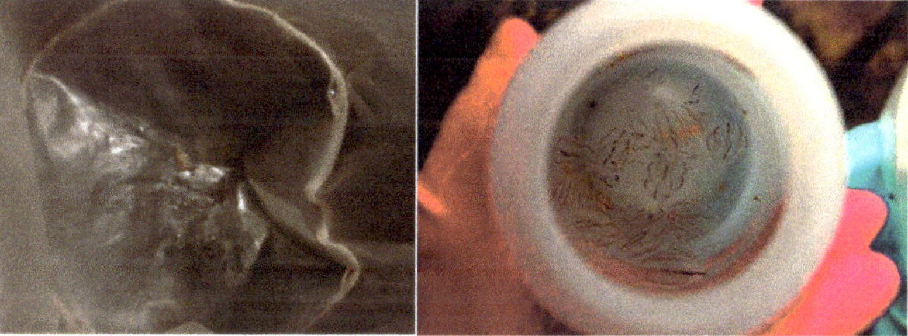

Figure 12. Samples in the trawl net and sample tube #3.

4. Discussion and Conclusions

4.1. Analysis of Failure to Close the Trawl Net Port

For this substantial sampling test, the sampling success rate of the two experiments was 100%, and the success rate of retaining the sample in the in-situ environment was 83.3%. The main reason that led to the failure of the No. 3 network in the first test was the failure of the No. 3 lever spring when it was reset. Every two levers control the opening and closing of a trawl, and each lever uses the contraction force of the spring to complete a 90° rotation to complete the opening and closing of the net port. As shown in Figure 13a, in the initial state, both pole 1 and pole 2 remain vertical, and the four nodes of the network port are fixed at four respective points: A, B, C, and D (the thick black line in Figure 13 represents the network port), where point A is fixed on the bottom plate, point B is on rod 1, and points C and D are on the upper and lower ends of rod 2. When the motor sends a rotation signal, the camshaft rotates by an angle to disengage the claw sleeve and the claw of pole 1. Rod 1 rotates 90° around the axis to reach the position shown in Figure 13b under the action of the spring force, so that net 1 is fully opened into a large "mouth" shape and enters the working state. When the motor gives a rotation signal again, the camshaft rotates by an angle again, which triggers pole 2 to rotate 90°, reaching the position shown in Figure 13c, and completing the closing of trawl 3. However, due to the failure of the spring reset, network port 3 is always kept open, so the temperature and pressure of the sample cannot be maintained.

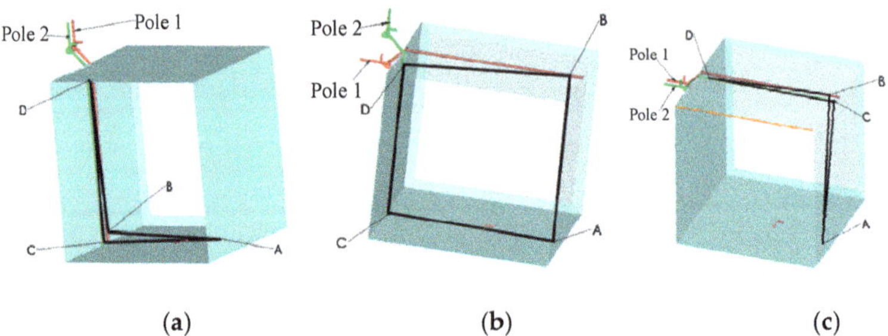

Figure 13. Samples in the trawl net and sample tube #3. (**a**) Initial state, (**b**) Working state, (**c**) Final state.

Since the work of the three trawls is relatively independent, their work sequence is determined by the movement of the levers. The bounce sequence and time interval of the levers can be freely controlled by arranging the position of the cam so that the opening and closing of the trawls can be controlled.

4.2. Conclusions

The multi-net visible zooplankton collector for near-bottom of the deep sea can carry out large-volume drag sampling and can maintain the pressure and temperature of the original environment of the sample. During the sampling process, the operation can be monitored by the camera, and environmental parameters can be collected simultaneously. In the Northwest Pacific experiment, five bottles of zooplankton samples were obtained, and the temperature and pressure of the samples were retained. The multi-net visible zooplankton collector provides a good start for obtaining near-bottom fidelity zooplankton and will be used for fidelity research of deeper-sea organisms.

Author Contributions: Conceptualization, J.X., J.C. and Z.T.; methodology, J.C.; software, J.X.; validation, Z.T., H.Z.; formal analysis, J.X., J.C.; investigation, H.Z.; resources, Z.T.; data curation, J.X.; writing—original draft preparation, J.X.; writing—review and editing, J.X., C.W.; visualization, J.Y.; supervision, Q.S.; project administration, D.Z.; funding acquisition, J.F. All authors have read and agreed to the published version of the manuscript.

Funding: The research was funded by the National Key R&D Program of China (grant number 2018YFC0310600) and Key Special Project for Introduced Talents Team of Southern Marine Science and Engineering Guangdong Laboratory (Guangzhou), China (GML2019ZD0506).

Institutional Review Board Statement: Not applicable.

Informed Consent Statement: Not applicable.

Data Availability Statement: Not applicable.

Acknowledgments: The study is based on the National Key R&D Program of China (2018YFC0310600) supported by the Ministry of Science and Technology of the People's Republic of China and the Key Special Project for Introduced Talents Team of Southern Marine Science and Engineering Guangdong Laboratory (Guangzhou), China (GML2019ZD0506).

Conflicts of Interest: We declare that we have no financial and personal relationships with other people or organizations that can inappropriately influence our work, and there is no professional or other personal interest of any nature or kind in any product, service and company. The manuscript entitled, "Visible fidelity collector of zooplankton sample from the near-bottom of the deep sea".

References

1. Danovaro, R.; Snelgrove, P.V.; Tyler, P. Challenging the paradigms of deep-sea ecology. *Trends Ecol. Evol.* **2014**, *29*, 465–475. [CrossRef] [PubMed]
2. Danovaro, R.; Corinaldesi, C.; Dell'Anno, A.; Snelgrove, P.V.R. The deep-sea under global change. *Curr. Biol.* **2017**, *27*, R1–R6. [CrossRef] [PubMed]
3. O'Dor, R.K.; Fennel, K.; Berghe, E.V. A one ocean model of biodiversity. *Deep-Sea Res. II* **2009**, *56*, 1816–1823. [CrossRef]
4. Danovaro, R.; Carugati, L.; Boldrin, A.; Calafat, A.; Canals, M.; Fabres, J.; Finlay, K.; Heussner, S.; Miserocchi, S.; Sanchez-Vidal, A. Deep-water zooplankton in the Mediterranean Sea: Results from a continuous, synchronous sampling over different regions using sediment traps. *Deep Sea Res. Part I Oceanogr. Res. Pap.* **2017**, *126*, 103–114. [CrossRef]
5. Brandt, A.; Gutt, J.; Hildebrandt, M.; Pawlowski, J.; Schwendner, J.; Soltwedel, T.; Thomsen, L. Cutting the umbilical: New technological perspectives in Benthic deep-sea research. *J. Mar. Sci. Eng.* **2016**, *4*, 36. [CrossRef]
6. Weikert, H.; Koppelmann, R.; Wiegratz, S. Evidence of episodic changes in deep-sea mesozooplankton abundance and composition in the Levantine sea Eastern Mediterranean. *Mar. Syst.* **2001**, *30*, 221–239. [CrossRef]
7. Bollens, S.M.; Rollwagen-Bollens, G.; Quenette, J.A.; Bochdansky, A.B. Cascading migrations and implications for vertical fluxes in pelagic ecosystems. *Plankton Res.* **2011**, *33*, 349–355. [CrossRef]
8. Siokou-Frangou, I.; Christaki, U.; Mazzocchi, M.G.; Montresor, M.; Ribera d'Alcalá, M.; Vaqué, D.; Zingone, A. Plankton in the open Mediterranean Sea: A review. *Biogeosciences* **2010**, *7*, 1543–1586. [CrossRef]
9. Sevadjian, J.C.; McManus, M.A.; Benoit-Bird, K.J.; Selph, K.E. Shoreward advection of phytoplankton and vertical re-distribution of zooplankton by episodic near-bottom water pulses on an insular shelf: Oahu, Hawaii. *Cont. Shelf Res.* **2012**, *50–51*, 1–15. [CrossRef]
10. Wiebe, P.; Beardsley, R.; Mountain, D.; Bucklin, A. U.S. GLOBEC Northwest Atlantic\Georges Bank program. *Oceanography* **2002**, *15*, 13–29. [CrossRef]
11. Johnson, D.L.; Fogarty, M.J. Intercalibration of MOCNESS and Bongo nets: Assessing relative efficiency for ichthyoplankton. *Prog. Oceanogr.* **2013**, *108*, 43–71. [CrossRef]
12. Huse, G.; Johannessen, A.; Fossum, P. A comparison of the length distributions of larval capelin (*Mallotus villosus*) taken by Gulf III, MIK and MOCNESS collectors. *J. Appl. Ichthyol.* **1996**, *12*, 135–136. [CrossRef]
13. Sameoto, D.D.; Jaroszynski, L.O.; Fraser, W.B. BIONESS, a New Design in Multiple Net Zooplankton Collector s. *Can. J. Fish. Aquat. Sci.* **1980**, *37*, 722–724. [CrossRef]
14. Guglielmo, L.; Arena, G.; Brugnano, C.; Guglielmo, R.; Granata, A.; Minutoli, R.; Sitran, R.; Zagami, G.; Bergamasco, A. MicroNESS: An innovative opening–closing multinet for under pack-ice zooplankton sampling. *Polar Biol.* **2015**, *38*, 2035–2046. [CrossRef]

Journal of
Marine Science and Engineering

Article

Parametric Hull Design with Rational Bézier Curves and Estimation of Performances

Tommaso Ingrassia [1], Antonio Mancuso [1], Vincenzo Nigrelli [1], Antonio Saporito [1,*] and Davide Tumino [2]

[1] Dipartimento di Ingegneria, Viale delle Scienze, Università degli Studi di Palermo, 90127 Palermo, Italy; tommaso.ingrassia@unipa.it (T.I.); antonio.mancuso@unipa.it (A.M.); vincenzo.nigrelli@unipa.it (V.N.)

[2] Facoltà di Ingegneria e Architettura, Università degli Studi di Enna Kore, Cittadella Universitaria, 94100 Enna, Italy; davide.tumino@unikore.it

* Correspondence: antonio.saporito@unipa.it

Abstract: In this paper, a tool able to support the sailing yacht designer during the early stage of the design process has been developed. Cubic Rational Bézier curves have been selected to describe the main curves defining the hull of a sailing yacht. The adopted approach is based upon the definition of a set of parameters, say the length of waterline, the beam of the waterline, canoe body draft and some dimensionless coefficients according to the traditional way of the yacht designer. Some geometrical constraints imposed on the curves (e.g., continuity, endpoint angles, curvature) have been conceived aimed to avoid unreasonable shapes. These curves can be imported into any commercial Computer Aided Design (CAD) software and used as a frame to fit with a surface. The resistance of the hull can be calculated and plotted in order to have a real time estimation of the performances. The algorithm and the related Graphical User Interface (GUI) have been written in Visual Basic for Excel. To test the usability and the precision of the tool, two existing sailboats with different characteristics have been successfully replicated and a new design, taking advantages of both the hulls, has been developed. The new design shows good performances in terms of resistance values in a wide range of Froude numbers with respect to the original hulls.

Keywords: sailing yacht design; rational Bézier curves; VBA; excel; CAD; VPP

Citation: Ingrassia, T.; Mancuso, A.; Nigrelli, V.; Saporito, A.; Tumino, D. Parametric Hull Design with Rational Bézier Curves and Estimation of Performances. *J. Mar. Sci. Eng.* **2021**, *9*, 360. https://doi.org/10.3390/jmse9040360

Academic Editor: Md Jahir Rizvi

Received: 3 March 2021
Accepted: 24 March 2021
Published: 27 March 2021

Publisher's Note: MDPI stays neutral with regard to jurisdictional claims in published maps and institutional affiliations.

1. Introduction

In the work of an engineer, the design is often the central and more important part of the entire process. In a wide range of industries, such as automobile, aircraft, and shipbuilding [1], the first step of the process consists of finding an existing well-designed geometry to be used as a benchmark for the new model. In this work, the interest of the authors is related to maritime applications. It is interesting to notice that, in this field, the design approach is mostly based on the traditional design techniques of trial-and-error. Consequently, the obtained results are highly dependent on the designer experience and knowledge [2,3]. To facilitate the design of hulls, naval engineers are investigating the possibility to define the so-called hull equation [4]. This equation should be able to describe, from a mathematical point of view, the hull of a sailboat, a motorboat, or a ship. Although an intensive effort in this sense, nowadays is not possible to describe the hull with one equation because the geometry of a hull depends on several parameters and most of them are related one each other [5]. Several works can be found in the literature where authors present methods to generate a quick but detailed preliminary design or, on the other hand, approaches to optimize the geometry. For instance, in [6] a design tool is developed using cubic polynomial expressions to define the control curves of a hull. In [7] cubic Bézier curves and the curve-plane intersection method are selected to properly design a submarine hull. Also, ref. [8] proposes a new design framework to generate the parametric design and modification of yacht hulls. In particular, the hull is split into three regions to assure better design flexibility. Splitting the whole hull domain into sub-domains is a common practice as

53

can be observed in [9] where the hull has two domains, one below the chine and one above the chine. Other authors were more focused on the optimization phase as in [10] where a novel simultaneous engineering design approach has been proposed or in [2] where an interactive design approach for hull forms optimization is developed. Concerning the evaluation of the performances of a given design, there are several tools that can be used to estimate the resistance generated by the hull. Velocity Prediction Programs (VPPs) are commonly used in the nautical field. These programs can calculate speed, heel, trim, forces of sails and of course the resistance of the hull among other important characteristics of a sailboat [5]. One of the most popular methods used to define a VPP is presented in [11] and an updated version in [12]. Both are based on the experimental campaign conducted at the Delft Ship Hydromechanics Laboratory of the Delft University of Technology; in these formulations the resistance of the hull is linked to its coefficients and parameters such as the prismatic coefficient, the length of the waterline, the beam of the waterline, the displacement and so on. In [11] the allowable range in terms of Froude number to estimate the value of the resistance of a sailboat is between 0.10 and 0.60; while in the updated version [12] the Froude number ranges between 0.15 and 0.75. Another popular approach is presented in [13], the focus of the authors in this study is related to the evaluation of the resistance for planing hulls in smooth and rough water. However, VPPs are not the only way to estimate the performances of a sailboat. There are more sophisticated softwares allowing the designer to learn more about his design like Computational Fluid Dynamics (CFD) software that is a method where Navier-Stokes equations are solved. CFD models are very powerful tools although more expensive and time consuming and need a rigorous process of verification and validation (V&V) [14,15]. In [16] the Least Square Root method (LSR) to define the validation uncertainty of the numerical model is described. Once the numerical model is verified and validated, the designer can explore different designs to investigate their performances as in [17] where planing hulls are studied. Other examples can be found in [18] where CFD is used to study the influence of the trim angle.

The design tool proposed in this paper, written in Visual Basic Application for Excel, is intended as a Computer Aided Design (CAD) software guiding the designer in creating an hull form according to the classical naval design methodology which is mostly based on shape coefficients and non-dimensional ratios. The user enters information in terms of control points coordinates, angles and weights; however, these data are strictly related to dimensions (e.g., length of waterline, maximum beam, draft), to tangency (e.g., deadrise, hard-chine, round bilge) and to shape (e.g., fullness, continuity). Since one of the most important aspects when developing a tool for a designer is to strongly link it to the experience, the developed software works with the most important coefficients and parameters of a sailboat by using specific properties of the rational Bézier curves. In Section 2, these curves are presented in terms of equations and properties, while in Section 3, an approach to estimate the resistance curve of a sailboat is presented. In Section 4, the applied approach to replicate and design a sailboat used in this work is showed. In Section 5 the method presented is validated replicating an existing sailboat and, finally, in Section 6 a new design is compared with two existing ones.

2. Rational Bézier Curves

The rational Bézier curves are a particular family of the Bézier curves as presented in [19]. They can be described with the following Equation (1).

$$C(t) = \frac{\sum_{i=0}^{n} w_i B_{i,n}(t) P_i}{\sum_{i=0}^{n} w_i B_{i,n}(t)}, \quad t = 0, \ldots, 1 \tag{1}$$

where n is the order of the curve, P_i and w_i are the control points and the weights respectively, while $B_{i,n}(t)$ are the Bernstein polynomials defined over the parametric abscissa t and described by the following Equation (2).

$$B_{i,n}(t) = \frac{n!}{i!(n-i)!} t^i (1-t)^{n-i}, \; i = 0, 1, \ldots, n, \tag{2}$$

The principal difference between a rational Bézier curve and a classic one consists in the possibility of modifying the shape of a given curve without moving the control points. This is possible thanks to the weights assigned to each control point. In this way the designer can manipulate the shape of the curve and maintain the order of continuity in the ending points in terms of tangency and curvature. Being these curves tangent to the control polygon at the endpoints, the designer can have direct control on the initial and final angle of tangency of the curve. An example of this property is shown in Figure 1.

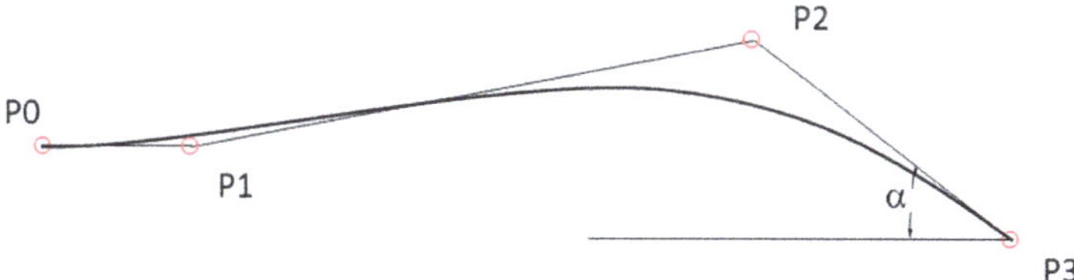

Figure 1. Tangency at the end points of a Bézier curve.

Regarding the control of the curvature (k) at the ending points there is another important property of the rational Bézier curves that allows the designer to link the position and weights of the control points to the value of the curvature. This relation is presented in the next Equation (3).

$$k(t_0) = \frac{w_0 w_2}{w_1^2} \frac{n-1}{n} \frac{h}{a^2}, \tag{3}$$

where n is the order of the curve, w_i are the weights of the control points while a and h are defined in Figure 2.

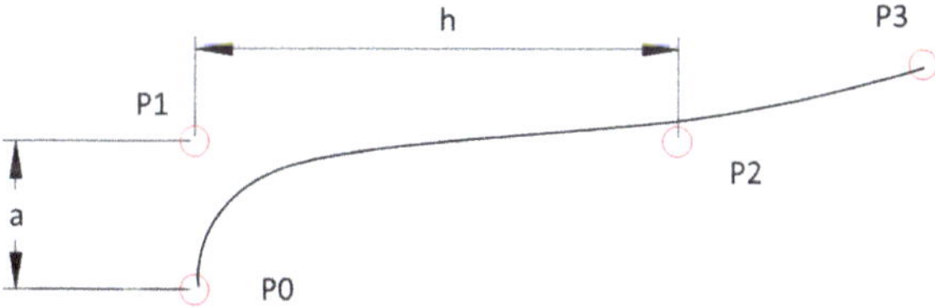

Figure 2. Curvature at the end points of a rational Bézier curve.

A practical application of these two properties can be seen in the following example. In Figure 3 two rational Bézier curves sharing the common point B are plotted.

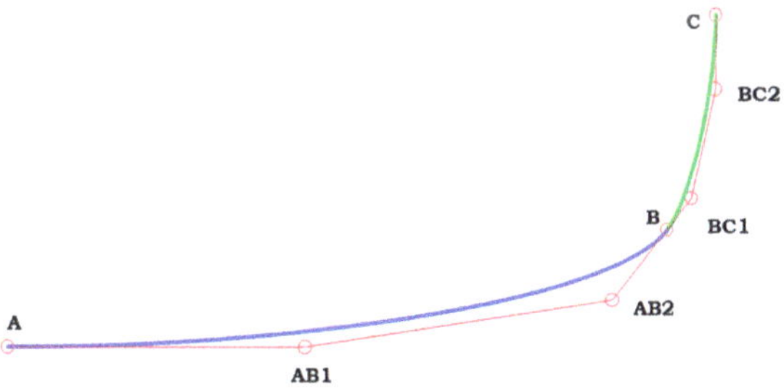

Figure 3. Two rational Bézier curves with a point in common.

The curve resulting from the union of these two curves presents a continuity of the second order. As it is possible to observe in Figure 3, the ending point of the blue curve is coincident with the starting point of the green curve (G0 = C0 continuity) and points AB2, B and BC1 are all aligned (G1 continuity). Regarding the curvature (G2 continuity), the designer can apply Equation (3) to establish that the curvature of the green curve is the same as the curvature of the blue curve calculated in point B (or vice-versa). To achieve this condition the designer can first calculate the value of the curvature in point B of the blue curve using Equation (3), then with the following Equation (4) calculate the value to assign to the weight of the control point BC1 to match the same value of the curvature of the blue curve and the green curve in their common point.

$$w_{BC1} = \sqrt{\frac{2}{3} \frac{w_B w_{BC2}}{k(B)} \frac{h}{a^2}}. \tag{4}$$

Another interesting property of the rational Bézier curves consists of the possibility of increasing the degree of the curve without modifying the shape of the curve itself. In Figure 4 is shown an example of this property.

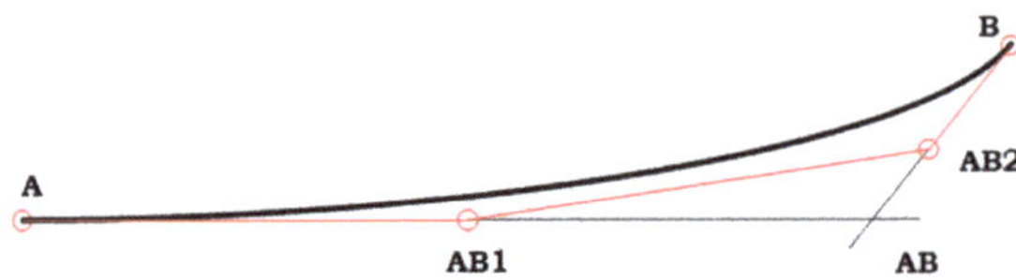

Figure 4. Degree elevation of a rational Bézier curve.

The original (quadratic) curve is defined by three points A, AB and B. To obtain the corresponding cubic curve the designer can apply the following procedure:

$$\begin{aligned}
X_A{}^* &= X_A \\
Y_A{}^* &= Y_A \\
X_{AB1}{}^* &= \frac{X_A}{3 + \frac{2}{3} X_{AB}} \\
Y_{AB1}{}^* &= \frac{Y_A}{3 + \frac{2}{3} Y_{AB}} \\
X_{AB2}{}^* &= \frac{2}{3} X_{AB} + \frac{X_B}{3} \\
Y_{AB2}{}^* &= \frac{2}{3} Y_{AB} + \frac{Y_B}{3} \\
X_B{}^* &= X_B \\
Y_B{}^* &= Y_B,
\end{aligned} \tag{5}$$

where the coordinates marked with * are representative of the new curve. The new curve is defined with four control points, A, AB1, AB2 and B, resulting in a higher degree curve compared to the starting one.

3. Estimation of Resistance for a Sailboat

Since 1975, researchers interested in sailing have been developing regression curves based on polynomial expressions with the aim of estimating the resistance of the hull of a sailboat. Nowadays there are several approaches to evaluate the resistance curve of a sailboat; one of the most common is the formulation proposed by [11] and the updated version presented in [12]. Both the formulations are based on the experimental campaign conducted at the Delft Ship Hydromechanics Laboratory of the Delft University of Technology. For the purpose of this work the older formulation has been preferred since in [12] typical coefficients concerning maxi yachts and higher Froude number are taken into account. Equation (6) shows the regression formulation for the estimation of the residuary resistance, Rrh, of the bare hull whose applicability ranges between Fr 0.10 and 0.60 stepped by 0.05.

$$\frac{Rrh}{\nabla c \cdot \rho \cdot g} = a_0 + \left(a_1 \cdot \frac{LCB_{fpp}}{Lwl} + a_2 \cdot Cp + a_3 \cdot \frac{\nabla c^{\frac{2}{3}}}{Aw} + a_4 \cdot \frac{Bwl}{Lwl} \right) \cdot \frac{\nabla c^{\frac{1}{3}}}{Lwl} + \left(a_5 \cdot \frac{\nabla c^{\frac{2}{3}}}{Sw} + a_6 \cdot \frac{LCB_{fpp}}{LCF_{fpp}} + a_7 \cdot \left(\frac{LCB_{fpp}}{Lwl} \right)^2 + a_8 \cdot Cp^2 \right) \cdot \frac{\nabla c^{\frac{1}{3}}}{Lwl}. \tag{6}$$

The coefficient a_i In Equation (6), can be found in reference [11], while in the Nomenclature section the meaning of the parameters are defined. To obtain the total hull resistance it is also necessary to compute the component of the resistance due to friction. In Equation (7) there is one of the possible empirical formulation widely used by the researchers [5] to define the friction coefficient.

$$c_f = \frac{0.075}{(\log(Re) - 2)^2}. \tag{7}$$

Consequently, the friction resistance can be expressed with the well-known equation:

$$R_f = \frac{1}{2} \rho v^2 S_w c_f. \tag{8}$$

The sum of the resistances estimated in Equations (6) and (8) permits to obtain the total resistance of the bare hull of a sailboat, as shown in Equation (9).

$$R = Rrh + R_f. \tag{9}$$

4. Design Approach

In this work, quadratic and cubic rational Bézier curves, whose formulation has been presented in Section 2, are used to design a sailboat. The shape of the hull is defined by three sections (fore, mid and aft), three longitudinal curves (sheer, chine and keel) and the right ahead. In this way, the whole domain is divided into three parts, as can be seen in Figure 5 where the red circles represent the intersection of the curves. In this way, the design variables of the problem are the Cartesian coordinates of the control points (the red circles) and the tangency of the curves at their ends.

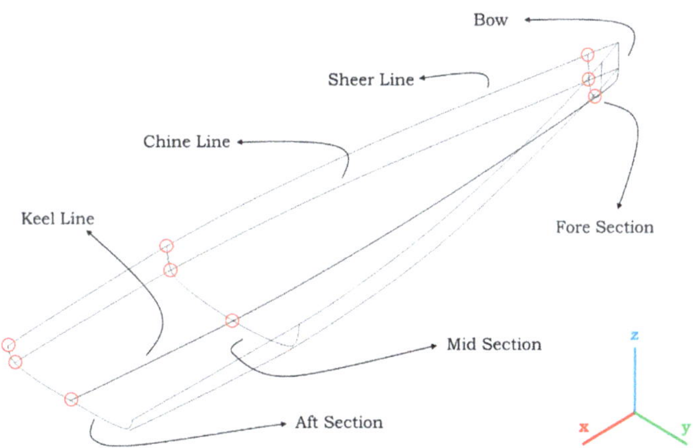

Figure 5. Curves frame used to define the hull surface.

Sections are defined with rational Bézier curves of second or third-degree depending on the type of the boat, respectively hard-chine or round-bilge. Each section is composed of two curves—one starting from the keel ending to the chine and the second one starting from the chine ending to the sheer. In this way, it is possible to generate a wide range of shapes. In Figure 6 the structure of the sections of a round-bilge hull is presented.

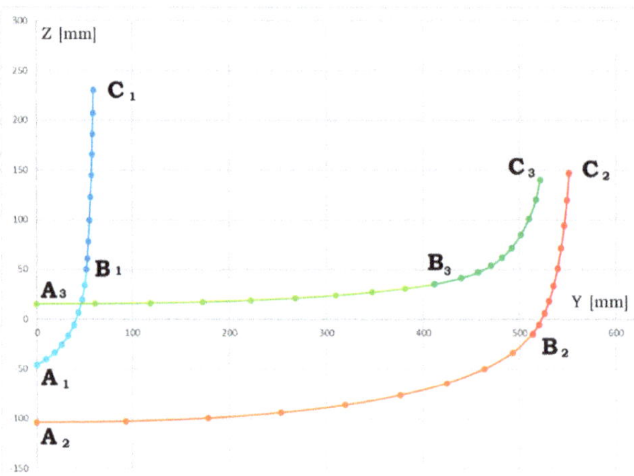

Figure 6. Structure of the curve of a section of the hull.

To properly define the shape of all the curves of Figure 5, the designer can insert the values of the coordinates of the control points of the curves with the help of a user-interface as shown in the following Figures 7 and 8. The definition of the curves by means of the rational Bézier formulation is particularly suitable for the design of the hull of a sailboat. In fact, once the main dimensions have been defined (e.g., max beam, max draught, length of water line), the designer can adjust the fullness of each curve without modifying the control polygon but just acting on the weights of the control points or the tangency at their ends.

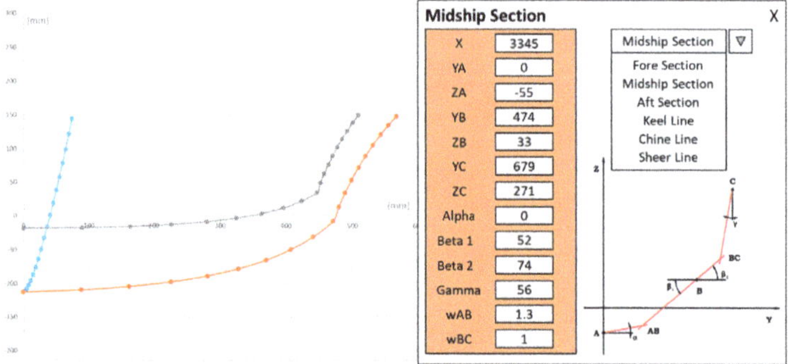

Figure 7. User-interface of a section.

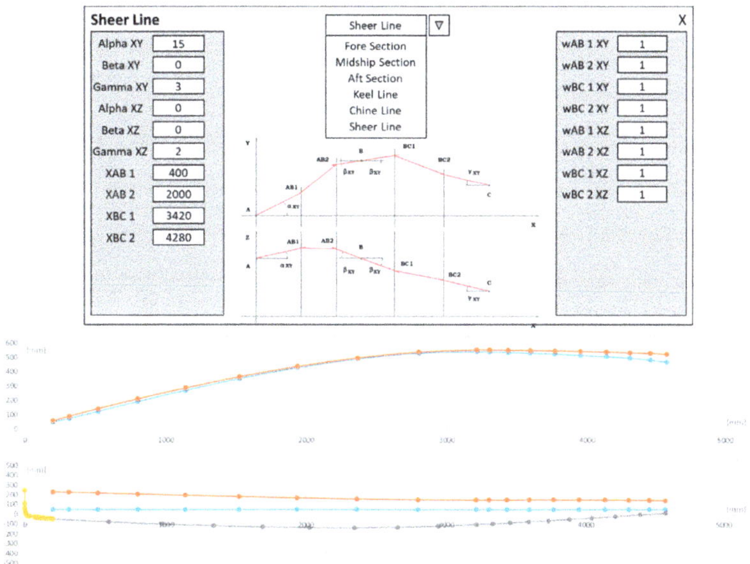

Figure 8. User-interface of the sheer, chine and keel lines.

To assure G0 continuity of the section, the position of the control points of the two curves at the chine have to be coincident. G0 continuity is enough for a hard-chine sailboat. While for a round-bilge sailboat, G1 and G2 continuity have to be imposed as well. G1 continuity can be obtained by controlling the tangency of the curves in the common point B lying on the chine (angles β_1 and β_2 in Figure 9). So, in the case of hard-chine $\beta_1 \neq \beta_2$, while in the case of round-bilge $\beta_1 = \beta_2$.

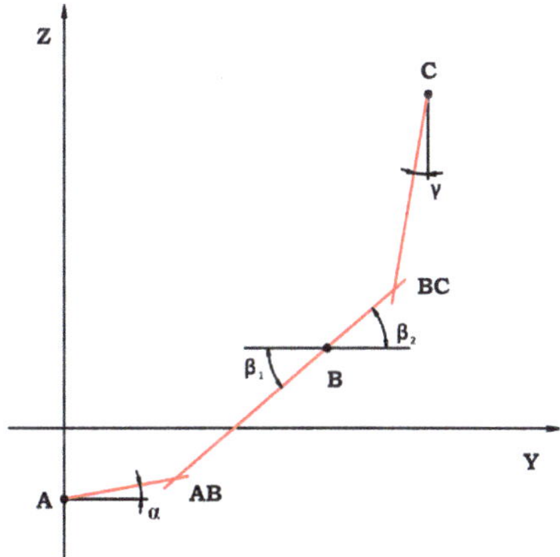

Figure 9. Control polygon of a G1 second-degree rational Bézier curve.

As can be seen the control points AB, B and BC are all lying in the same line so the angle at the end of the first curve is the same as the starting angle of the second curve and G1 continuity is respected. To assure the G2 continuity, the second-degree curve is not sufficient so, as presented in Section 2, the curve is automatically modified to obtain a third-degree curve using Equation (5), then Equations (3) and (4) are applied to impose the same curvature at the common point of the two curves (point B). In Figure 10 the new control polygon assuring G2 continuity of a section is shown.

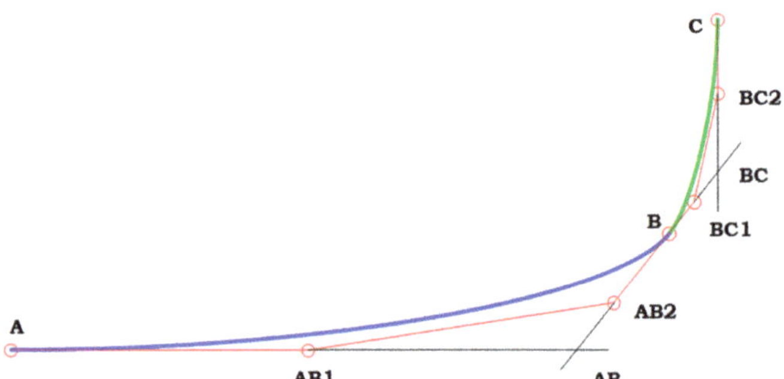

Figure 10. Section modeled with a G2 third-degree rational Bézier curve.

Once all the curves have been defined, the Visual Basic for Application (VBA) tool prints several exchange files with the information of each curve following two different strategies. In the first one each curve is sampled in a fixed number of points and their coordinates (in ASCII format) can be imported in several CAD software that re-creates the curves by interpolating these points. In the second format, each curve is defined with the syntax form of an IGES file [20] preserving, in this way, its mathematics. According to the latter method, a macro has been set up in the parametric software CREO 4.0 from PTC aimed to import all the curves and automatically generate the sweep surface of the whole hull. Figure 11 shows the surface of a round-bilge sailboat.

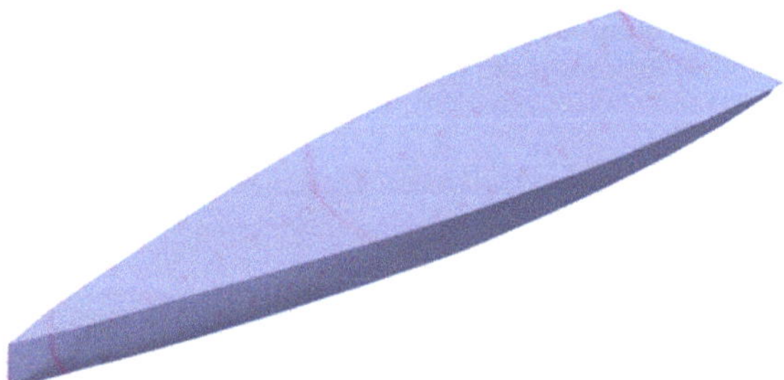

Figure 11. Surface model of a sailboat designed in CREO 4.0 with the information from the VBA tool.

Now the designer can have a closer look to the curves and surface using all the feature of a commercial CAD software. For instance, Figure 12 shows a screenshot of the gaussian curvature of a hard-chine hull.

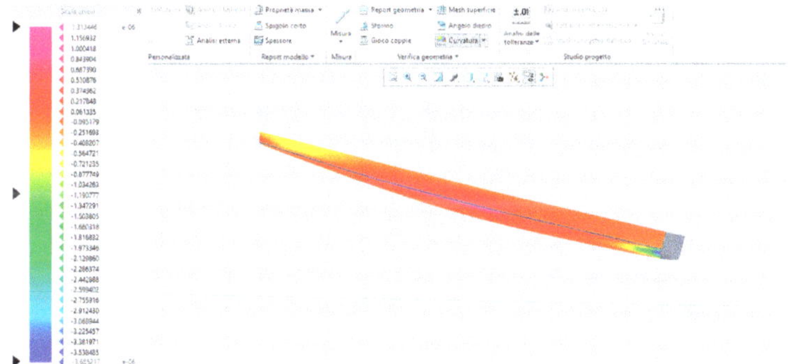

Figure 12. Curvature analysis of the surface of the hull.

Still inside CREO 4.0 the coefficients used in Equation (6), whose definition is given in the Nomenclature section, are automatically calculated. The VBA tool grasps these information and evaluates the resistance curve of the specific hull vs. the Froude number [11], plotting the results as shown in Figure 13.

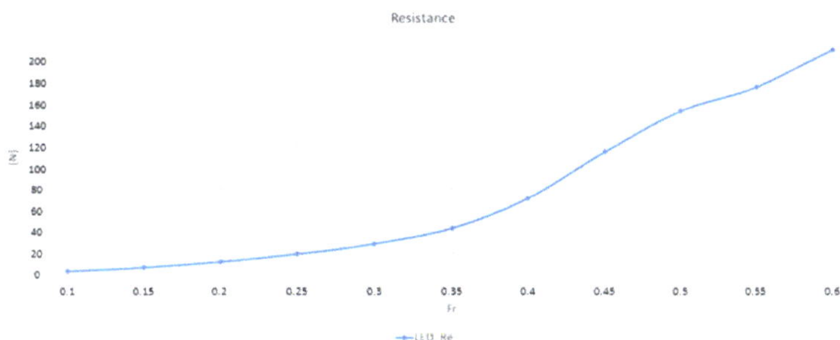

Figure 13. Resistance curve of the hull of a sailboat.

5. Design an Existing Sailboat

In this section two existing sailboats are replicated to validate the method presented in Section 4. The type of sailboat selected are the so called SKIFF (Sail Keep It Flat and Fast) shown in Figure 14 that take part to the international annual regatta 1001Velacup®held in Italy every year in September.

Figure 14. The Sail Keep It Flat and Fast (SKIFF) LED and TryAgain during the regattas of 2013 in La Spezia, Italy.

These two boats, although quite similar in terms of displacement, length and main parameters, are characterized by a different hull shape. LED is a classical round-bilge, designed to sail at a low Froude number while TryAgain is a hard-chine hull, mainly designed for sail at a higher Froude number. In Figure 15 the designs of both the hulls are shown (LED on the top and TryAgain on the bottom).

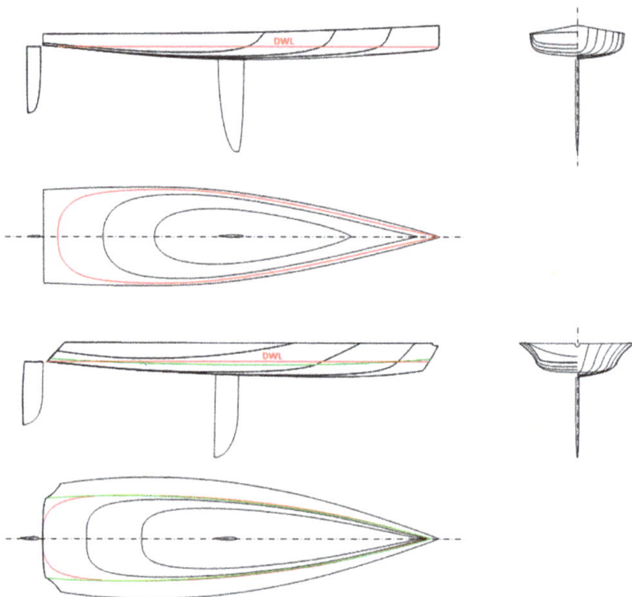

Figure 15. Hull design of LED (**top**) and TryAgain (**bottom**).

Applying the method presented in Section 4, the two hulls were replicated. In particular, Figure 16 shows the overlap of the original and rebuilt curves, while Figure 17 shows the cut-off of the rebuilt surfaces modelled as previously said in CREO 4.0 with transversal planes to show the sections (red curves), with horizontal planes to show the waterlines (blue curves) and with longitudinal planes to show the buttocks (green curves). A well faired curves frame has been obtained without undesired changes in slope or curvature confirming the goodness of the proposed approach.

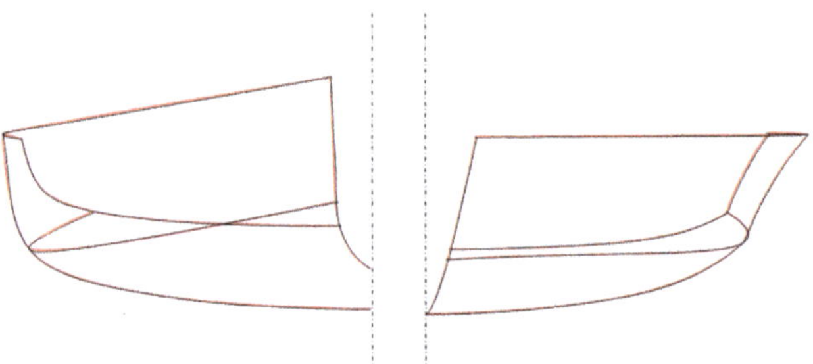

Figure 16. Lines of the original hulls (black) and rebuilt ones (red). LED on the left part and TryAgain on the right (part).

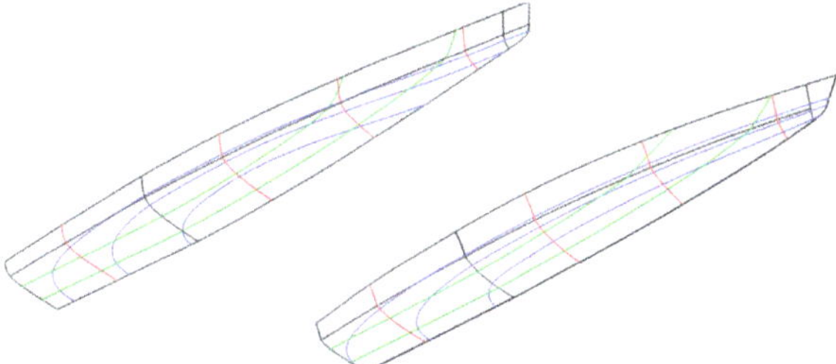

Figure 17. Rebuilt hull surfaces: LED (**left**) and TryAgain (**right**).

Once the geometry of the hulls has been defined, the main characteristics of the two sailboats are automatically calculated. Table 1 shows the comparison of the main hull characteristics whose definitions can be found in [5].

Table 1. Coefficients of the original and rebuild hulls.

Entity	Symbol	Unit	LED		TryAgain	
			Original	Rebuilt	Original	Rebuilt
Displacement	∇	m^3	0.257	0.258	0.262	0.263
Length Overall	LOA	m	4.60	4.60	4.60	4.60
Length Water Line	LWL	m	4.46	4.46	4.49	4.49
Max Beam Water Line	BWL	m	1.05	1.05	0.95	0.95
Wetted Surface	S_W	m^2	3.49	3.48	3.46	3.50
Water Plane Area	A_W	m^2	3.21	3.20	3.10	3.14
Max Transversal Area	A_X	m^2	0.107	0.107	0.094	0.093
Long. Centre of Buoyancy	LCB	m	2.48	2.52	2.25	2.26
Long. Centre of Flotation	LCF	m	2.69	2.70	2.60	2.60
Max Draught	Tc	m	0.14	0.14	0.17	0.17
Prismatic Coefficient	Cp		0.539	0.540	0.621	0.629
Midship Section Coefficient	Cm		0.728	0.728	0.582	0.576

As can be seen in the previous table, also the differences in terms of coefficients are very narrow so the method results effectively to replicate a sailboat. In addition, the resistance curves have been calculated with the procedure presented in Section 3 and in Figure 18 the original and rebuilt hulls are compared. It is evident that the rebuild process does not affect the performance prediction: the resistance curves of both the hulls are completely overlapped, and no appreciable differences can be observed. Some basic comments about this plot can be done as follows: differences between the two hulls can be appreciated in the range Fr = [0.25–0.4] where LED has lower resistance than TryAgain and in the range Fr = [0.4–0.55] where the opposite happens. This is mainly due to the performance characteristic of the round bilge (LED) and of the hard chine (TryAgain): at low speed, faired streamlines are favored by the smoothness of the round bilge reducing the friction resistance; at higher speed, lifting forces generated by the hard chine enhance pre-planing phenomena and, as a consequence, a reduced wave resistance [5].

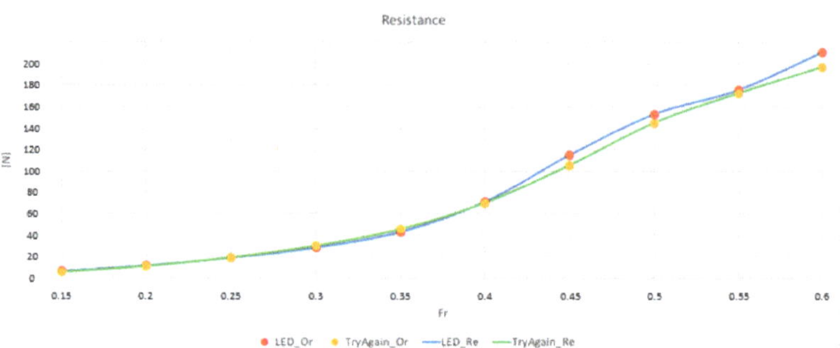

Figure 18. Resistance curves of the original and rebuild hulls.

6. Design of a New Sailboat

As can be seen in Figure 18, depending on the sailing conditions, LED or TryAgain is better than the other one. To understand the reasons for these differences in terms of the resistance of the two hulls, an investigation on the effect of the coefficients on the computed resistance was carried on. In Table 2 the coefficients of LED and TryAgain are compared to see which one is the most different. The last column reports the difference ($\text{Coeff}_{\text{TryAgain}} - \text{Coeff}_{\text{LED}})/\text{Coeff}_{\text{LED}}$.

Table 2. Comparison between the coefficients of LED and TryAgain.

Coeff.	LED	TryAgain	Diff. (%)
V_C	0.258	0.263	+1.90
LCB	2.47	2.26	−9.29
S_W	3.48	3.50	+0.57
L_{WL}	4.46	4.49	+0.67
B_{WL}	1.05	0.95	−10.53
A_W	3.20	3.14	−1.91
LCF	2.70	2.60	−3.85
T_C	0.14	0.17	+17.65
A_X	0.107	0.093	−15.05
C_P	0.540	0.629	+14.15
C_M	0.728	0.576	−26.39

As can be seen in the previous table, the largest differences in terms of coefficients of the two sailboats are the Longitudinal Centre of Buoyancy (LCB), the Beam of the Water Line (Bwl), the Max Draught (Tc) and the Max Transversal Area (Ax).

VPP is very sensitive to slight changes of the coefficients. In particular, there are appreciable differences in terms of resistance starting from Froude 0.4, since for lower ranges of Froude number the biggest quote to the resistance is related to friction. Reducing the value of LCB causes better performances for medium speeds, reducing BWL generates a sailboat that performs in a better way at high Froude numbers, reducing T_C and A_X have a not optimal resistance trend for low values of Froude numbers but on the other hand good one for high Froude numbers.

The goal is to obtain a new shape that has the advantages of the two hulls LED and TryAgain. To reach this objective the new hull should have a low B_{WL} and a low C_P in order to preserve the performances of LED at a low Froude number; at the same time, this new hull should need a hard-chine to preserve the performances of TryAgain at higher Froude number. As a matter of course, reducing the value of B_{WL} will lead to having a higher T_C to match the total displacement of the sailboat.

With all this information, the authors started an intensive campaign to obtain a new sailboat. The approach can be considered heuristic and knowledge-based with the goal of obtaining a new hull design as similar as possible to LED, with good performances at low values of Froude numbers, but with a chine to take advantage of its positive effect at high values of Froude numbers. After several attempts using the approach here presented a new design was defined and the obtained lines are shown in Figure 19 while in Table 3 the coefficients of the three sailboats are compared.

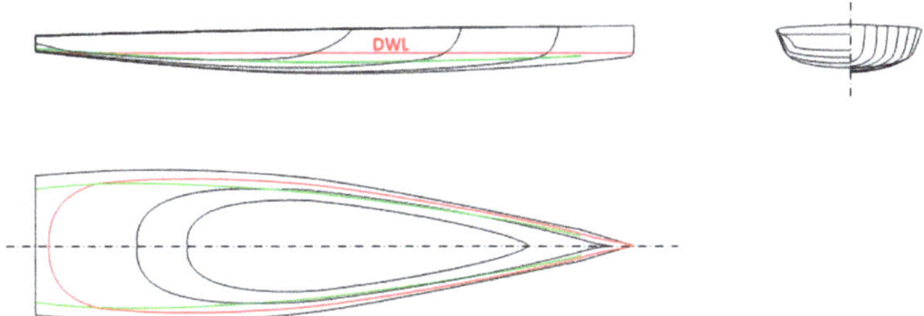

Figure 19. The lines of the new design.

Table 3. Comparison of the coefficients of LED, TryAgain and LED_UP_06.

Coeff.	LED	TryAgain	LED_UP_06
V_C	0.258	0.263	0.259
LCB	2.47	2.26	2.48
S_W	3.48	3.50	3.39
L_{WL}	4.46	4.49	4.48
B_{WL}	1.05	0.95	0.995
A_W	3.20	3.14	3.08
LCF	2.70	2.60	2.71
T_C	0.14	0.17	0.16
A_X	0.107	0.093	0.107
C_P	0.540	0.629	0.540
C_M	0.728	0.576	0.689

In Figure 20 the surface of the new hull is generated and cut off with transversal planes to show the sections (red curves), with horizontal planes to show the waterlines (blue curves) and with longitudinal planes to show the buttocks (green curves).

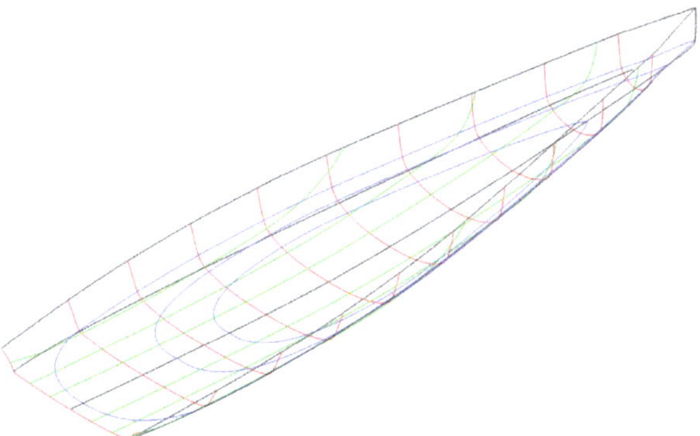

Figure 20. Surface of LED_UP_06.

Finally, applying the method presented in Section 3, the resistance curves of the three hulls are compared and the result plotted in Figure 21.

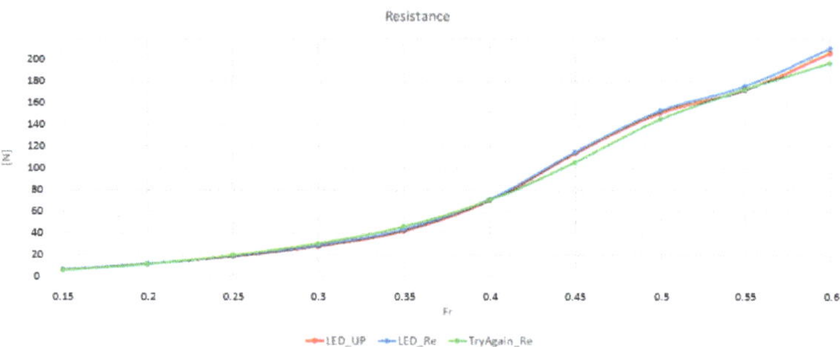

Figure 21. Comparison between the resistance curves of LED, TryAgain and LED_UP_06.

As can be seen in the previous figure, the VPP indicates that LED_UP_06 is the best hull in light wind condition (low Froude number), then TryAgain gains advantage for medium ranges of speed, followed by LED_UP_06.

To resume, in Figure 22, the resistances calculated for LED and TryAgain have been compared to the ones calculated for LED_UP_06. The blue areas indicate where and how much LED_UP_06 is better than the original hulls, while the red areas indicate the opposite. The green line compares LED with LED_UP_06 and the area below this line is always blue in the analyzed Fr range. This means that the new hull guarantees better performances with respect to LED at any Fr. The red line compares TryAgain with LED_UP_06 and the areas below this line are red or blue depending on the value of Fr. Basically, LED_UP_06 behaves better than TryAgain in the range Fr = [0.21–0.4] and the improvements arrive at the 9% of the resistance. TryAgain has lower values of resistance for medium-high values of Froude number (Fr > 0.4), which is a consolidated result in literature since a higher value of C_P leads to better performances in this range of speeds. Nevertheless, the better performance due to the pre-planing attitude shown by TryAgain at high Fr can become significant in the choice between the two hulls only when high speed is assumed during regattas. Usually these competitions are held in light breeze conditions and rarely boats are fully planing. Moreover, frequent restarts of the boat due to maneuvering require a boat able to quickly accelerate and in this sense, the low resistance shown by LED_UP_06 at medium Fr range could be of great importance.

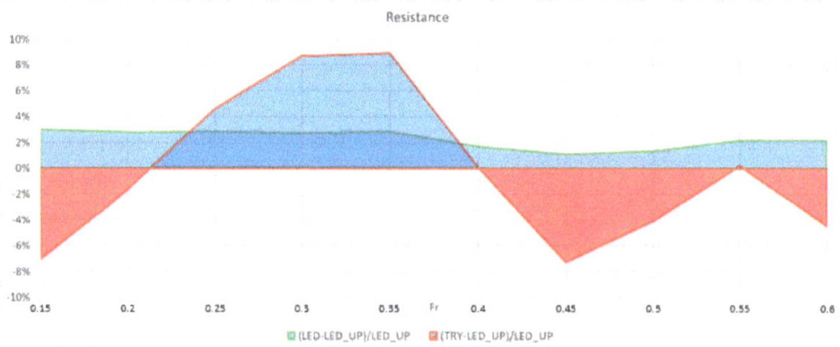

Figure 22. Difference in percentage between LED and LED_UP_06 and TryAgain and LED_UP_06.

7. Conclusions

In this work, a tool and the relative methodology to design and evaluate the performances of hulls of sailing boats is presented. The algorithm and the related Graphical User Interface (GUI) have been written in Visual Basic for Excel. A total of seven rational Bézier curves of the third-degree are selected to define the geometry of the hull. To prove the validity of the tool and the applied approach, two existing sailboat hulls have been successfully replicated and a new design of a sailboat is presented.

This procedure has been used to rapidly design a new hull that includes benefits given by a round bilge and a hard chine hull. It has been demonstrated that a hybrid solution between these two opposite shapes ensures better performance especially in the mid-range of Fr, before lifting effects on the hull due to dynamic pressure prevail.

In future works the new hull will be investigated by means of CFD simulations, in a wider range of real sailing conditions in terms of Froude number, trim and leeway angles, allowing the designer to understand in a deeper way why the hulls present differences in terms of performances.

Author Contributions: Conceptualization, T.I., A.M., V.N., A.S. and D.T.; methodology, A.M., A.S. and D.T.; software, A.S.; validation, A.M. and A.S.; formal analysis, T.I., A.M., V.N., A.S. and D.T.; investigation, T.I., A.M., V.N., A.S. and D.T.; resources, T.I., A.M., V.N., A.S. and D.T.; data curation, A.S.; writing—original draft preparation, A.S.; writing—review and editing, T.I., A.M., V.N., A.S. and D.T.; visualization, A.M., A.S. and D.T.; supervision, T.I., A.M., V.N. and D.T.; project administration, T.I., A.M., V.N. and D.T.; funding acquisition, NONE. All authors have read and agreed to the published version of the manuscript.

Funding: This research received no external funding.

Conflicts of Interest: The authors declare no conflict of interest.

Nomenclature

R_{rh}	Residuary Resistance	N
∇c	Displacement	m^3
ρ	Density of Water	kg/m^3
g	Gravity Acceleration	m/s^2
L_{OA}	Length Overall	m
S_W	Wetted Surface	m^2
A_X	Max Transversal Area	m^2
L_{wl}	Length of Water Line	m
B_{wl}	Beam of Water Line	m
T_c	Canoe Body Draft	m
LCB_{fpp}	Center of Buoyancy	m
LCF_{fpp}	Center of Flotation	m
C_p	Prismatic Coefficient ($C_P = \nabla c / L_{WL} Ax$)	-
A_w	Water Plane Area	m^2
a_i	Coefficients	-
P_i	Coordinate of control points	m
w_i	Weight of control points	-
$B_{i,n}$	Bernstein polynomials	-
Re	Reynolds number	-

References

1. Nam, J.; Bang, N.S. A curve based hull form variation with geometric constraints of area and centroid. *Ocean Eng.* **2017**. [CrossRef]
2. Khan, S.; Gunpinar, E.; Sener, B. GenYacht: An interactive generative design system for computer-aided yacht hull design. *Ocean Eng.* **2019**. [CrossRef]
3. Cirello, A.; Cucinotta, F.; Ingrassia, T.; Nigrelli, V.; Sfavara, F. Fluid-structure interaction of downwind sails: A new computational method. *J. Mar. Sci. Technol.* **2019**, *24*, 86–97. [CrossRef]
4. Mancuso, A. Parametric design of sailing hull shapes. *Ocean Eng.* **2006**, *33*, 234–246. [CrossRef]
5. Larsson, L.; Eliasson, R.E.; Orych, M. *Principles of Yacht Design*; Adlard Coles Nautical: London, UK, 2014.
6. Calkins, D.E.; Schachter, R.D.; Oliveira, L.T. An automated computational method for planning hull form definition in concept design. *Ocean Eng.* **2001**, *28*, 297–327. [CrossRef]
7. Chrismianto, D.; Zakki, A.F.; Arswendo, B.; Kim, D.J. Development of Cubic Bezier Curve and Curve-Plane Intersection Method for Parametric Submarine Hull Form Design to Optimize Hull Resistance Using CFD. *J. Mar. Sci. Technol.* **2015**, *14*, 399–405. [CrossRef]

8. Khan, S.; Gunpinar, E.; Dogan, K.M. A novel design framework for generation and parametric modification of yacht hull surfaces. *Ocean Eng.* **2017**. [CrossRef]

9. Pérez-Arribas, F. Parametric generation of planing hulls. *Ocean Eng.* **2014**, *81*, 89–104. [CrossRef]

10. Ingrassia, T.; Mancuso, A.; Nigrelli, V.; Tumino, D. A multi-technique simultaneous approach for the design of a sailing yacht. *Int. J. Interact. Des. Manuf.* **2017**, *11*, 19–30. [CrossRef]

11. Keuning, J.A.; Sonnenberg, U.B. Approximation of the hydrodynamic forces on a sailing yacht based on the Delft Systematic Yacht Hull Series. In Proceedings of the 15th International Symposium on "Yacht Design and Yacht Construction", Amsterdam, The Netherlands, 16 November 1998; ISBN 902700171-8.

12. Keuning, J.A.; Katgert, M. A bare hull resistance prediction method derived from the results of the Delft systematic yacht hull series extended to higher speeds. In Proceedings of the International Conference on Innovation in High Performance Sailing Yachts, Lorient, France, 29–30 May 2008.

13. Savitsky, D.; Brown, P.W. Procedures for Hydrodynamic Evaluation od Planing Hulls in Smooth and Rough Water. *Mar. Technol.* **1976**, *13*, 381–400.

14. ITTC. Practical Guidelines for Ship CFD Applications. In Proceedings of the 26th International Towing Tank Conference, Rio de Janeiro, Brazil, 28 August–3 September 2011.

15. ITTC. Uncertainty Analysis in CFD Verification and Validation Methodology and Procedures. In Proceedings of the Recommended Procedures and Guidelines of the International Towing Tank Conference, 2017. Available online: https://www.ittc.info/media/7691/75-04-01-011.pdf (accessed on 26 March 2021).

16. Eca, L.; Hoekstra, M. A procedure for the estimation of the numerical uncertainty of CFD calculations based on grid refinement studies. *J. Comput. Phys.* **2014**, *262*. [CrossRef]

17. Raymond, J.; Cudby, K. Using Parametric Modelling, CFD, and Historical Data, to Estimate Planing Hull Performance on a Laptop. In Proceedings of the 4th High Performance Yacht Design Conference, Auckland, New Zealand, 12–14 March 2012.

18. Viola, I.M.; Enlander, J.; Adamson, H. Trim effect on the resistance of sailing planing hulls. *Ocean Eng.* **2014**, *88*, 187–193. [CrossRef]

19. Sederberg, T.W. *Computer Aided Geometric Design. Computer Aided Geometric Design Course Notes*; BYU Scholars Archive, 2012. Available online: https://scholarsarchive.byu.edu/cgi/viewcontent.cgi?article=1000&context=facpub (accessed on 26 March 2021).

20. ANSI Standard. *Initial Graphics Exchange Specification IGES 5.3*; U.S. Product Data Association: Charleston, SC, USA, 1996.

Journal of
Marine Science and Engineering

Article

The Effect of Longitudinal Rails on an Air Cavity Stepped Planing Hull

Filippo Cucinotta [1,*], Dario Mancini [2], Felice Sfravara [1] and Francesco Tamburrino [3]

[1] Department of Engineering, University of Messina, Contrada Di Dio (S. Agata), 98166 Messina, Italy; fsfravara@unime.it
[2] Istituto Nazionale di Astrofisica-Osservatorio Astronomico di Capodimonte, 80131 Naples, Italy; dario.mancini@inaf.it
[3] Department of Civil and Industrial Engineering, University of Pisa, Largo Lucio Lazzarino 1, 56122 Pisa, Italy; francesco.tamburrino@ing.unipi.it
* Correspondence: filippo.cucinotta@unime.it; Tel.: +39-0906765292

Abstract: The use of ventilated hulls is rapidly expanding. However, experimental and numerical analyses are still very limited, particularly for high-speed vessels and for stepped planing hulls. In this work, the authors present a comparison between towing tank tests and CFD analyses carried out on a single-stepped planing hull provided with forced ventilation on the bottom. The boat has identical geometries to those presented by the authors in other works, but with the addition of longitudinal rails. In particular, the study addresses the effect of the rails on the bottom of the hull, in terms of drag, and the wetted surface assessment. The computational methodology is based on URANS equation with multiphase models for high-resolution interface capture between air and water. The tests have been performed varying seven velocities and six airflow rates and the no-air injection condition. Compared to flat-bottomed hulls, a higher incidence of numerical ventilation and air–water mixing effects was observed. At the same time, no major differences were noted in terms of the ability to drag the flow aft at low speeds. Results in terms of drag reduction, wetted surface, and its shape are discussed.

Keywords: computational fluid dynamics; hull design; air cavity ships; hull ventilation; stepped planing hull

Citation: Cucinotta, F.; Mancini, D.; Sfravara, F.; Tamburrino, F. The Effect of Longitudinal Rails on an Air Cavity Stepped Planing Hull. *J. Mar. Sci. Eng.* **2021**, *9*, 470. https://doi.org/10.3390/jmse9050470

Academic Editor: Apostolos Papanikolaou

Received: 22 March 2021
Accepted: 25 April 2021
Published: 27 April 2021

Publisher's Note: MDPI stays neutral with regard to jurisdictional claims in published maps and institutional affiliations.

1. Introduction

The drag reduction is the main issue in limiting fuel consumption or enhancing the performance of High Speed Vessels (HSV). Traditionally the global drag is divided into two principal components: wave-making drag and viscous drag [1]. Under the name of viscous drag are included all the effects due to the fluid viscosity, i.e., all the effects that would be zero if the fluid is inviscid. The viscous part affects several phenomena, including the so-called form effect, the friction effect, the roughness, and the flat plate friction. According to the ship typology and its Froude number, the resistance can change from about 40% to 95% of the global resistance [2]. Over time, there have been numerous attempts to reduce the drag's viscous component, including the use of multihull, the Surface Effect Ships, the use of foils (to reduce the wet surface), or the use of hull ventilation [3]. From the first experiments to today, there have been many efforts to exploit the effect of the ventilation for drag reduction. Hull ventilation can be addressed both in a natural or forced manner. The presence of an air pocket can also have a lift role or only a lubricating effect. The shape or the mass of air can be in the form of cushion [4], layer [5], bubbles, or microbubbles [6].

In order to exploit natural ventilation of planing hulls, wedge boats and stepped bottom were theorized and studied, both experimentally and numerically, since the fore-runners' work of Savitsky [7]. In fact, the planing hulls behave in a similar way to an

69

airfoil. In this way, the natural under pressure generated behind a step can be exploited to facilitate the entry and diffusion of air for ventilating the bottom.

Dashtimanesh et al. dealt with multistep hulls suggesting empirical methods [8] and by the CFD approach [9,10]. De Marco et al. tried the Large Eddy Simulations (LES) approach, with different moving mesh techniques [11]. Niazmand Bilandi et al. applied the 2D+ T theory with success, extending 2D theory to tridimensional problems at double-stepped planing hulls [12,13]. The results have been furthermore applied to laboratory experiments and mathematical modeling [14]. In the range of experimental studies, the studies of systematic variations in a series of stepped hull are particularly interesting, such as the works of Lee et al. [15] and of Taunton et al. [16].

Butuzov et al. were pioneers in the concept of modern forced ventilated hulls [17]. Since then, many authors have carried out numerical and experimental studies on Air Cavity Ships (ACS) and Air Cushion Vehicles (ACV). Mäkiharju et al. studied the problem of scaling air behavior from model to real scale [18]. A great contribution to this topic comes from Matveev's research who studied a two-dimensional [19] and three-dimensional [20] approach. In 2015, Butterworth carried out experimental campaigns on a container ship [21], while in 2017, Cucinotta et al. performed an experimental campaign on three different planning models and a mother hull [22]. Wang et al. carried out experimental tests obtaining similar results and finding excellent performances in terms of porpoising [23]. Barbaca et al. [24] and Qin et al. [25] focused their attention on the cavity flow, taking advantage of cavity growth, shedding, and closure characteristics.

A tool that has made it possible to achieve significant research progress in this field is Computational Fluid Dynamic (CFD). This made it possible to reduce research times and costs, overcome scale problems, and better monitor the phenomena under the hull [26]. Hull [27] and foils [28] fluid dynamics can be addressed successfully by implementing CFD analyses. Furthermore, it is possible to implement the calculus with Fluid Structure Interaction analysis (FSI) [29] and optimization analysis [30].

By means of CFD, Cucinotta et al. simulated the hydrodynamics of two models, [31] and [32], validating them through experimental data. In these analyses, the authors showed good results regarding drag reduction and numerical convergence and reliability. Since air, as has been widely observed, tends to escape from the hull sides, especially at low speeds, the attempt to use longitudinal rails to contain the air flow has been proposed and assessed in this work. Therefore, the authors present the results of experimental tests and CFD analyses carried out on a model identical with that presented in [22] but with the addition of longitudinal rails within the stepped part of the hull and assess their evaluation.

The article is structured in a first part, which describes in depth the used methods, firstly of the experimental campaign and secondly of the CFD campaigns, and in a second part that shows the results and the relative comments. Finally, after the conclusions, a list of nomenclature is reported.

2. Methods

The chapter describes the geometric characteristics of the hull and the position of the step for the injection of the air and the geometric dimensions of the rails. In order to have a general idea of the line of the hull, a lines plan is proposed and the trend of the transversal section area is reported for the sake of completeness. The second part of this chapter introduces the simulation settings. The mesh and the refinement zones are presented and a complete description of the sequence of simulations conducted allows us to understand the approach used during this campaign of investigation. Particular attention concerns the methodology used for the wall-treatment inside the commercial CFD software and the initial boundary conditions.

2.1. The Model

As reported in [33], the original yacht is 18 m in length and whose principal dimensions are reported by Cucinotta et al. [22].

The bottom part of the hull was modified in order to have a step for injection of the air. A series of rails from the step to the transom was added. The model is fully described by [22] and for the sake of readability, the main dimensions are reported in Table 1.

Table 1. Main dimensions.

Dimensions	Ship	Model	Unit
LWL	14.884	2.481	m
LP	16.2	2.7	m
BWL	4.314	0.719	m
BPX	3.72	0.62	m
BPA	3.2	0.54	m
BPT	3.66	0.61	m
T	1.000	0.167	m
Δ	34	0.153	t
S	78.6	2.182	m^2
x_G (% of LOA)	35.92%	35.92%	-
AP	52.2	1.45	m^2
N_{ST}	1	1	-
L_{ST}	8.00	1.333	m
S_{ST}	0.304	0.0084	m^2
N_{IN}	10	10	-
$B \times H$	0.31×0.09	0.051×0.015	$m \times m$
S_{IN}	0.279	0.0077	m^2
Longitudinal Rails Dimensions			
N	8	8	-
$B_R \times H_R$	0.041×0.059	0.007×0.01	$m \times m$
TD_R	0.340	0.056	m

Figure 1 shows the linesplane, with the presence of the step and its position with respect to the transom. The step is located 8 m away from the transom (1.33 m for the scale model) and the rails starting from this step until the end of the hull. The presence of the step produces a discontinuity in the transversal section area of about 0.3 m².

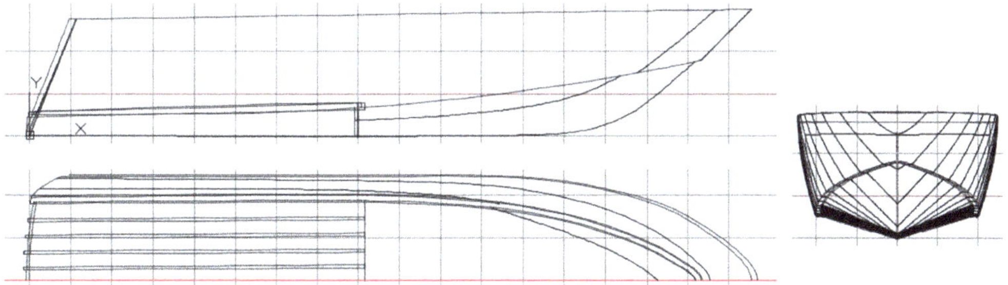

Figure 1. Multi-Step hull–Position of rails.

The experimental campaign was carried out in the University of Naples facilities and the results are reported in [22]. Figures 2 and 3 and show, respectively, the bottom of the model and a picture from the tests in the towing tank.

Figure 2. The bottom of the hull.

Figure 3. A test in towing tank.

2.2. Simulation Settings

The approach used for solving the fluid dynamics around the hull is the Computational Fluid Dynamic (CFD in this paper). The main goal of simulation is to solve the pressure and velocity field around the hull, the volume fraction (the interface between the air and the water), and, finally, the relative position of the boat in terms of trim and sinkage. In order to capture all characteristics of the fluid field in terms of pressure and velocity, a URANSE (Unsteady RANS) approach was chosen. This approach allows to use the Navier–Stokes equations to solve all the unknowns of fluid. The finite volume method is implemented in the commercial software STAR CCM+ [34] used for this campaign of simulations. A crucial parameter is the time-step and the discretization order of the scheme used (in this simulation 2nd order) for the time-marching solution. The ITTC [35] suggests using a time-step in a range given by a function of the length of the hull and velocity of the boat, according to the formula

$$\Delta t = 0.01 \div 0.05 \frac{l}{v}$$

in which l is the length and v the velocity of the boat. A complete uncertainty analysis that comprises the evaluation of time step size, Courant number, grid dimension, and convergence ratio was carried out according to [35]. The detailed results are reported in [31].

The complexity of the simulation is due to the interaction between water and air in two different ways (interface of the sea, injection of the air in the bottom of the hull). Both fluids are treated as incompressible.

The multiphase simulation, with the VoF (Volume of Fluid) scheme, is used and thanks to the High-resolution interface capturing scheme (HRIC), a clear interface is captured during the entire duration of the simulation. A surface tension equal to 0.072 N/m between the two fluids is imposed. All the solver settings are the same as the ones used by Cucinotta et al. [31]. However, for the sake of clarity, Table 2 reports all the settings used during the simulation.

Table 2. Settings.

Discretization Method	Finite Volume Method
Solver	Implicit Unsteady
Approach	Segregated Flow
Continuity and Momentum Equation coupling	SIMPLE-Algorithm
Convection Term	2nd Order
Turbulence Model	k-Omega Menter
Surface tension	CSF model
Temporal Discretization	2nd order
Iteration for Time Step	10
Time Step	Equation (1)
Gradient Discretization	Hybrid Gauss-LSQ
Algebraic system of Equations solver	AGM-Algebraic Multigrid Solver
Interface	VoF-Volume of Fluid
Convection Scheme for VoF	HRIC-High-Resolution Interface Capturing
Ship hull motion	DBI-Dynamic Fluid Body Interaction
Inner Iterations for Ship Motion	10
Mesh motion	Overset Mesh
Interpolation for Overset	Linear

In addition to the multiphase problem, the simulation also solves the motion of the boat. The approach used is the overset mesh. It helps to keep the quality of mesh very high. In order to use this procedure, the model was divided into two different regions: the background region (the fixed region in the space) and the overset region. The latter is free to move along the Z-axis (axis perpendicular to initial free-surface) and it is free to rotate itself along the Y-axis. The method used for solving the motion of the boat is the DFBI (Dynamic Fluid Body Interaction). It allows to solve the flow around a rigid body and simultaneously the motion of the rigid body caused by external forces and the forces induced by the flow (viscous forces, pressure forces). The equations of motion are the classic rigid body equations. For each step, the Navier–Stokes equations are solved considering the volume fraction under the hull by means VoF model. The viscous effects and the pressure field are integrated over the vessel in order to obtain the value of forces and moments. In function of these quantities, the solver applies the rigid body equations considering as center of motion the center of gravity of the boat. Finally, considering all the effects solved thanks to Navier–Stokes equations and Vof method (turbulence, air under the hull, waves, pressure field around the hull, viscous effects), the position, velocity, and acceleration of the boat is updated.

The geometry was defined with the commercial CAD software and all lines of the hull were defined with the higher precision possible (NURBS modeling). The ITTC [35] defines a series of guidelines in order to do a virtual towing tank. All these dimensions are shown in Figure 4.

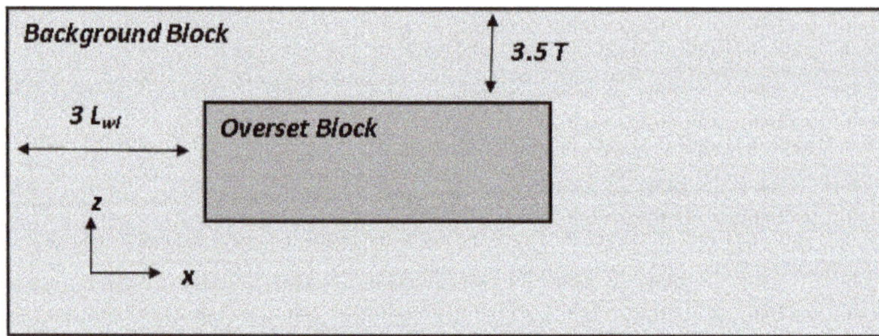

Figure 4. Overset and Background regions.

2.3. Boundary Conditions and Mesh

Figure 5 shows the background block with all the boundary conditions defined inside the solver. Each boundary condition is a field function that involves the velocity and pressure component of the fluid. The inlet has a Velocity Field function, it considers the position of the free surface and the velocity of the boat. The free surface is an unknown quantity that is solved each time step. The hidden surface is defined as symmetry in order to simulate only half of the domain. The outlet surface has a pressure boundary condition. It helps to update the free surface also in the outlet.

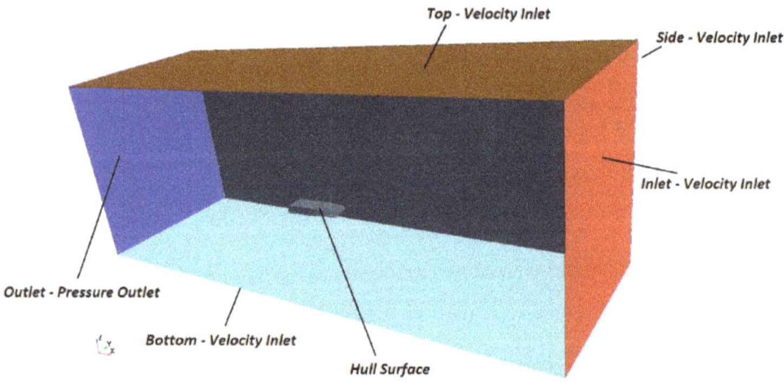

Figure 5. Boundary conditions.

In the overset region, the external surfaces are the ones linked with the background region. A zone where it is possible to have problems during the simulation is the overlapping zone between overset and background. There is a gap of elements where the solver exchanges the results of the corresponding zones; it is crucial to have a size of mesh very similar in this gap to avoid overflow errors. The blue surface is the symmetry area and the hull in grey is a wall no-slip surface. The step for injection of the air is a wall in no-slip condition (simulation is without air injection) and Inlet mass flow when the simulation involves the air injection.

The trimmed cartesian mesh process is used for this simulation. This approach allows defining different blocks for refinement. As shown in Figure 6, the refinement is conducted to keep a low aspect ratio and capture physics phenomena of different zones of interest. It is evident that a clear refinement is inside the overset block, in the free-surface zone, and the area around the boat with the Kelvin triangle (Figure 7).

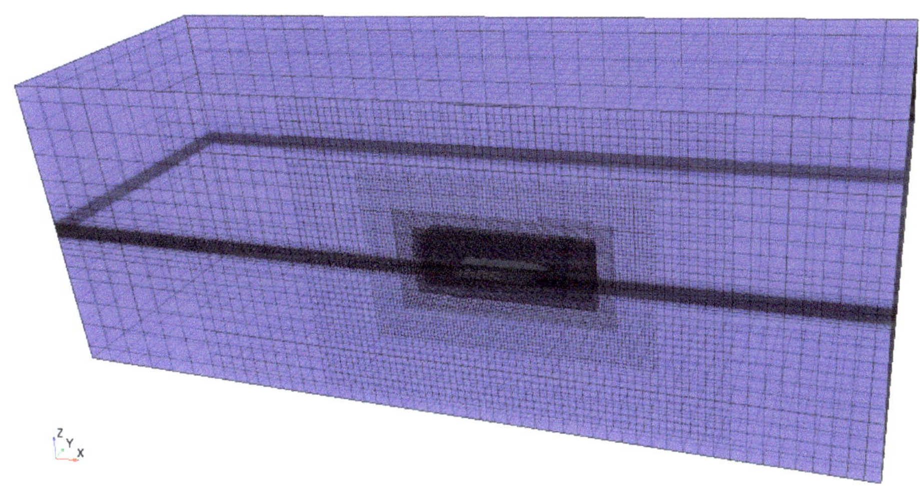

Figure 6. Mesh in all regions of the simulation. It is possible to notice the overset refinement.

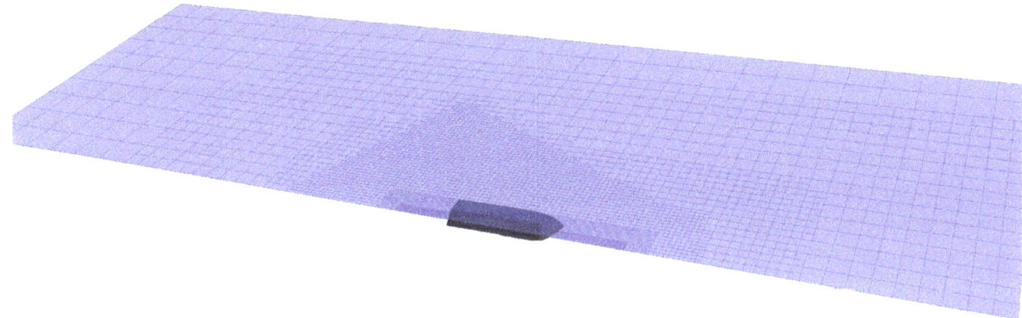

Figure 7. Kelvin refinement.

The hull has meshed with particular attention to the immersed zone (Figure 8). The rails were captured by the mesh with a local refinement, this increases the number of elements of the mesh. The total number of cells for the overset zone is 4.57 Mln and for the background zone is 0.49 Mln.

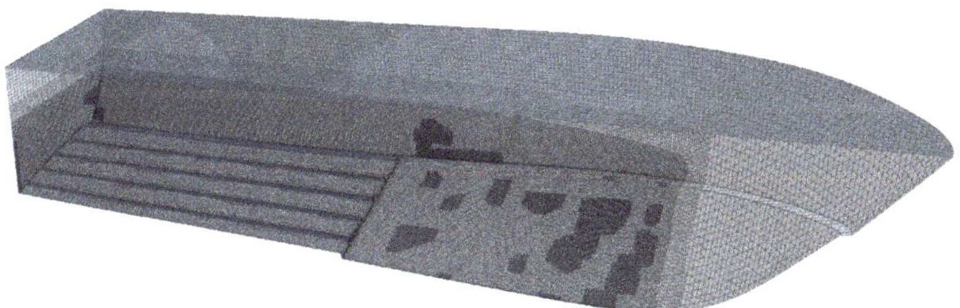

Figure 8. Hull refinement.

The wall surface of the hull has meshed with a prism layer of elements that allows capturing the boundary layer appropriately. The parameter that the authors keep under control is the Wall y+ that for all simulations is always between 30 and 200 (Figure 9).

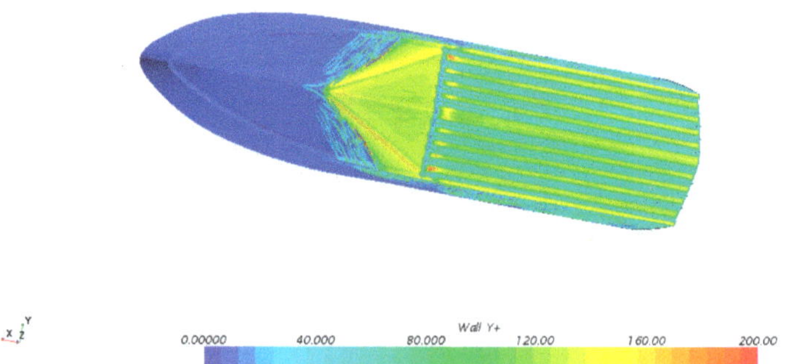

Figure 9. Wall y+ at full speed.

Finally, thanks to CFD, it is possible to know the area of hull wetted by the water or by the air. It is possible to solve this quantity considering a cut-off method on the volume fraction. The wetted surface by the water is calculated defining an interval of volume fraction where the hull is considered wet by the water. In this work, the chosen threshold was a volume fraction comprises between 0.5 and 1. The solver integrates the area where there are these values of water fraction and so evaluate the amount of hull wetted by the water.

2.4. Simulations Campaign

The simulations were conducted in two different parts. The first part involves the construction of the resistance curve in function of the velocity without the injection of the air. These results will be compared with experimental tests. During the simulation, the trend of the main physic quantities (drag, trim, and sinkage) was plotted and when these curves reached a stable condition, the simulation was over.

The velocities under investigation are the same as the experimental tests conducted and reported by Cucinotta et al. [22]. From the first campaign, the recorded pressure and velocity fields of the fluid and the relative position of the yacht in terms of trim and sinkage to each Froude number were recorded. These steady states became the initial conditions of the transient simulation of the second campaign. This methodology helps the convergence and the speed of the calculus.

During the first campaign of simulation, an uncertainty analysis has been conducted following the methodology proposed by ITTC [35]. The total uncertainty can be divided in three different terms: iterative (U_I), grid (U_G), and time-step uncertainties (U_T). Stern et al. [36] suggested that the value of uncertainty caused by the grid is an order of magnitude greater than other ones and it is recommended for each simulation to evaluate this value of uncertainty. The method consists in evaluating several n-th physic quantities (S_{n1}, S_{n2}, and S_{n3}) to three different meshes (1 fine mesh, 2 medium mesh, 3 coarse mesh) with a constant ratio between number of cells (r_k) (see Table 3). In this case, the physic quantities assessed are Drag, Trim, Sinkage, and Wetted Surface. For each parameter chosen, it can be calculated the difference between medium-fine mesh (2) and between coarse-medium mesh (3). Thanks to these values, it is possible to evaluate the convergence ratio R_G (4) and the order of accuracy P_G (5).

Table 3. Convergence ratio.

Value of Convergence Ratio	Type of Convergence
$0 < R_G < 1$	Monotonic
$R_G < 0 \ \& \ |R_G| < 1$	Oscillatory
$R_G > 1$	Monotonic divergence
$R_G < 0 \ \& \ |R_G| > 1$	Oscillatory divergence

Finally, with these two values it is possible to obtain the uncertainty with the Richardson formulation (6).

$$\varepsilon_{n21} = S_{n2} - S_{n1}, \tag{1}$$

$$\varepsilon_{n32} = S_{n3} - S_{n2}, \tag{2}$$

$$R_G = \frac{\varepsilon_{n21}}{\varepsilon_{n32}}, \tag{3}$$

$$P_G = \frac{ln\left(\frac{\varepsilon_{n32}}{\varepsilon_{n21}}\right)}{\ln r_k}. \tag{4}$$

$$U_n = F_S \left| \frac{\varepsilon_{n21}}{R_G^{P_G} - 1} \right| \tag{5}$$

The main objective of this part is to define the reduction of resistance with the injection of the air. The flow rates analyzed were generally ranging from 5500 L/min (liter per minute) to 10,500 L/min, with a step of 1000 L/min. The Froude number studied were between 0.64 and 0.89 and from 5500 to 10,500 L/min for the Froude number between 1.02 and 1.36. Only for low Froude numbers the same high flowrate was not tested, since the difference between low and high flowrate became negligible.

The total number of conditions analyzed is 28. Another important objective of this campaign of simulation is the distribution of the air under the hull during the the air injection. The conditions that were simulated are briefly summarized in Table 4.

Table 4. Velocities and time step used for the simulations.

	First Part of Simulation			Second Part of Simulation			
V_M [m/s]	Fn	Δ_t [s]	Two-Steps	Interval of Flow Rate [L/min]	Step Flow Rate [L/min]	Δ_t [s]	Two-Steps
3.15	0.64	0.01	Wall-No slip	5500–8500	1000	0.005	Air Flow Inlet
3.78	0.76	0.01	Wall-No slip	5500–8500	1000	0.005	Air Flow Inlet
4.41	0.89	0.01	Wall-No slip	5500–8500	1000	0.005	Air Flow Inlet
5.04	1.02	0.01	Wall-No slip	5500–10,500	1000	0.005	Air Flow Inlet
5.67	1.15	0.01	Wall-No slip	5500–10,500	1000	0.005	Air Flow Inlet
6.30	1.27	0.01	Wall-No slip	5500–10,500	1000	0.005	Air Flow Inlet
6.72	1.36	0.01	Wall-No slip	5500–10,500	1000	0.005	Air Flow Inlet

2.5. Hardware

All the operational phases described were conducted on a workstation with an Intel Xeon 2 GHz, 2 CPU with 16 core, 36 GB of memory, and a Nvidia Quadro M5000 8 GB.

3. Results

3.1. Results of the First Part of the Simulation

The first part of the simulation concerns the model without air injection. The curves of resistance, trim, and sinkage were compared with the ones of experimental tests. In adding to this information, the other purpose of this block of simulations is to have a starting point for the next campaign of simulation. Figure 10 shows the comparison between experimental and numerical simulation. The maximum difference between the two curves is to the maximum velocity.

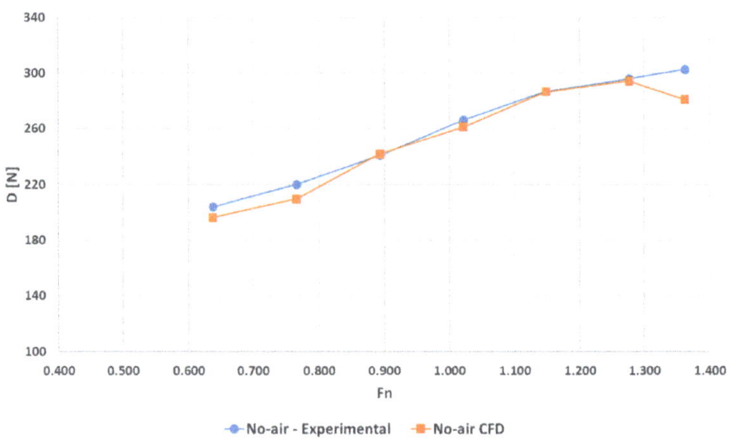

Figure 10. Resistance curves comparison.

Usually, the increase of velocity could lead to a higher probability of numerical ventilation effect. This effect is caused by two main reason: the VoF method, at high Fn and with an overset mesh [37], and the planing hull characteristics (the acute angle caused by the intersection between the hull and the free-surface) [38]. Even if this phenomenon is known and studied, there are very few specific studies on a problem as complex as that relating to a cavity of air injected at high speed under a surface equipped with dynamic motions and with a mesh overset. This effect produces localized areas under the hull with the presence of the air which in reality does not exist. This presence of the air could lead to a reduction of the drag with respect to experimental tests. In this case the effect is localized and produces a difference lower than 7%.

Another two crucial parameters compared are the trim and the sinkage. Figure 11 shows the behavior of the trim of the model during the different velocities. The trend of the curves is similar, with peaks of the CFD model more marked than the experimental test. Trim prediction is always the most challenging issue because it depends not only on the pressure but also on its distribution and, in particular, in this case, it is strictly related to the wetted length. However, even if the trend tends to change about Fn 0.8, the maximum difference is lower than 10%. In Figure 12 the sinkage comparison is reported, the two curves have the same trend.

The CFD results of this campaign confirm that the settings used for the Virtual Towing Tank allow having results very similar to experimental tests. The comparison highlights the same trend for the three different quantities reported. Another important physic quantity is the wetted surface of the boat. This quantity is not a result of the experimental tests, but it can be evaluated with CFD simulations. Figure 13 shows the ratio between the wetted surface area in static condition and during the corresponding speed. The graph highlights a drastic decrease of the wetted surface between 0.7 and 1 of the Froude number. In this condition, the wetted surface is about 65% of the static one. At the Froude number of 0.7, the hull starts to planing so there is a decreasing of the wetted surface. The complete planing is reached at a Froude number of about 1.0.

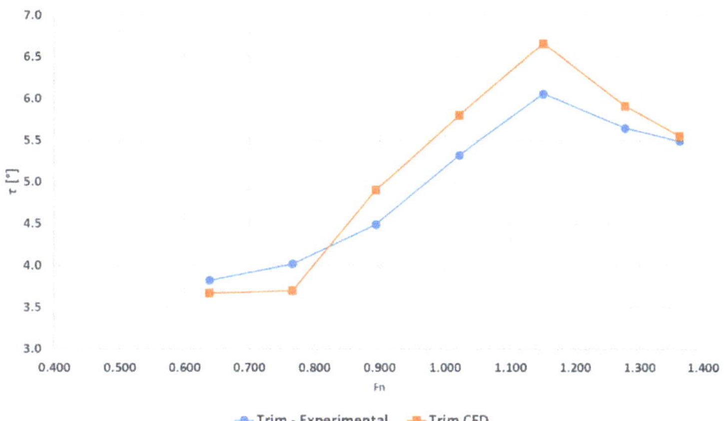

Figure 11. Trim comparison.

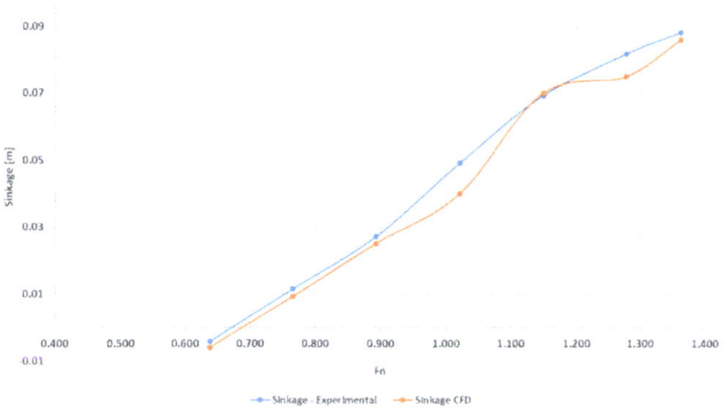

Figure 12. Sinkage comparison.

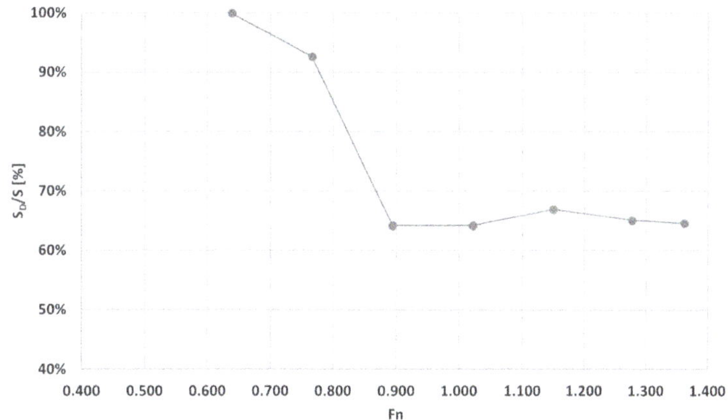

Figure 13. Wetted surface ratio–Dynamic/Static surface.

At the last value of this campaign of velocity, as mentioned above, an uncertainty analysis was conducted in order to evaluate the influence of the grid to the final result. At the value of Fn of 1.36, three different meshes were investigated with the number of cells in ratio of 1.4 (as suggested by ITTC [35]). Table 5 shows the results of the physical quantities investigated to the three different meshes. For all the parameters under investigation, the method used suggests a monotonic convergence (Table 6). Thanks to the monotonic convergence, the safety factor of the Richardson expression (6) as suggested by Stern [36] can be equal to 1. The same table shows that the maximum uncertainty is about 5.1% for the mesh used in this campaign of simulations. The maximum uncertainty is for the value of drag.

Table 5. Number of elements and different result for different meshes.

Parameter	Fine	Medium	Coarse
Elements	2,345,723	1,663,633	1,184,081
D	281	293	315
τ_{CFD} [deg]	5.55	5.72	6.01
$DTBow_{CFD}$ [m]	0.086	0.091	0.102
S_{DYN} [m^2]	1.408	1.443	1.502

Table 6. Uncertainty analysis.

Parameter	Medium/Fine ε_{n21}	Coarse/Medium ε_{n32}	R_G	P_G	U [%]
D	12	22	0.55	1.75	5.1
τ_{CFD}	0.17	0.29	0.59	1.54	4.3
$DTBow_{CFD}$	−0.005	−0.011	0.45	2.28	4.8
S_{DYN}	−0.035	−0.06	0.59	1.51	3.6

3.2. The Second Campaign of Simulation

In this part of the simulation, for each number of Froude simulated in the first campaign, the virtual tests were performed adding the air injection. Also in this case, the numerical simulations were compared with experimental tests and in adding, thanks to the Virtual Towing tank, the wetted surface was reported for each condition of air flow.

Figure 14 shows the first two flow rates with the relative curves of experimental tests. In both cases, the CFD results of air injection have a trend lower than experimental tests. In general, the air injection to these flow rates produces a decrease of resistance in the magnitude of 1% on average for the experimental tests. For CFD results, this decrease of resistance, on average, is in the magnitude of 5%. The difference between experimental tests and numerical simulations is lower than 15% in all the points. The difference between experimental tests and numerical simulations could be caused by the great mixture of air and water to these values of flow rate. The VoF scheme, with the HRIC model tends to underestimate the mix of air and water in the rails and underestimate the wetted surface, introducing a more positive effect on drag than experimental tests.

Thanks to CFD it is possible to evaluate the differences in terms of reduction of the wetted surface with the air injection. The reduction of the wetted surface allows decreasing the frictional resistance of the model. On average, the reduction of wetted surface for 5500 L/min and 6500 L/min is about 60%. Figure 15 shows the trend to different velocities for both the flow rates. The first two flow rates investigated suggest that no great difference there is between 5500 L/min and 6500 L/min. The air under the hull probably is not entirely developed and the effect in terms of drag is very low in both flow rates.

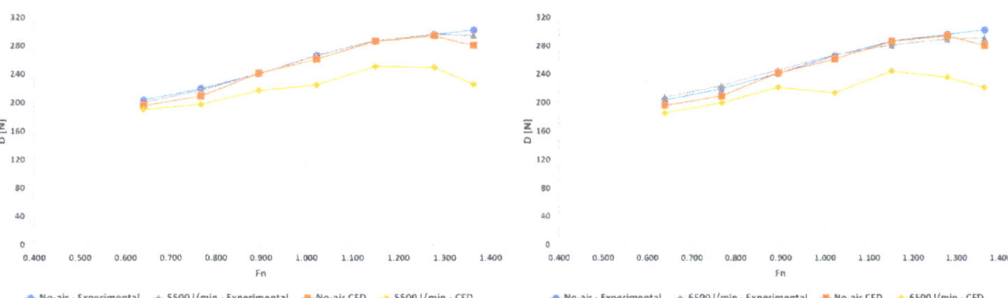

Figure 14. Experimental and CFD results to air injection 5500 L/min and 6500 L/min.

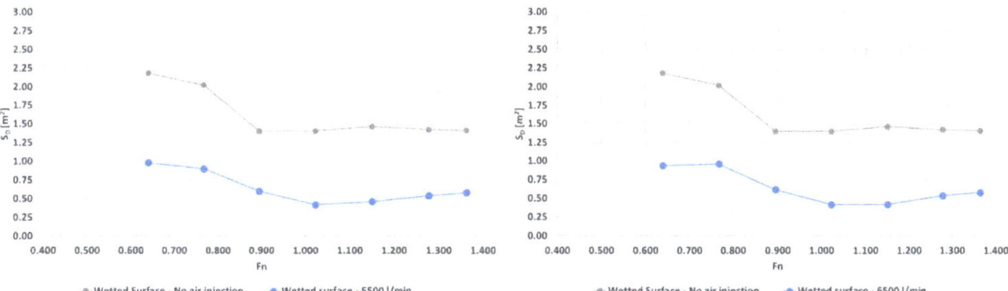

Figure 15. Wetted surface comparison–Air (5500 L/min and 6500 L/min) and No air.

Figure 16 shows the drag curves for 7500 L/min and 8500 L/min flow rates and the relative ones without air injection. In both cases, the CFD results of air injection have a trend lower than experimental tests. In general, the air injection to flow rate of 7500 L/min produces a decrease of resistance in the magnitude of 2% on average for the experimental tests. For CFD results, this decrease is about 7%. In the case of 8500 L/min, the reduction is about 8% for experimental tests and 11% for CFD results. The difference between experimental tests and numerical simulations is lower than 15% in all the points.

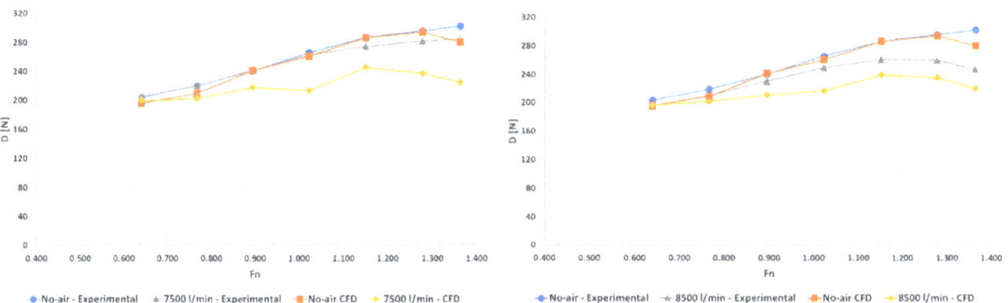

Figure 16. Experimental and CFD results to air injection 7500 L/min and 8500 L/min.

Figure 17 shows the wetted surface at different Fn for flow rates of 7500 and 8500 L/min. The decrease of the wetted surface area is respectively 62% and 65%. To the flow rate of 8500 L/min the decrease of drag is appreciable in experimental tests and in CFD simulations. The flow of air under the hull starts to cover a higher percentage of bottom of the hull. This value of flow rate seems to be the turning point, lower than this value the effects are negligible.

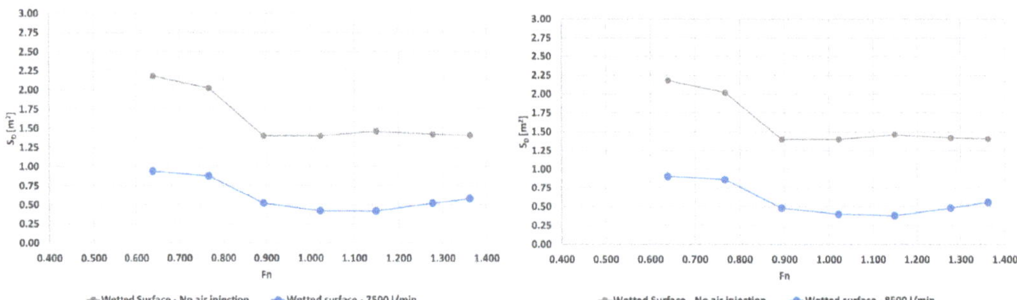

Figure 17. Wetted surface comparison–Air (7500 L/min and 8500 L/min) and No air.

Figure 18 shows the drag curves for 9500 L/min and 10,500 L/min flow rates and the relative ones without air injection. In this case, the CFD curves have a trend lower than the experimental ones, but the difference is reduced to lower than 10%. In general, the air injection to flow rate of 9500 L/min produces a decrease of resistance in the magnitude of 11% on average for the experimental tests. For CFD results, this decrease is about 15%. For a flow rate of 10,500 L/min the decrease is 13% for the experimental test and 17% for CFD results.

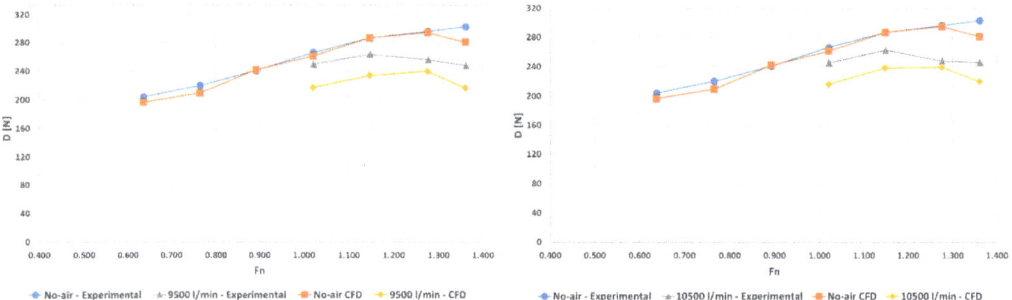

Figure 18. Experimental and CFD results to air injection 9500 L/min and 10,500 L/min.

Comparing to similar experiences carried out on the same hull but without the presence of the rails (i.e., [31,32]), in this study, the agreement between experimental and numerical data is lower for all the ventilated cases. This result probably depends on the difficulty of the interface models used to predict the real ventilated hull surface. In particular, the HRIC scheme tends to underestimate the air-water mixing in correspondence of hull discontinuity, and consequently to overestimate the ventilated surface, leading a reduction in the drag prediction. This numerical phenomenon must be properly evaluated as it was not a problem for flat bottom hulls.

Figure 19 shows the trend to different velocities for flow rates to 9500 and 10,500 L/min. The decrease of the wetted surface area is respectively 68% and 69%. The flow rates are the higher ones and the air under the hull cover almost the entire bottom.

Figure 20 shows the trend of drag reduction to different flow rates at the same Fn. The curves show that with the increase of the Froude number the curves tend to an high level of reduction of resistance. The curves show that to Fn number of 0.639 and 0.766, the decrease of resistance is almost equal to zero and also with the increase of the flow rates this behaviour remains the same.

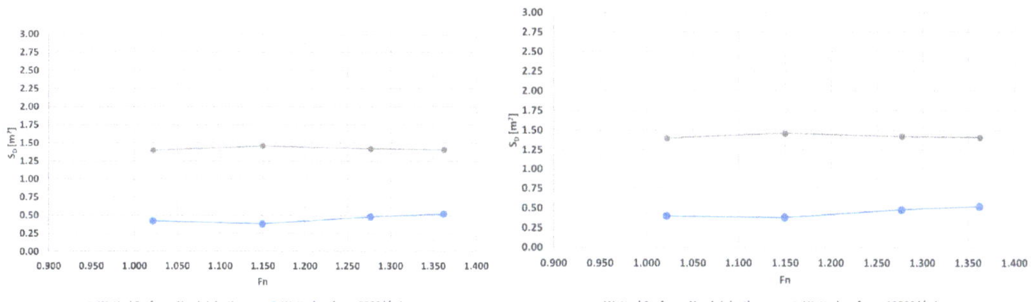

Figure 19. Wetted surface comparison–Air (9500 L/min and 10,500 L/min) and No air.

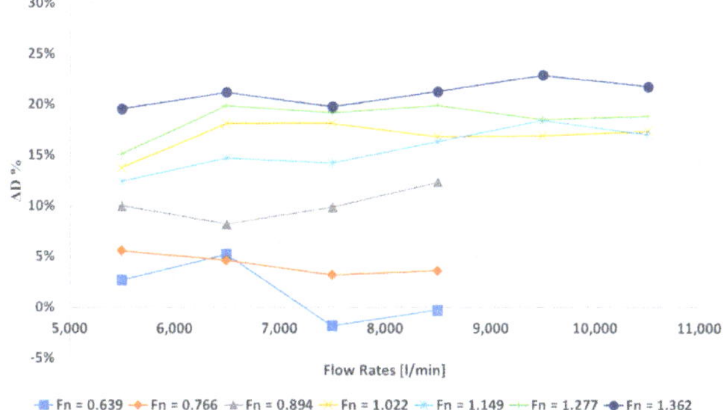

Figure 20. Reduction of resistance vs flow rate to the same Fn.

To the Fn number of 0.894 the starting point of increase of percentage reduction is the flow rate of 6500 L/min. This starting point it is also visible for the Froude number equal to 1.149 but it is translated to the flow rate of 7500 L/min. To the highest values of Froude number, the curves seem flat and the flow rates does not affect the reduction of drag. To these values of Fn, there is not great need to push the flow rate until to 10,500 in order to obtain the desidered effects.

The advantage of this hull grows with increasing speed, up to a peak reduction of 24% (at Fn 1.362 and 9500 L/min). For a speed of Fn < 1, the ventilation is not particularly effective and tends not to lead to benefits. For Fn $\geq$ 1, on the other hand, the benefits grow rapidly and the most advantageous air flow rates are those at 9500 L/min, above which the gain is reduced again.

In Figure 21 are reported all the wetted surfaces, for each Fn, in three different conditions, respectively, the no-air condition and the minimum and maximum tested flow rates. In the pictures, the water is represented by the color red and the air by the color blue. The figure highlights that to all the Froude numbers, the flow rate of 5500 L/min is not enough to wet the bottom of the hull completely. At this flow rate, for little Froude numbers, the air tends to escape from the sides, while for high Froude numbers there is a tendency of the air to be dragged up to the stern, remaining at the center of the boat. At the same time, at high speeds, air and water are less mixed. For this reason, for the Froude number higher than 1.27, the air does not escape from the sides of the boat and on the bilge area there is a channel of a stable water. From the same figure, it is possible to see that the ideal condition seems to be reached with the condition of 10,500 L/min at the Froude number of 1.15. In this case, the bottom of the hull and the area around the bilge are almost wholly wetted by the air without mixing with water.

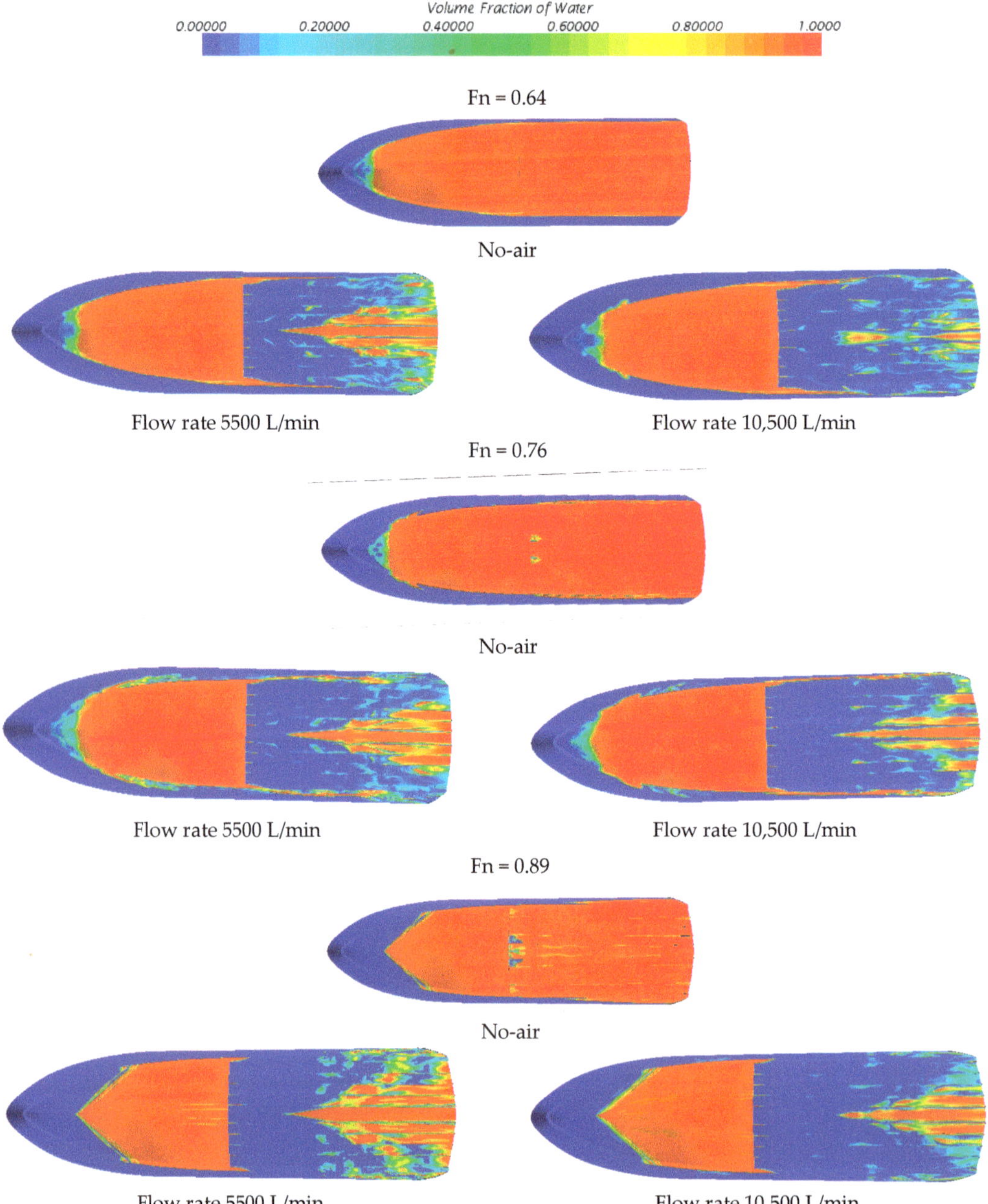

Figure 21. *Cont.*

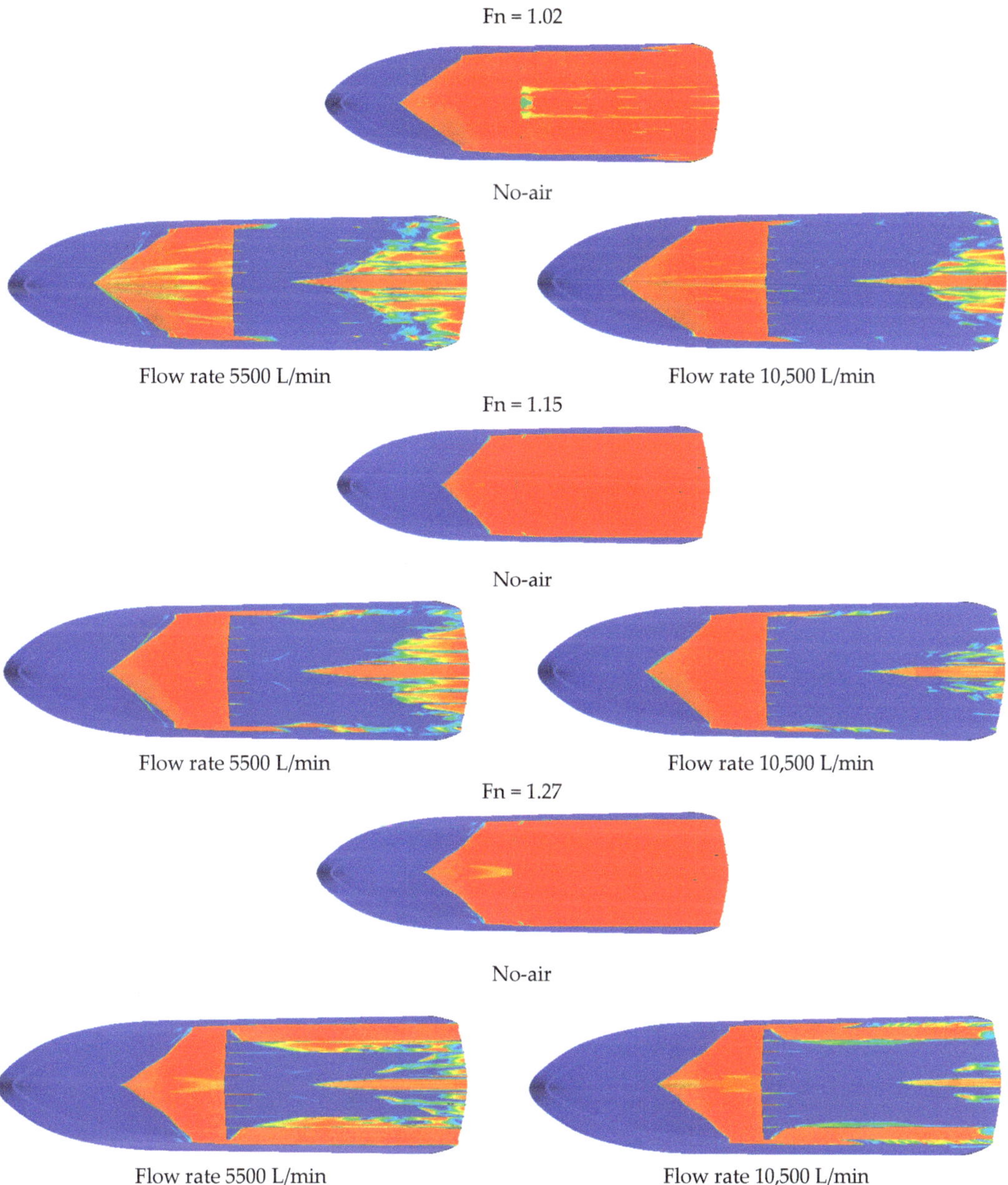

Figure 21. *Cont.*

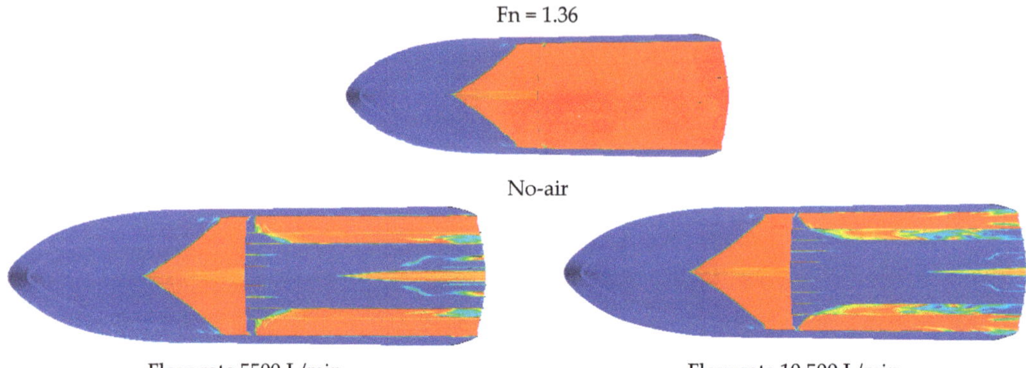

Figure 21. Wetted surface in three different conditions: no-air and injection of the air to the minimum and maximum flow rate.

The longitudinal rails at low flow rates cannot retain the air and consequently are a disadvantage more than an advantage since they increase the wetted surface.

4. Conclusions

The experimental and numerical campaigns carried out on an ACS showed the high potentiality of the hull ventilation, also for planing hulls. The natural under pressure generates under the step, as in an airfoil lower surface, help the air to ventilate the hull without escaping too fast.

The hull reach a greatest advantage of about 24% of drag reduction. In the best conditions, it is clearly visible that the chain is wet (a sign that the air does not escape laterally) and that consequently the air channel completely fills the cavity up to the transom.

Event if it can be considered a good result, the rails under the hull have not been shown to give a great benefit compared to the hull without them. The idea for which they were designed, was that of channeling the flow of air towards the stern, limiting the lateral escaping, has proved to be ineffective. The airflow towards the stern in fact depends mainly on the Froude number. As Fn increases, the air tends to follow the travel direction, in the same way as the hull with the flat bottom [31]. On the contrary, the rails considerably increase the wet surface. The overall balance is therefore negative.

Even if the accordance between experimental and CFD results is good, the hull equipped with rails, compared to the flat ones showed in other papers, reveals a more significant error. This is probably due to greater difficulty in estimating the actual wetted surface in the area of the rails.

In this sense, the VoF scheme with the HRIC model, which well describes the wave surface into the air–water interface, tends to underestimate the mix of air and water in the rails and underestimate the wetted surface. Indeed, the real effect showed by experiments leads to the generation of microbubbles, instead of big, well-formed bubbles. The CFD campaign, therefore, showed that on this kind of boat, when the discontinuities under the hull increase, not only the hull has not great benefits in terms of drag reduction, but also that the numerical error increases.

The obtained results indicate interesting possibilities for future development in the study of biphasic phenomena under the hull, particularly in the presence of abrupt discontinuities. Different models than HRIC can be, in future, developed and applied to search for a better matching between experimental and numerical outcomes.

Author Contributions: Conceptualization, F.C. and F.S.; methodology, F.C.; software, F.S.; validation, F.S., F.T., and D.M.; data curation, F.T. and D.M.; writing—original draft preparation, F.S.; writing—review and editing, F.C.; supervision, F.C. All authors have read and agreed to the published version of the manuscript.

Funding: This work was partially supported by European Union (PON R&C 2007–2013 and PON 2015–2020 ARS01_00334) funding. Authors wish to thank the personnel of Naples Towing Tank for the support during the experimental tests.

Institutional Review Board Statement: Not applicable.

Informed Consent Statement: Not applicable.

Data Availability Statement: The data presented in this study are available on request from the corresponding author.

Conflicts of Interest: The authors declare no potential conflicts of interest with respect to the research, authorship, and/or publication of this article.

Nomenclature

Definition	Symbol	Unit
Overall Length	LOA	m
Waterline Length	LWL	m
Projected Chine Length	LP	m
Waterline Beam	BWL	m
Projected Maximum Beam	BPX	m
Projected Beam at generic X position	BPC	m
Projected Beam Transom	BPT	m
Deadrise Angle	β	$^\circ$
Height of medium buttock line	HLM	m
Draught	T	m
Displacement	Δ	t
Total Resistance	R_T	N
Weight force	W	N
Longitudinal centre of gravity	x_G	m
Wetted surface Area	S	m^2
Dynamic Wetted Surface Without Air Injection	S_D	m^2
Dynamic Wetted Surface With Air Injection	S_{DAir}	m^2
Projected Area	AP	m^2
Velocity of Ship	V_S	m/s
Velocity of Model	V_M	m/s
Froude number	Fn	-
Length of Ship	L_S	m
Length of Model	L_M	m
Scale	λ	-
Number of Steps	N_{ST}	-
Position of the step relative to the transom	L_{ST}	m
Refinement ratio of the three meshes	r_K	
Area Step	S_{ST}	m^2
Number of nozzles	N_{IN}	-
Dimensions-Basis × Height	$B_{IN} \times H_{IN}$	m × m
Area of nozzles	S_{IN}	m^2
Volumetric Flow Rate	Q	m^3/s

References

1. Lewis, E.V. *Principles of Naval Architecture*; The Society of Naval Architects and Marine Engineers: Jersey City, NJ, USA, 1988; ISBN 0-939773-01-5.
2. Larsson, L.; Raven, H. *Ship Resistance and Flow*; Paulling, J.R., Ed.; The Society of Naval Architects and Marine Engineers: Jersey City, NJ, USA, 2010; ISBN 978-0-939773-76-3.

3.   Papanikolaou, A. Developments and Potential of Advanced Marine Vehicles Concepts. *Bull. Kansai Soc. Nav. Achitects* **2002**, *55*, 50–54.

4.   Matveev, K. On the Limiting Parameters of Artificial Cavitation. *Ocean Eng.* **2003**, *30*, 1179–1190. [CrossRef]

5.   Cucinotta, F.; Nigrelli, V.; Sfravara, F. A Preliminary Method for the Numerical Prediction of the Behavior of Air Bubbles in the Design of Air Cavity Ships. In *Advances in Mechanical Engineering*; J.B. Metzler: Stuttgart, Germany, 2017; pp. 509–516.

6.   Kodama, Y.; Kakugawa, A.; Takahashi, T.; Sugiyama, T. Drag Reduction of Ships by Microbubbles. In Proceedings of the 24th Symposium on Naval Hydrodynamics, Fukuoka, Japan, 8–13 July 2002.

7.   Savitsky, D. Hydrodynamic Design of Planing Hulls. *Mar. Technol. SNAME News* **1964**, *1*, 71–95. [CrossRef]

8.   Dashtimanesh, A.; Tavakoli, S.; Sahoo, P. A Simplified Method to Calculate Trim and Resistance of a Two-Stepped Planing Hull. *Ships Offshore Struct.* **2017**, *12*, S317–S329. [CrossRef]

9.   Dashtimanesh, A.; Roshan, F.; Tavakoli, S.; Kohansal, A.; Barmala, B. Effects of Step Configuration on Hydrodynamic Performance of One- and Doubled-Stepped Planing Flat Plates: A Numerical Simulation. *Proc. Inst. Mech. Eng. Part M J. Eng. Marit. Environ.* **2019**, *234*, 181–195. [CrossRef]

10.   Dashtimanesh, A.; Tavakoli, S.; Kohansal, A.; Khosravani, R.; Ghassemzadeh, A. Numerical Study on a Heeled One-Stepped Boat Moving Forward in Planing Regime. *Appl. Ocean Res.* **2020**, *96*, 102057. [CrossRef]

11.   De Marco, A.; Mancini, S.; Miranda, S.; Scognamiglio, R.; Vitiello, L. Experimental and Numerical Hydrodynamic Analysis of a Stepped Planing Hull. *Appl. Ocean Res.* **2017**, *64*, 135–154. [CrossRef]

12.   Bilandi, R.N.; Dashtimanesh, A.; Tavakoli, S. Development of a 2D+T Theory for Performance Prediction of Double-Stepped Planing Hulls in Calm Water. *Proc. Inst. Mech. Eng. Part M J. Eng. Marit. Environ.* **2018**, *233*, 886–904. [CrossRef]

13.   Bilandi, R.N.; Dashtimanesh, A.; Tavakoli, S. Hydrodynamic Study of Heeled Double-Stepped Planing Hulls using CFD and 2D+T Method. *Ocean Eng.* **2020**, *196*, 106813. [CrossRef]

14.   Bilandi, R.N.; Vitiello, L.; Mancini, S.; Nappo, V.; Roshan, F.; Tavakoli, S.; Dashtimanesh, A. Calm-Water Performance of a Boat with two Swept Steps at High-Speeds: Laboratory Measurements and Mathematical Modeling. *Procedia Manuf.* **2020**, *42*, 467–474. [CrossRef]

15.   Lee, E.; Pavkov, M.; McCue-Weil, L. The Systematic Variation of Step Configuration and Displacement for a Double-Step Planing Craft. *J. Ship Prod. Des.* **2014**, *30*, 89–97. [CrossRef]

16.   Taunton, D.J.; Hudson, D.A.; Shenoi, R.A. Characteristics of a Series of High Speed Hard Chine Planing Hulls-Part 1: Performance in Calm Water. *Trans. R. Inst. Nav. Archit. Part B Int. J. Small Cr. Technol.* **2010**, *152*. [CrossRef]

17.   Butuzov, A.A.; Vasin, A.I.; Drozdov, A.L.; Ivanov, A.N.; Kalyuzhny, V.G.; Matveev, I.I.; Ruzanov, V.E. Full-Scale Trials of a Boat with an Air Cavity. *Shipbuild. Probl.* **1988**, *28*, 45–51.

18.   Mäkiharju, S.A.; Elbing, B.R.; Wiggins, A.; Schinasi, S.; Vanden-Broeck, J.-M.; Perlin, M.; Dowling, D.R.; Ceccio, S.L. On the Scaling of Air Entrainment from a Ventilated Partial Cavity. *J. Fluid Mech.* **2013**, *732*, 47–76. [CrossRef]

19.   Matveev, K.I. Two-Dimensional Modeling of Stepped Planing Hulls with Open and Pressurized Air Cavities. *Int. J. Nav. Arch. Ocean Eng.* **2012**, *4*, 162–171. [CrossRef]

20.   Matveev, K.I. Three-Dimensional Wave Patterns in Long Air Cavities on a Horizontal Plane. *Ocean Eng.* **2007**, *34*, 1882–1891. [CrossRef]

21.   Butterworth, J.; Atlar, M.; Shi, W. Experimental Analysis of an Air Cavity Concept Applied on a Ship Hull to Improve the Hull Resistance. *Ocean Eng.* **2015**, *110*, 2–10. [CrossRef]

22.   Cucinotta, F.; Guglielmino, E.; Sfravara, F. An Experimental Comparison between Different Artificial Air Cavity Designs for a Planing Hull. *Ocean Eng.* **2017**, *140*, 233–243. [CrossRef]

23.   Wang, L.; Huang, B.; Qin, S.; Cao, L.; Fang, H.; Wu, D.; Li, C. Experimental Investigation on Ventilated Cavity Flow of a Model Ship. *Ocean Eng.* **2020**, *214*, 107546. [CrossRef]

24.   Barbaca, L.; Pearce, B.W.; Ganesh, H.; Ceccio, S.L.; Brandner, P.A. On the Unsteady Behaviour of Cavity Flow over a Two-Dimensional Wall-Mounted Fence. *J. Fluid Mech.* **2019**, *874*, 483–525. [CrossRef]

25.   Qin, S.; Wu, Y.; Wu, D.; Hong, J. Experimental Investigation of Ventilated Partial Cavitation. *Int. J. Multiph. Flow* **2019**, *113*, 153–164. [CrossRef]

26.   Cao, L.; Karn, A.; Arndt, R.E.A.; Wang, Z.; Hong, J. Numerical Investigations of Pressure Distribution Inside a Ventilated Supercavity. *J. Fluids Eng.* **2016**, *139*, 021301. [CrossRef]

27.   Begovic, E.; Bertorello, C.; Mancuso, A.; Saporito, A. Sailing Dinghy Hydrodynamic Resistance by Experimental and Numerical Assessments. *Ocean Eng.* **2020**, *214*, 107458. [CrossRef]

28.   Saporito, A.; Persson, A.; Larsson, L.; Mancuso, A. A New Systematic Series of Foil Sections with Parallel Sides. *J. Mar. Sci. Eng.* **2020**, *8*, 677. [CrossRef]

29.   Cirello, A.; Cucinotta, F.; Ingrassia, T.; Nigrelli, V.; Sfravara, F. Fluid–Structure Interaction of Downwind Sails: A New Computational Method. *J. Mar. Sci. Technol.* **2019**, *24*, 86–97. [CrossRef]

30.   Cella, U.; Cucinotta, F.; Sfravara, F. *Sail Plan Parametric CAD Model for an A-Class Catamaran Numerical Optimization Procedure using Open Source Tools*; Springer: Cham, Switzerland, Basel, Switzerland; 2017.

31.   Cucinotta, F.; Guglielmino, E.; Sfravara, F.; Strasser, C. Numerical and Experimental Investigation of a Planing Air Cavity Ship and its Air Layer Evolution. *Ocean Eng.* **2018**, *152*, 130–144. [CrossRef]

32. Cucinotta, F.; Guglielmino, E.; Sfravara, F. A Critical CAE Analysis of the Bottom Shape of a Multi Stepped Air Cavity Planing Hull. *Appl. Ocean Res.* **2019**, *82*, 130–142. [CrossRef]
33. Cucinotta, F.; Guglielmino, E.; Sfravara, F. Life Cycle Assessment in Yacht Industry: A Case Study of Comparison between Hand Lay-Up and Vacuum Infusion. *J. Clean. Prod.* **2017**, *142*, 3822–3833. [CrossRef]
34. Siemens. *Simcenter STAR-CCM+ Documentation*; Granite Park One; Siemens: Plano, TX, USA, 2020.
35. ITTC. *Uncertainty Analysis in CFD Verification and Validation Methodology and Procedures*; Ocean Technology Laboratory COPPE/UFRJ-Federal University of Rio de Janeiro: Rio de Janeiro, Brazil, 2011.
36. Stern, F.; Wilson, R.V.; Coleman, H.W.; Paterson, E.G. Comprehensive Approach to Verification and Validation of CFD Simulations—Part 1: Methodology and Procedures. *J. Fluids Eng.* **2001**, *123*, 793–802. [CrossRef]
37. Samuel; Kim, D.J.; Fathuddiin, A.; Zakki, A.F. A Numerical Ventilation Problem on Fridsma Hull Form Using an Overset Grid System. *IOP Conf. Series Mater. Sci. Eng.* **2021**, *1096*, 012041. [CrossRef]
38. Wang, H.; Zhu, R.; Huang, S.; Zha, L. Hydrodynamic Analysis of a Planing Hull in Calm Water Using Overset Mesh and Rigid Body Motion Method. In Proceedings of the 30th International Ocean and Polar Engineering Conference 2020 Sep 2. International Society of Offshore and Polar Engineers, Shanghai, China, 11–16 October 2020.

Journal of
Marine Science and Engineering

Article

Cartesian Mesh Generation with Local Refinement for Immersed Boundary Approaches

Luca Di Angelo *, Francesco Duronio, Angelo De Vita and Andrea Di Mascio

Department of Industrial and Information Engineering and Economics, University of L'Aquila, Piazzale E. Pontieri, Monteluco di Roio, 67100 L'Aquila, Italy; francesco.duronio@graduate.univaq.it (F.D.); angelo.devita@univaq.it (A.D.V.); andrea.dimascio@univaq.it (A.D.M.)
* Correspondence: luca.diangelo@univaq.it

Abstract: In this paper, an efficient and robust Cartesian Mesh Generation with Local Refinement for an Immersed Boundary Approach is proposed, whose key feature is the capability of high Reynolds number simulations by the use of wall function models, bypassing the need for accurate boundary layer discretization. Starting from the discrete manifold model of the object to be analyzed, the proposed model generates Cartesian adaptive grids for a CFD simulation, with minimal user interactions; the most innovative aspect of this approach is that the automatic generation is based on the segmentation of the surfaces enveloping the object to be analyzed. The aim of this paper is to show that this automatic workflow is robust and enables to get quantitative results on geometrically complex configurations such as marine vehicles. To this purpose, the proposed methodology has been applied to the simulation of the flow past a BB2 submarine, discretized by non-uniform grid density. The obtained results are comparable with those obtained by classical body-fitted approaches but with a significant reduction of the time required for the mesh generation.

Keywords: Cartesian adaptive grids; immersed boundaries; LES simulation

Citation: Di Angelo, L.; Duronio, F.; De Vita, A.; Di Mascio, A. Cartesian Mesh Generation with Local Refinement for Immersed Boundary Approaches. *J. Mar. Sci. Eng.* **2021**, *9*, 572. https://doi.org/10.3390/jmse9060572

Academic Editors: Antonio Mancuso and Davide Tumino

Received: 28 April 2021
Accepted: 20 May 2021
Published: 25 May 2021

Publisher's Note: MDPI stays neutral with regard to jurisdictional claims in published maps and institutional affiliations.

1. Introduction

The increasing popularity of Computational Fluid Dynamics (CFD) in marine engineering sciences, observed in the last few decades, is to be ascribed to the growth of computational power, in combination with the increase of robustness and accuracy of CFD solvers. Today, Reynolds averaged Navier–Stokes simulations on body-fitted meshes are commonly performed in naval architecture, in order to save time in the design process and make it less expensive; conventional towing tank or water channel tests are usually limited to a few shapes obtained in the final design. Nevertheless, the bottleneck of the whole simulation procedure remains mesh generation; in order to obtain a mesh that satisfies the requirements of smoothness and proper clustering, particularly in boundary layers and wakes, its generation still requires lengthy human interaction and relevant expertise by the user.

Nowadays, the most used method for geometry discretization is the body-fitted approach, particularly for high Reynolds number flows, and most solvers handle unstructured or block-structured grids, possibly with partial overlapping: their generation remains the most demanding task in the total effort and time for the complete simulation [1,2]. The major reason for this last aspect is that the process is never completely automatic, except in those cases where the geometry is so simple that it is possible to parametrize its shape. This is even more complicated in optimization algorithms, where only minor model changes in shape (and not in topology) are allowed.

In the two last decades, the Immersed Boundary (IB) method has emerged as a valid alternative to the body-fitted meshes-based CFD methods. The key idea of IB methods is to locally modify the governing equations in order to enforce the boundary conditions without a body-fitted mesh: this avoids the complex and time-consuming body-fitted meshes generation, by allowing the discretization with a simple structured Cartesian mesh. The most important advantages of this technique are the easy grid generation, also when dealing with moving boundaries. The IB idea can be attributed to Peskin et al. [3], who used it to simulate cardiac mechanics and the associated blood flow. Nowadays, many variants and procedures of the IB original ideas are proposed: in order to enforce the boundary condition on the interfaces, some exploit a continuous forcing term in the field equations, others explicitly locally modify the equations; excellent reviews can be found, for instance, in [4,5]. With the introduction of wall models [6,7], the use of IB methods was applied also to resolve high Reynolds number flows, mitigating the need for accurate resolution within the boundary layer.

This paper proposes an efficient and robust Cartesian Mesh Generation with Local Refinement for Immersed Boundary Approaches. In particular, the proposed methodology was developed to be suited for the method proposed in [8], which couples the Immersed Boundary approach to the level-set method, and also makes use of wall functions at rigid walls. Although several methods have been published in the literature [8], none of them completely satisfies the requirements of the specific IB considered here; furthermore, for the authors' knowledge, none considers the differential geometric properties of the model surfaces to be analyzed to define an optimized geometry-based discretization. The proposed method aims to overcome these limitations by introducing strategies of diversification of the mesh dimensions in the different parts of the model, based on automatic segmentation of the surfaces enveloping the object.

The proposed methodology, i.e., the Cartesian Mesh Generation with Local Refinement and the IB method with wall functions, is applied to study the flow past a submarine at a high Reynolds number. The obtained results show how the use of the proposed CFD tools is extremely helpful to capture the main flow characteristics in the wake of all the appendages, although the details in the boundary layers are lost because of the adoption of wall functions. The reported results suggest that the use of immersed boundary approach is mature enough to be used as an investigation tool in naval architecture.

2. The Cartesian Mesh Generation Method

The proposed Cartesian mesh generation method is specifically developed for the IB method developed in [1]; the approach produced a Cartesian grid with the following features:

- a signed distance from the wall (positive inside the body and negative outside) is defined at each point;
- the mesh can be easily refined close to the boundary and where the solution requires finer discretization (typically in the wake);
- it can consist of block structured Cartesian blocks with possible partial overlapping.

Particular attention is given to the data structure, in order to optimize the storage and minimize interfaces, in view of parallel calculation.

In the related literature, many methods have been published for the generation of Cartesian grids (the interested reader can find details in [8]), although none completely suited to the needs of the developed flow solver. Therefore, a specific grid generation algorithm, whose flow-chart is shown in Figure 1, is implemented.

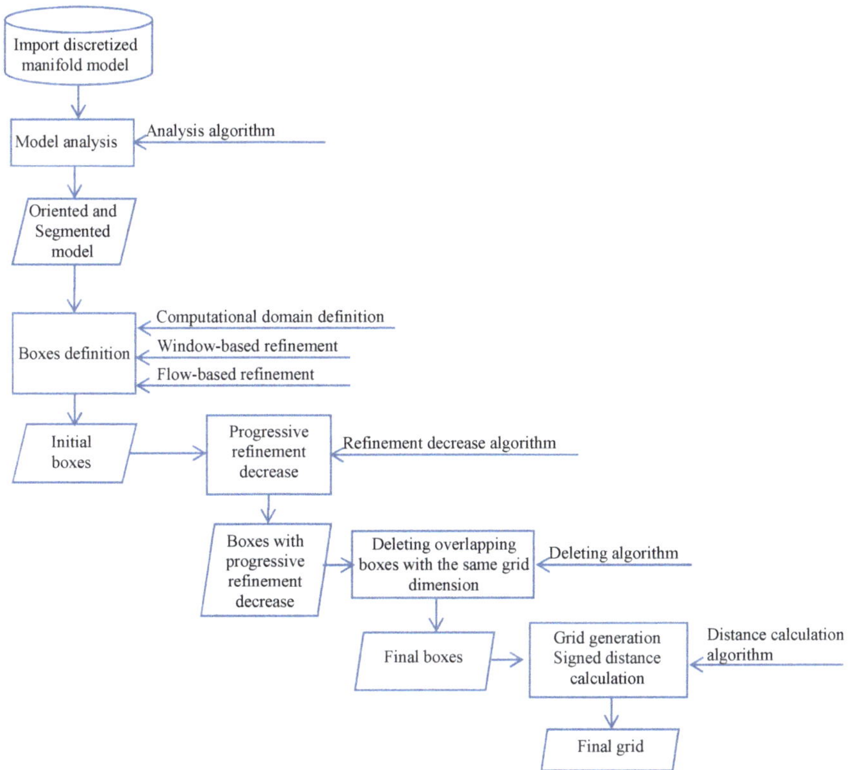

Figure 1. Flow-chart of the proposed method.

2.1. Model Analysis

The proposed algorithm starts from a discrete manifold model (in STL format) and store it into two tables (named as "Points" and "Triangles") containing the information about the planar triangular facets (Figure 2a):

- **Points** (x_i, y_i, z_i) for $i = i, \ldots, np$: where the coordinates of the np unique points are stored;
- **Triangles**: where three pointers to **Points** are stored for each triangle.

The structure of the two tables avoids redundancy of information. The model is then positioned by rigid roto-translation operations in the Global Reference System (O, x, y, z) of the computational domain so that:

$$\begin{cases} x^M_{\min} = 0 \\ y^M_G = 0 \\ z^M_G = 0 \\ \text{flow direction} \; // \; x\text{-axis} \end{cases}$$

Here, $\left\{ X^M_G, Y^M_G, Z^M_G \right\}$ are coordinates of the model centroids where the origin of the new reference frame is translated; all the coordinates of the **Points** are then recomputed in this reference system (Figure 2b). The model processing includes an automatic surfaces segmentation, based on a fuzzy analysis of the discrete differential properties, according to the method proposed in [9] (Figure 2c).

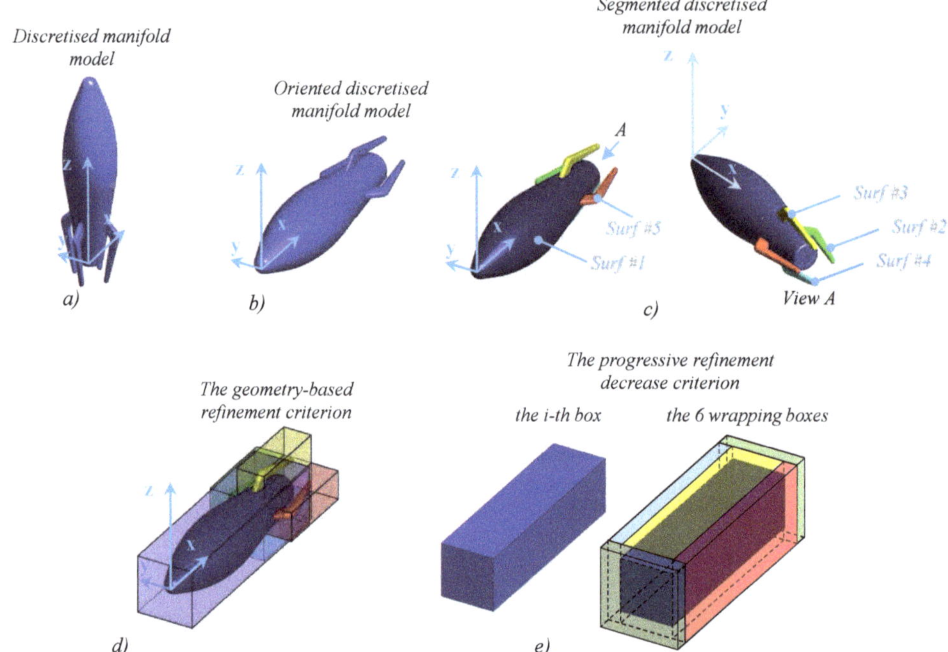

Figure 2. Key steps of the proposed grid generation method: (**a**) the imported discretised manifold model; (**b**) the oriented discretised manifold model; (**c**) the results of the discretised manifold model segmentation; (**d**) the boxes generated with the geometry-based refinement criterion; (**e**) the boxes generated with the progressive refinement decrease criterion.

2.2. Boxes Definition

The sole explicit operation from the user is the definition of the computational domain where local refinement is required on the basis of the expected flow structure (i.e., wakes); this is done by the definition of a box, identified by the two extreme vertices $\{X_{min}, Y_{min}, Z_{min}\}$ and $\{X_{max}, Y_{max}, Z_{max}\}$ in the Global Reference System and by the size of the far-boundary cells $\left\{\Delta x^{far}, \Delta y^{far}, \Delta z^{far}\right\}$, where $\Delta x^{far} = \Delta y^{far} = \Delta z^{far} = h \times 2^{k_G}$, where k_G is an integer and h is the minimum cell size. The first generated grid is a set of hexahedra having face normal oriented along in the three axes directions of the Global Reference System. The number of Voxels along the three directions is defined as follows:

$$\begin{cases} N_x = \max\left(1, int\left(\frac{X_{max}-X_{min}}{\Delta x^{far}}\right)\right) \\\\ N_y = \max\left(1, int\left(\frac{Y_{max}-Y_{min}}{\Delta y^{far}}\right)\right) \\\\ N_z = \max\left(1, int\left(\frac{Z_{max}-Z_{min}}{\Delta z^{far}}\right)\right) \end{cases} \tag{1}$$

Each of the $N_v = N_x \cdot N_y \cdot N_z$ cells is identified by three sets of generalized indices, defined in the following:

- Equivalent structure cell indices $G_{ijk} = (G_i, G_j, G_h)$ where $1 \leq G_i \leq N_x, 1 \leq G_j \leq N_y, 1 \leq G_h \leq N_z$;
- Three coordinates of the center $C_{ijk} = (C_{ijk,x}, C_{ijk,y}, C_{ijk,z})$;
- Refinement level index $K_{ijk} = k_G$.

The complete grid with proper local size is generated by implementing the following refinement criteria, applied to the initial grid:

- geometry-based criterion: refinement based on the distance from the wall surface of the model;
- flow-based criterion: refinement defined on the bases of flow features, in regions with relevant variations of the fluid-dynamic quantities (for example, pressure gradient and vorticity);
- explicit window-based criterion: any other refinement in regions of interest.

The first criterion is automatic; the second and the third ones require that the operator defines each of the N_{ROI} region of interest by its two extreme vertices $\left\{ X^w_{\min}, Y^w_{\min}, Z^w_{\min} \right\}_l$ and $\left\{ X^w_{\max}, Y^w_{\max}, Z^w_{\max} \right\}_l$ (where $1 \leq l \leq N_{ROI}$) coincident with two vertices of the initial grid and the refinement level as 2^{k_W} (with the integer $k_W < k_G$). In the sequel of this research activity, the last two will also be rendered fully automatic.

The geometry-based refinement criterion is based on the surfaces segmentation (Figure 2d): for the *i*-th segmented surface, the algorithm:

- creates a box (whose extreme vertices $\left\{ X^M_{\min}, Y^M_{\min}, Z^M_{\min} \right\}$ and $\left\{ X^M_{\max}, Y^M_{\max}, Z^M_{\max} \right\}$ coincide with the initial grid) that contains it;
- calculates a value of the grid dimension ($2^{k_{M,i}}$ with the integer $k_M, i < k_G$) on the basis of the surface minimum characteristic dimension.

Regardless of the refinement criteria, the boxes are generated with the extreme points coincident with grid nodes of the computational domain. This allows for keeping the consistency of the discretization schemes also with local refinements.

2.3. Progressive Coarsening

To guarantee the smoothness of the refinement level in each direction between a cell and its neighborhoods, an isotropic recursive algorithm working is implemented. At each iteration, six wrapping boxes are generated around each box for which $k_i < \frac{k_G}{2}$ (Figure 2e) with grid dimension $k_{i,new} = 2 * k_i$. The algorithm stops when all the outermost boxes within the computational domain have grid dimension $k_i < \frac{k_G}{2}$.

2.4. Removal of Overlapping Boxes with the Same Grid Dimension

The above processes can generate prisms with the same grid dimension that overlaps: to reduce this redundancy of information, a voxelization-based method with subsequent clustering of adjacent prisms is implemented.

2.5. Grid Generation with Signed Distance Calculation

For each Voxel of each box, the vertices are generated according to the scheme of Figure 3. In order to minimize redundancy, all this information is stored in two tables:

- **nodes** (x_i, y_i, z_i) for $i = 1, \dots, N_n$, where the coordinates of the N_n unique vertices are stored;
- **Voxels** where, for each Voxel, the eight pointers to **nodes** are stored.

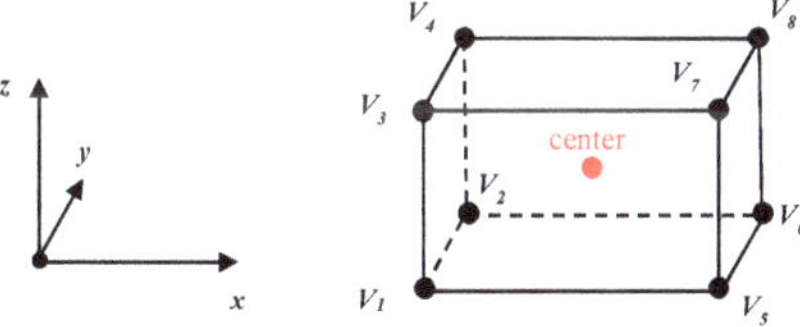

Figure 3. The scheme generation of the vertices of each voxel.

The fourth column of the table **nodes** is the signed distance between the corresponding node and the model, whose sign encodes whether the point is *inside* (negative) or *outside* (positive) to the watertight surface. The value of the distance is evaluated by searching the minimum distance between each node and some points generated parametrically on the triangular faces of the model. The distance sign is defined, for its simplicity and computational efficiency, by using a ray-tracing technique [8].

3. Case Study

The case study analyzed in this paper is a BB2 submarine with casing and appendage taken from https://www.marin.nl/markets/defence/naval-subsurface-vessel-hydrodynamic-design-services, accessed on the 24 February 2021 (Figure 4). The model has foreplanes on the sail and tailplanes; no propulsion systems and mobile appendages or rudders are present.

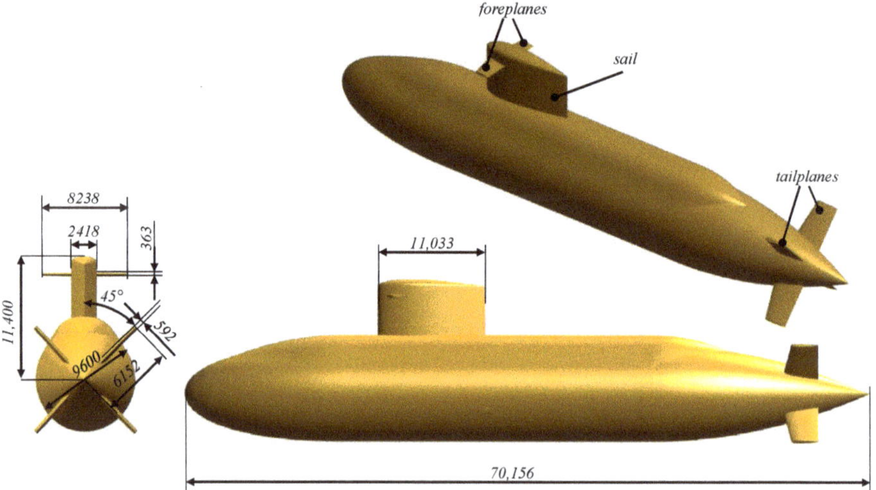

Figure 4. The analyzed BB2 submarine with some of the characteristic dimensions.

The choice of this model was done to apply the proposed method to a practical geometry of interest for naval architecture. The Cartesian grid generation of this model is critical because of the different characteristic dimensions and shape of the appendages.

3.1. The Refined Cartesian Meshes' Generation

The original solid model was transformed into a discrete model defined by triangular flat faces. Figure 5a shows the oriented model with the bounding box dimension and points' density measures. The clear non-uniformity of points distribution permits to analyze the robustness of the proposed generation method. In Figure 5b, the fundamental characteristics of the computational domain defined for the CFD analysis are depicted. First of all, the far boundaries are placed far enough to minimize blockage effects; then, once grid dimension (k_G) is defined, its extreme points ($\{X_{min}, Y_{min}, Z_{min}\}$ and $\{X_{max}, Y_{max}, Z_{max}\}$) were recomputed in order to obtain integer numbers of cells according to (1). Figure 6 shows the results of the surfaces segmentation method superimposing the auto-generated geometry-based boxes refinement. The algorithm correctly recognizes and segments eight surfaces and calculates the $k_{M,i}$ integers on the base of the corresponding surface minimum characteristic dimension.

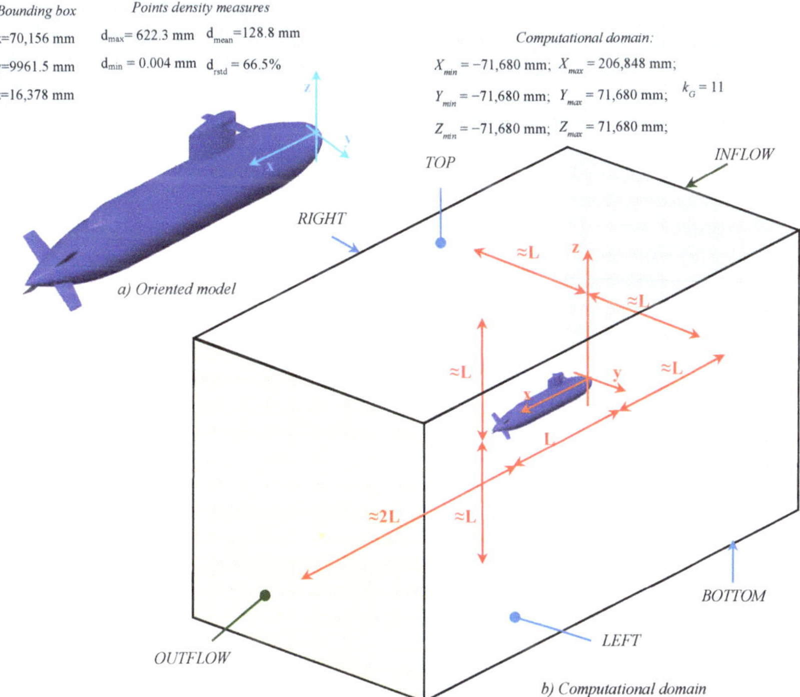

Figure 5. The oriented model (*a*) and the computational domain (*b*).

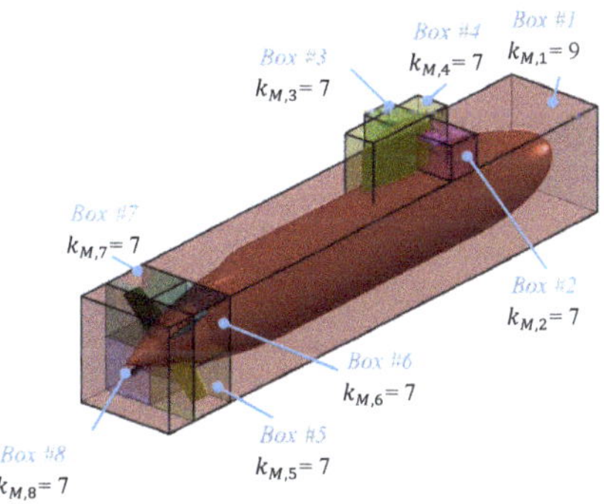

Figure 6. The segmented model with geometry-based boxes refinement.

For the application of the proposed algorithm to a realistic CFD analysis, three refinement boxes are introduced to properly capture the expected wakes behind the sail, foreplanes, and tailplanes (Figure 7).

Once the computational domain and the refinement windows are defined, the grid is generated according to the operations shown in Figure 1 and discussed in Section *Grid generation*; for the test case under consideration, the generated grid consists of 12.8 million cells. This value is about 20 times smaller than that used in [10] to analyze the same geometry, simplified by eliminating the two foreplanes. Figure 8a shows the zero level of the signed distance function with superimposing the grid sections on three perpendicular planes, whereas Figure 8b highlights the difference between the original shape and the one obtained by interpolation from the signed distance function. From the figure, it can be seen that the larger errors are close to the sharp trailing edges of the profiles which, anyhow, are between 60 mm and 100 mm with a maximum error of always <0.14%.

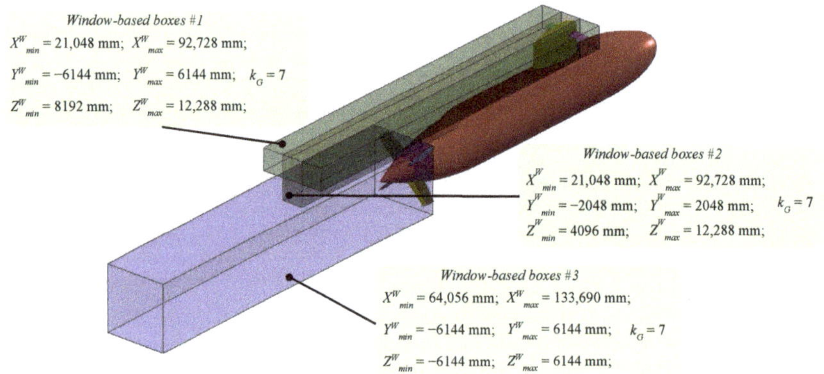

Figure 7. The segmented model with window-based boxes refinement.

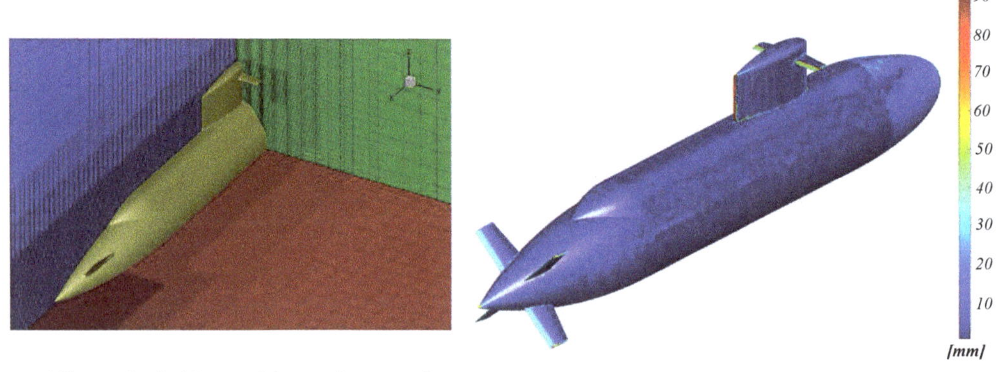

a) The zero level of the signed distance function with superimposed the grid sections in three perpendicular planes

b) The distance map between the original model and the zero level of the signed distance function

Figure 8. The grid generation results.

3.2. Mathematical Models and Numerical Algorithms

The CFD simulation was carried out by the immersed boundary algorithm described in [8], to which the reader is referred for details. For the sake of completeness, the key elements of the algorithm are summarized here.

The immersed boundary approach is applied to the solution of the Navier–Stokes equations for incompressible flows. The governing equations that are solved by numerical approximations are here reported with index notation (the repeated index convention is used):

$$\begin{cases} \dfrac{\partial u_i}{\partial x_i} = 0 \\[2mm] \dfrac{\partial u_i}{\partial t} + \dfrac{\partial u_i u_j}{\partial x_j} + \dfrac{1}{\rho}\dfrac{\partial p}{\partial x_i} = \dfrac{\partial \tau_{ij}}{\partial x_j} \end{cases} \tag{2}$$

The symbols adopted for physical quantities are:

- t for time;
- $\mathbf{e}_i$, $i = 1, 2, 3$ for the base unit vectors;

- x_i, $i = 1, 2, 3$ for spatial coordinates;
- $\mathbf{x} = x_i \mathbf{e}_i$ for the position vector;
- ρ for the density;
- u_i, $i = 1, 2, 3$ for the i-th velocity component;
- $\mathbf{u} = u_i \mathbf{e}_i$ for the velocity vector;
- p for the pressure;
- μ for the dynamic viscosity;
- $\nu = \mu/\rho$ for kinematic viscosity;
- ν_T for the turbulent viscosity;
- and $\tau_{ij} = (\nu + \nu_T)(u_{i,j} + u_{j,i})$ for the stress tensor divided by ρ (the Boussinesq hypothesis was adopted).

Detached Eddy Simulation [11–13] was used for the computation of ν_t required to model the turbulent stresses. The above equations hold in the fluid domain. On the solid wall, no-slip conditions were applied (i.e., $u_i = 0$), whereas, on the fictitious boundary in the far-field, the velocity was enforced on the inlet boundary, and ambient pressure was fixed on the outlet. As initial conditions, the flow was started from a resting position and accelerated to the final value during a transient of time length given by L/U_∞, L being the body length, and U_∞ the velocity in the far-field.

The equations are discretized by a finite difference approach, where the convective and pressure fluxes are discretized by a fifth-order WENO scheme [14], while the viscous terms are approximated with second-order centered approximation. Time integration was performed by a second-order fully implicit scheme, with a constant time step equal to $U_\infty \Delta t / L = 5 \times 10^{-3}$.

3.3. Numerical Set-Up

In all the simulations, the adopted Reynolds number was $Re = 2.7 \times 10^6$, as in the experiments reported in [10].

To enforce the boundary conditions on the submarine walls, the Immersed Boundary procedure described in detail in [8] was applied; the algorithm can be summarized as follows:

1. at the beginning of each time step, the solution is extrapolated inside the body in the normal direction to the body surface;
2. the solution at internal points is then modified in order to get null velocity on the rigid walls;
3. the discrete equations are locally modified to retain at least second order accuracy in the neighborhood of the wall.

Given the high value of the Reynolds number, the wall stresses were evaluated by the use of wall functions, as described in the referenced paper; of course, by this approximation, the details of the boundary layers on the hull are lost, and only the wall stress exerted on the external flow is represented in the model. With the adopted grid, cell size on the walls in terms of wall units is $y^+ = du_\tau/\nu = O\,(200\sim300)$, d being the cell thickness, $u_\tau = \sqrt{\tau_w/\rho}$ the friction velocity, and τ_w the tangential stress on the wall. Nevertheless, vorticity production on the solid walls and the following evolution in the wakes are very well represented, as shown in the next section.

3.4. Results

The computed pressure on the submarine hull is reported in Figure 9, whereas the instantaneous vortex structures are reported by the Q-criterion [15] with $Q = -50$ in Figure 10. From this figure, it can be seen that the grid is able to capture the details of the large vortical structure; in particular, the tip vortices from the sail wings are very well captured, together with their interaction with the vortex structures in the wake of the sail. Similarly, all the details of the large eddies in the wake of the main body and of the tail appendages are captured, and their evolution is very well represented in all the refined regions.

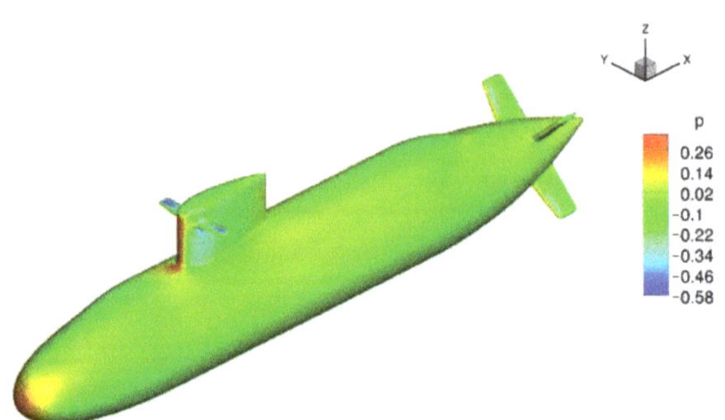

Figure 9. Non-dimensional pressure contours of the submarine surface.

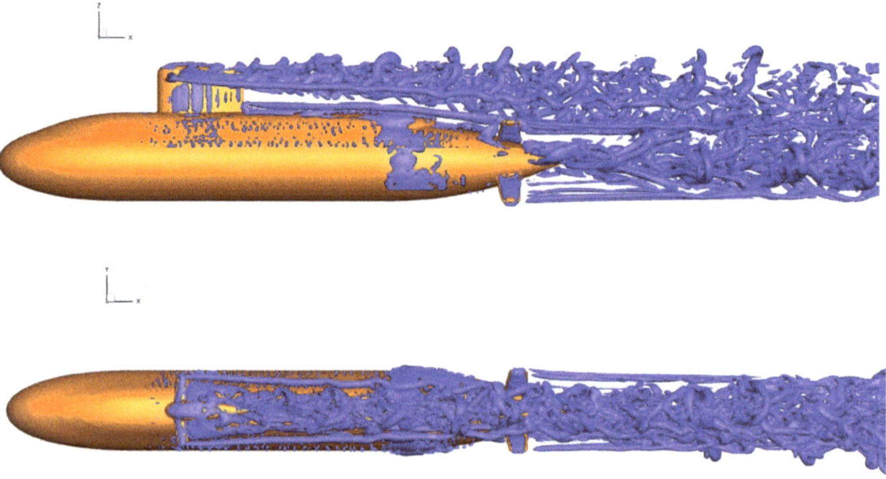

Figure 10. Instantaneous vortex structures visualized by the Q criterion (Q = −50).

The time average of the computed solution is reported in the lower part of Figure 11 in terms of axial velocity on the symmetry plane; in the top part of the same figure, the instantaneous contours of the same variable are also reported. In Figure 12, the averaged axial velocity is reported on six cross planes downstream the hull.

The numerical uncertainty was evaluated by following the procedure described in [16], as recommended by most international engineering associations (e.g., International Towing Tank Conference ITTC and American Institute of Aeronautics and Astronautics AIAA). A first level of coarser grid was generated from the finest one by removing every other point in each direction. A third level of coarsening was impossible because the grid would have been too coarse to capture some basic element of both geometry and flow characteristics. Therefore, we adopted the two–grid verification procedure in [16], where the uncertainty U is evaluated as

$$U = F \frac{||u^f - u^c||_1}{r^2 - 1} \frac{1}{|u^f||_1} \times 100 \tag{3}$$

where $r = 2$ is the adopted refinement ratio in each direction, u^f is the solution computed on the fine grid, u^c is the solution on the coarse grid, and F is a safety factor that, as suggested in [16] for the two-grid uncertainty verification, was chosen to be equal to 3. The quantity $||u^f - u^c||_1$ is the L_1-norm of the difference between the two solution, whereas $|u^f||_1$ is the L_1-norm of the field computed on the fine grid. The uncertainty, computed on the averaged velocity field, was $U = 3.08\%$ for the case considered in the reported example.

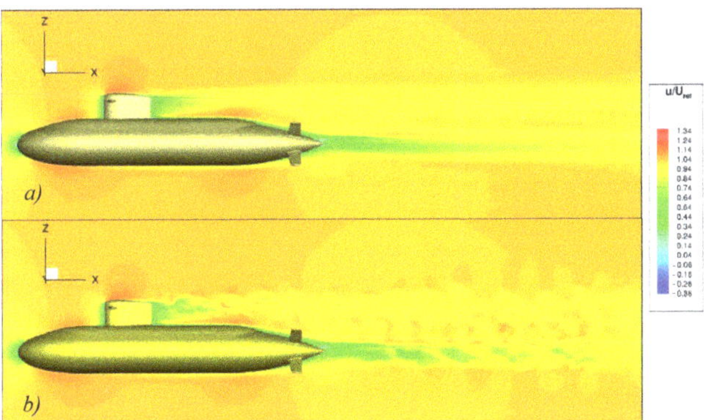

Figure 11. Non–dimensional instantaneous (*b*) and averaged (*a*) axial velocity component on the plane $y = 0$.

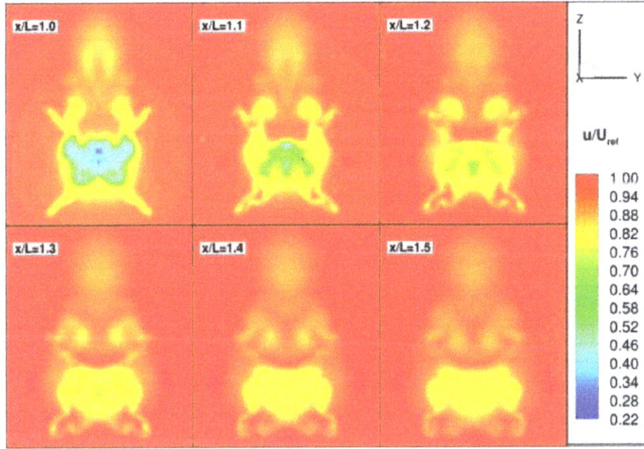

Figure 12. Averaged non-dimensional axial velocity on six downstream sections.

Finally, in Figure 13, the contours of the resolved and modeled kinetic energy are reported on both the symmetry plane and on several cross-sections. From this figure, it can be seen that, according to the Pope criterion [17], the grid is adequate for a correct LES simulation, and the ratio between the modeled to the total turbulent kinetic energy being always below 0.2.

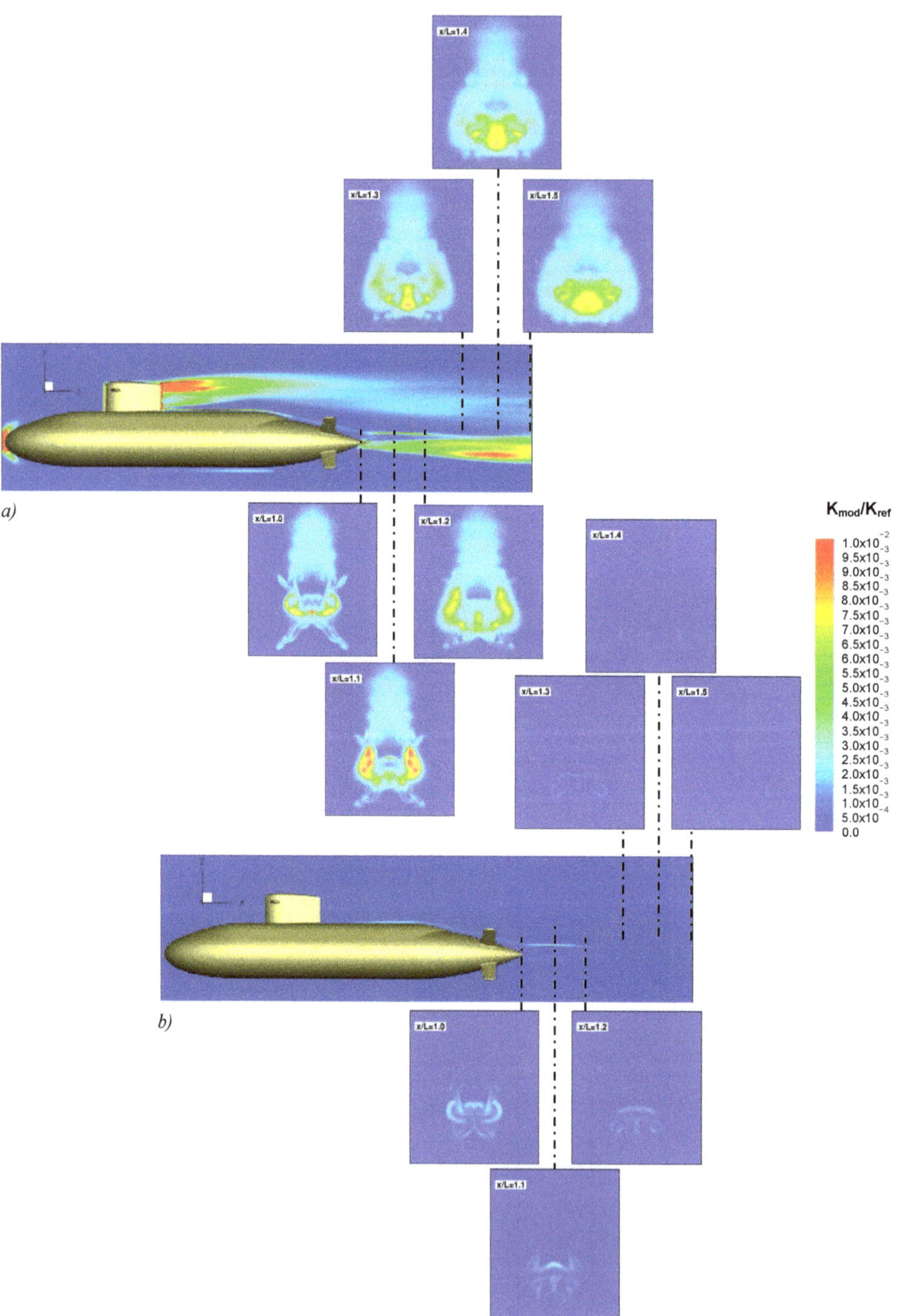

Figure 13. Resolved (**a**) and modeled (**b**) turbulent kinetic energy.

4. Conclusions

In this paper, an almost fully automatic methodology for CFD analysis for high Reynolds number flows is also presented. The proposed workflow includes a new method for Cartesian Mesh Generation with Local Refinement devised and applied to the IB method developed in [8]. The innovative aspects in the present Cartesian adaptive grid method can be found in the strategies of diversification of the mesh dimensions in the different parts of the model, based on automatic segmentation of the surfaces enveloping the object. In addition, grid refinement can be explicitly controlled in regions where the flow is expected to exhibit high gradients. This, together with the use of the IB method and of the wall functions described in [8], allows the simulation of high Reynolds number flows, with limited grid requirements of the boundary layers. The whole methodology, starting from a discrete manifold model of the object to be analyzed and from the following input:

- definition of the computational domain and regions of interest with the dimensions of the corresponding grid;
- the key information for the CFD simulation (expected high flow gradients);

automatically produce the grid for the CFD analysis. The aim of this paper is to show that this automatic workflow is robust and enables to obtain quantitative results on geometrically complex configurations such as marine vehicles. For this purpose, the proposed methodology was applied to study the flow past a BB2 submarine. The grid was able to capture the details of the large vortical structures from the sail wings and from the tailplanes, as well as their interaction with the wakes emanating from the sail and from the main body. Furthermore, the grid proved to be adequate for a correct LES simulation in the wake.

The present research activity will be extended to include the development of an automated mesh refinement strategy, able to capture flow details without explicit input from the user. Moreover, other operating conditions (underwater maneuvers, surfacing, diving) will be addressed, and the results of the fluid dynamic simulations will be verified and validated against available experimental data. In particular, in future research, the capability of the present Immersed Boundary approach coupled with automated mesh refinement will be checked for free surface flows around surface piercing vessels, like ship hull or submarine vehicles operating at snorkeling depth.

Author Contributions: Conceptualization, L.D.A. and A.D.M.; methodology, L.D.A. and A.D.M.; software, L.D.A. and A.D.M.; validation, L.D.A. and A.D.M.; investigation, F.D.; resources, F.D.; data curation, L.D.A. and A.D.M.; writing—original draft preparation, L.D.A., A.D.M., and F.D.; writing—review and editing, A.D.V.; visualization, L.D.A. and A.D.M.; supervision, A.D.M.; project administration, A.D.V. All authors have read and agreed to the published version of the manuscript.

Funding: This research received no external funding.

Institutional Review Board Statement: Not applicable.

Informed Consent Statement: Not applicable.

Conflicts of Interest: The authors declare no conflict of interest.

Abbreviations

The following abbreviations are used in this manuscript:

CFD	Computational Fluid Dynamics
IB	Immersed Boundary
LES	Large Eddy Simulation
$\{X_G^M, Y_G^M, Z_G^M\}$	coordinates of the model centroids
ROI	Region of Interest

$\{X_{\min}, Y_{\min}, Z_{\min}\}$ and $\{X_{\max}, Y_{\max}, Z_{\max}\}$	two extreme vertices of the computational domain
2^{k_G}	the computational domain grid dimension
$\{X_{\min}^{M,i}, Y_{\min}^{M,i}, Z_{\min}^{M,i}\}$ and $\{X_{\max}^{M,i}, Y_{\max}^{M,i}, Z_{\max}^{M,i}\}$	two extreme vertices of the box containing i-th segmented surface
$2^{k_{M,i}}$	the grid dimension of the box containing i-th segmented surface
$\{X_{\min}^{w}, Y_{\min}^{w}, Z_{\min}^{w}\}_l$ and $\{X_{\max}^{w}, Y_{\max}^{w}, Z_{\max}^{w}\}_l$	two extreme vertices of the l-th ROI defined by the operator
$2^{k_{w,l}}$	the grid dimension of the l-th box defined by the operator

References

1. Di Mascio, A.; Zaghi, S. An immersed boundary approach for high order weighted essentially non-oscillatory schemes. *Comput. Fluids* **2021**, *222*, 104931. [CrossRef]
2. Montanaro, A.; Allocca, L.; De Vita, A.; Ranieri, S.; Duronio, F.; Meccariello, G. Experimental and Numerical Characterization of High-Pressure Methane Jets for Direct Injection in Internal Combustion Engines. In Proceedings of the SAE Powertrains, Fuels & Lubricants Meeting, Kraków, Poland, 22–24 September 2020; SAE International: Warrendale, PA, USA, 2020;.
3. Peskin, C.S. Flow patterns around heart valves: A numerical method. *J. Comput. Phys.* **1972**, *10*, 252–271. [CrossRef]
4. Mittal, R.; Iaccarino, G. Immersed boundary methods. *Annu. Rev. Fluid Mech.* **2005**, *37*, 239–261. [CrossRef]
5. Kim, W.; Choi, H. Immersed boundary methods for fluid-structure interaction: A review. *Int. J. Heat Fluid Flow* **2019**, *75*, 301–309. [CrossRef]
6. Berger, M.J.; Aftosmis, M.J. An ODE-based wall model for turbulent flow simulations. *AIAA J.* **2018**, *56*, 700–714. [CrossRef]
7. Capizzano, F. Turbulent wall model for immersed boundary methods. *AIAA J.* **2011**, *49*, 2367–2381. [CrossRef]
8. Capizzano, F. Automatic generation of locally refined Cartesian meshes: Data management and algorithms. *Int. J. Numer. Methods Eng.* **2018**, *113*, 789–813. [CrossRef]
9. Di Angelo, L.; Di Stefano, P. Geometric segmentation of 3D scanned surfaces. *Comput.-Aided Des.* **2015**, *62*, 44–56. [CrossRef]
10. Fureby, C.; Anderson, B.; Clarke, D.; Erm, L.; Henbest, S.; Giacobello, M.; Jones, D.; Nguyen, M.; Johansson, M.; Jones, M.; et al. Experimental and numerical study of a generic conventional submarine at 10 yaw. *Ocean Eng.* **2016**, *116*, 1–20. [CrossRef]
11. Spalart, P.R. Detached-eddy simulation. *Annu. Rev. Fluid Mech.* **2009**, *41*, 181–202. [CrossRef]
12. Pena, B; Muk-Pavic, E.; Thomas, G. Fitzsimmons, P. An approach for the accurate investigation of full-scale ship boundary layers and wakes. *Ocean Eng.* **2020**, *214*, 107854. [CrossRef]
13. Hosseini, A.; Tavakoli, S.; Dashtimanesh, A.;Sahoo, P. K.; K orgesaar, M. Performance Prediction of a Hard-Chine Planing Hull by Employing Different CFD Models. *J. Mar. Sci. Eng.* **2021**, *9*, 481. [CrossRef]
14. Jiang, G.S.; Shu, C.W. Efficient implementation of weighted ENO schemes. *J. Comput. Phys.* **1996**, *126*, 202–228. [CrossRef]
15. Jeong, J.; Hussain, F. On the identification of a vortex. *J. Fluid Mech.* **1995**, *285*, 69–94. [CrossRef]
16. Roache, P.J. Quantification of uncertainty in computational fluid dynamics. *Annu. Rev. Fluid Mech.* **1997**, *29*, 123–160. [CrossRef]
17. Pope, S.B. *Turbulent Flows*; IOP Publishing Ltd.: Bristol, UK, 2001.

Journal of
Marine Science and Engineering

Article

VPP Coupling High-Fidelity Analyses and Analytical Formulations for Multihulls Sails and Appendages Optimization

Ubaldo Cella [1,2,*], Francesco Salvadore [3], Raffaele Ponzini [3] and Marco Evangelos Biancolini [1]

[1] Department of Enterprise Engineering, University of Rome "Tor Vergata", 00133 Rome, Italy; biancolini@ing.uniroma2.it
[2] Design Methods—Aerospace Engineering, 98121 Messina, Italy
[3] HPC Department, CINECA, 40033 Casalecchio di Reno, Italy; f.salvadore@cineca.it (F.S.); r.ponzini@cineca.it (R.P.)
* Correspondence: ubaldo.cella@designmethods.aero

Abstract: A procedure for the optimization of a catamaran's sail plan able to provide a preliminary optimal appendages configuration is described. The method integrates a sail parametric CAD model, an automatic computational domain generator and a Velocity Prediction Program (VPP) based on a combination of sail RANS computations and analytical models. The sailing speed and course angle are obtained, with an iterative process, solving the forces and moment equilibrium system of equations. Analytical formulations for the hull forces were developed and tuned against a matrix of CFD solutions. The appendages aerodynamic polars are estimated by applying preliminary design criteria from aerospace literature. The procedure permits us to find the combination of appendages configuration, rudders setting, sail planform, shape and trim that maximise the VMG (Velocity Made Good). Two versions of the sail analysis module were implemented: one adopting commercial software and one based on the use of only Open-Source codes. The solutions of the two modules were compared to evaluate advantages and limitations of the two approaches.

Keywords: velocity prediction program; numerical optimization; High-Fidelity analysis; geometric parameterization; multihull design

Citation: Cella, U.; Salvadore, F.; Ponzini, R.; Biancolini, M.E. VPP Coupling High-Fidelity Analyses and Analytical Formulations for Multihulls Sails and Appendages Optimization. *J. Mar. Sci. Eng.* **2021**, *9*, 607. https://doi.org/10.3390/jmse9060607

Academic Editors: Antonio Mancuso and Davide Tumino

Received: 10 May 2021
Accepted: 28 May 2021
Published: 1 June 2021

Publisher's Note: MDPI stays neutral with regard to jurisdictional claims in published maps and institutional affiliations.

1. Introduction

A sailing boat is a mechanical system in which several forces and moments act in a complex environment whose static and dynamic equilibrium affect the overall performance. The sailing speed depends on the boat characteristics and on the sails performances with a mechanism that requires several aspects of physics involved to be opportunely modelled. For this reason, yacht design should be faced within so-called Velocity Prediction Program (VPP) environments [1]. VPPs are procedures that evaluate the global equilibrium of the system balancing hull and sail forces. Their accuracy is strongly related to the accuracy of the model adopted to estimate the forces [2]. The typical approaches to feed the models are the generation of experimental databases [3,4] or the integration of CFD (Computational Fluid Dynamic) computations [5]. Adopting numerical flow solutions, however, can significantly increase the cost of computations leading to procedures that are not compatible with the practical requirements of design especially in the preliminary sizing phases. The identification of the most opportune compromise between accuracy and computational requirements is the winning strategy to develop efficient design tools. The adoption of a combination of surrogate models and available sails data to reduce the calculation requirements in VPPs is proposed in [6]. In [7] analytical formulations for sail aerodynamics have also been described. In [8], the data deriving from experiments and computations are used to develop a VPP for the Olympic Nacra 17 foiling catamaran. In some cases, as for instance in foiling vessels, the sailing dynamic stability plays a crucial role. In [9], the dynamic stability is included in a VPP developed for the International Moth dinghy. In [10], such aspect is studied for a 16-foot foiling catamaran.

The method here proposed integrates a parametric geometry model of the sail plan, an automatic computational domain generator and a VPP based on a combination of CFD computations and analytical models. Sailing speed and course angle are obtained, with an iterative process, solving the forces and moments equilibrium system of equations. The hull forces are modelled by empirical analytical formulations whose coefficients are tuned against a matrix of known solutions of the isolated demihull. This model provides a very fast evaluation of the forces at a given velocity, displacement and leeway angle. Dagger boards and rudders are modelled as wings. Their aerodynamic polars are estimated applying preliminary design criteria from the aerospace literature. The closure of the equations system is assured by the sail forces and the position of the aerodynamic centre of effort provided by RANS (Reynolds Averaged Navier–Stokes) computations. The VPP is integrated in a numerical optimization environment which permits to find the sail planform, shape and trim that, in combination to a preliminary evaluation of the optimal appendages configuration and rudders setting, maximise the boat VMG (Velocity Made Good). The test case used for the development of the method is a Classic A-Class catamaran.

Part of the work reported in this paper was funded within the 4th EU PRACE's SHAPE programme. At the end of the project a technical report was produced and made available online [11]. The present work was inspired by that report from which it differs for a deeper description of the implemented modules with particular focus on the VPP.

The paper is divided into three parts: the first introduces the VPP that couples the analytical model of the boat to the sail RANS solution, the second one describes the development of the modules of the optimization environment and the last one compares the solutions of two versions of the sail analysis model applied to a simple A-Cat sail optimization problem. The two proposed CFD modules differ in the numerical tools used to implement them. One was developed adopting well-known commercial software while the other is based on Open-Source codes. The integration of the latter model allows to propose a design tool completely free from commercial licenses (the VPP and the parametric CAD model of the sail are already based on Open-Source software) with a clear advantage in economic terms. Nevertheless, its accuracy and the complexity related to its implementation in an optimization environment should be verified case by case. The comparison here reported was planned with this purpose.

2. Performance Prediction Model

A boat model, fully based on analytical formulations, is proposed. The objective is to provide a very fast and versatile tool able to estimate the boat characteristics with an accuracy acceptable in the preliminary development phase and suitable for an optimization environment. The strength of this approach is the capability to easily parametrise several aspects of the boat components providing the possibility to involve in the optimization a wide range of design variables as chord, draft, twist, setting, airfoil and planform of the appendages. The boat model is developed in form of a *Scilab* function [12] able to interact, in a comparative iterative process within the equilibrium equations system, with the sail RANS aerodynamic solution to constitute a VPP.

2.1. Boat Global Forces and Moment Equilibrium

Figure 1 reports the orientation of the adopted reference frame and summarises the forces acting on the boat.

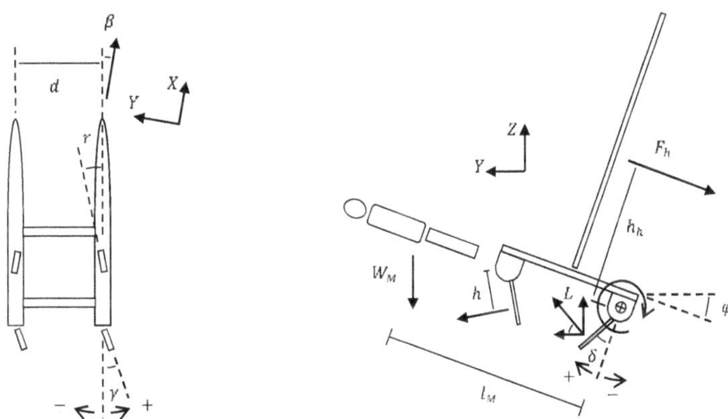

Figure 1. Scheme of forces acting on the boat.

The forces equilibrium equations of the complete boat, referred to a frame with the X axis aligned with the sailing direction and the Z axis perpendicular to the water plane, are:

X equilibrium

$$D_{TOT} = D_{D_D} + D_{D_U} + D_{R_D} + D_{R_U} + D_H + D_{M_x} + D_{B_x} = F_t \tag{1}$$

D refers to drag, the subscripts $*_{D_D}$, $*_{D_U}$, $*_{R_D}$ and $*_{R_U}$ refer, respectively, to downwind dagger board, upwind dagger board, downwind rudder and upwind rudder. D_H is the demihull resistance, D_{M_x} and D_{B_x} are the aerodynamic drag component along the X direction, respectively, of the crew and the boat. F_t is the thrust force of the sail.

Y equilibrium (assuming to neglect the lift generated by the boat)

$$F_h\cos\varphi + D_{M_y} + D_{B_y} = L_{D_D}\cos(\varphi + \delta_D) + L_{D_U}\cos(\varphi - \delta_D) + L_{R_D}\cos(\varphi + \delta_R) + L_{R_U}\cos(\varphi - \delta_R) + L_H \tag{2}$$

F_h is the sail heeling force, φ is the heeling angle, D_{M_y} and D_{B_y} are the aerodynamic drag component along the Y direction, respectively, of the crew and the boat. L refers to lift force, δ refers to the appendages dihedral angle and L_H is the side force generated by the demihull. The subscripts $*_D$ and $*_R$ refer to dagger boards and rudders.

Z equilibrium

$$W_M + W_{BE} + F_h\sin\varphi = W_{BO} + L_{D_D}\sin(\varphi + \delta_D) + L_{D_U}\sin(\varphi - \delta_D) + L_{R_D}\sin(\varphi + \delta_R) + L_{R_U}\sin(\varphi - \delta_R) \tag{3}$$

W_M is the crew weight, W_{BE} is the boat empty weight and W_{BO} is the boat operative weight.

The moment equilibrium along the X axis and around the centre of buoyancy of the downwind hull gives:

M_X equilibrium

$$W_{BE}\frac{d}{2}\cos\varphi + W_M l_M\cos\varphi = W_{BE}h_g\sin\varphi + F_h h_h + L_{D_D}h_{D_D} + L_{D_U}(h_{D_U} - d\sin\delta_D) + L_{R_D}h_{R_D} + L_{R_U}(h_{R_U} - d\sin\delta_R) + W_{BE}h_B\sin\varphi \tag{4}$$

where the left-hand side of the equation represents the maximum possible righting moment with the helmsman at trapeze. The term h_h is the height of the sail centre of effort and h_g is the height of the boat centre of gravity. The other terms h refer to the arm of the resulting force relative to the element identified by the subscripts.

It was decided to not involve the yaw and pitching moment equilibrium in the system. The first in general impacts the rudder angle while the latter mainly influences the hull longitudinal setting. Both parameters can be controlled with an opportune sail rig/appendage centring and crew position. In more detail, the idea is to avoid including the yaw equilibrium, for which the variables would be the relative mast/daggerboard positions, and to include the rudder setting as variable of design. Its most opportune setting can be obtained as output of the optimization and operatively achieved by positioning the mast and the appendages in a location that allows sailing with the required rudder setting. The option to avoid solving the pitching moment equilibrium might appear not opportune in case of very slender hulls such as the catamaran's ones. In our case, nevertheless, the extremely light empty weight of the boat (comparable with the weight of the crew) allows to force the longitudinal attitude in all sailing conditions controlling the position of the crew without the requirement of including the longitudinal degree of freedom in the system. As it was done for the rudder, the longitudinal setting can be imposed as variable of design without deriving it as consequence of the equilibrium. In the view of developing a static VPP, the assumption to neglect the two additional equilibrium equations is considered acceptable for the presented application.

2.1.1. Hull Forces Modelling

A large amount of literature is available, and several strategies are offered to model a traditional monohull sailing yachts, from databases solutions to accurate regression based polynomial expressions [13,14]. Such methods are not valid in case of fast catamarans hulls although experiments on slender bodies, both in calm water and head waves, are available to support the development of analytical models [15]. A significant contribution on this topic is provided by the Molland's work [16]. Nevertheless, the documented correlations inspired the develop of simplified formulations customised to a typical A-Class cat hull shape whose coefficients are to be tuned knowing a limited set of forces data of the hull to be modelled. The formulations were developed by a comparison with a wide matrix of data of a reference hull at several attitudes and leeway angles. The reference database was obtained by RANS analyses in place of experimental measurements. The literature confirms, in fact, the confidence in the accuracy of CFD solutions for both displacing and planning hulls [17,18]. The adopted base of validation of the analytical model is the demihull of a *Flyer S* A-Class catamaran (Figure 2).

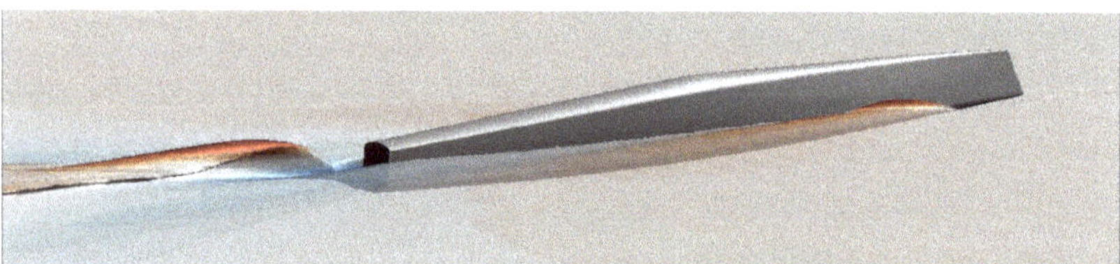

Figure 2. Reference demihull adopted to develop the analytical hull forces model.

Two formulations were developed for hull side force and for drag. To estimate the hull side force (force laying in a plane parallel to the water plane and normal to the sailing direction), the bare hull is modelled as a lifting body as follows:

$$L_H = \frac{1}{2}\rho_w V^2 S_H \frac{\partial C_{L_H}}{\partial \beta}\beta \tag{5}$$

where ρ_w is the sea water density, V is the boat velocity and β is the leeway angle. The reference surface S_H is the side projection, on the symmetry plane, of the submerged part of the demihull and changes with displacement. It is modelled, for a typical A-Class hull shape, by two formulations approximating a set of values computed by a CAD system and valid before and after a defined operative weight W_{BO_0}:

$$S_H = \begin{cases} \left(k_{S_{H1}} W_{BO_0} + k_{S_{H2}}\right)\left(\frac{W_{BO}}{W_{BO_0}}\right)^{\tau_{S_H}}, & W_{BO} < W_{BO_0} \\ k_{S_{H1}} W_{BO} + k_{S_{H2}}, & W_{BO} \geq W_{BO_0} \end{cases} \tag{6}$$

The coefficients $k_{S_{H1}}$ and $k_{S_{H2}}$ are computed knowing two surface values in the linear region. τ_{S_H} is estimated adding a known value at an operative weight lower than W_{BO_0}.

According to the matrix of CFD solutions of the reference demihull, the slope of the side force curve $\frac{\partial C_{L_H}}{\partial \beta}$ linearly change with the displacement. Nevertheless, also a non-linear relation with the velocity (Reynolds effect) and the leeway angle was observed. To account for those effects the developed formulation contains exponential expressions of the two parameters:

$$\frac{\partial C_{L_H}}{\partial \beta} = V^{\tau_{H1}} \beta^{\tau_{H2}} (k_{H1} W_{BO} + k_{H2}) \tag{7}$$

The parameters k_{H1}, k_{H2}, τ_{H_1} and τ_{H_2} are tuned against the known hull data. The graphs in Figure 3 compare the analytical solutions at two velocities and three leeway angles with the reference CFD database.

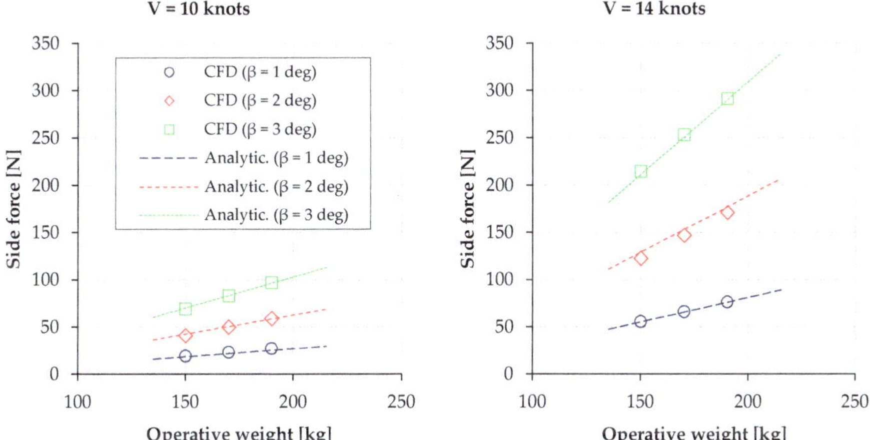

Figure 3. CFD and analytical solutions of demihull side force for the reference hull.

The bare demihull total resistance is expressed by:

$$D_H = \frac{1}{2} \rho_w V^2 S_{wet} C_T \tag{8}$$

The reference wet surface S_{wet} depends on the operative weight and is modelled with two formulations similar to the ones adopted to model S_H. The values to be approximated are computed by CAD. The model estimates the surface value starting from zero displacement in order to provide the procedure the capability to analyse also configurations in which the lifting contributions of the foils are predominant to the hull forces.

$$S_{wet} = \begin{cases} (k_{S_{w1}} W_{BO_0} + k_{S_{w2}}) \left(\frac{W_{BO}}{W_{BO_0}}\right)^{\tau_{Sw}}, & W_{BO} < W_{BO_0} \\ k_{S_{w1}} W_{BO} + k_{S_{w2}}, & W_{BO} \geq W_{BO_0} \end{cases} \tag{9}$$

The terms $k_{S_{w1}}$, $k_{S_{w2}}$ and τ_{Sw} are found knowing the hull wet surface at two displacement values in the linear and one in the non-linear region. Figure 4 reports the comparison between the wet surface modelled by the developed analytical formulation and the values of the reference hull computed by CAD.

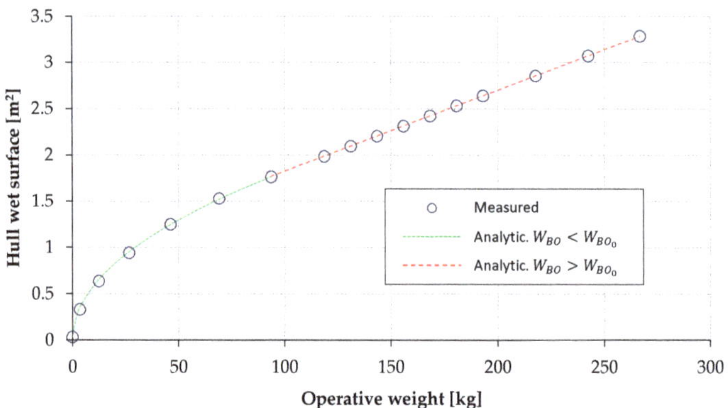

Figure 4. Hull wet surface of the reference demihull.

The total resistance coefficient is modelled as a combination of a friction and a residuary component [19]:

$$C_T = (1+k)C_f + C_w \tag{10}$$

The form factor $(1+k)$ accounts for the over velocity generated by the thick shape of the body [20]. Its value is evaluated from literature or from a known bare hull drag value. The skin friction coefficient is estimated according to the ITTC-57 friction line expression [21]:

$$C_f = \frac{0.075}{(log R_N - 2)^2} \tag{11}$$

where $R_N = \frac{V L_{wl}}{\nu}$ is the Reynolds number referred to the hull waterline length. Good agreement with CFD computations was observed (Figure 5) adopting as L_{wl} the full length of the hull (modern A-Cat hulls have an inversed bow shape).

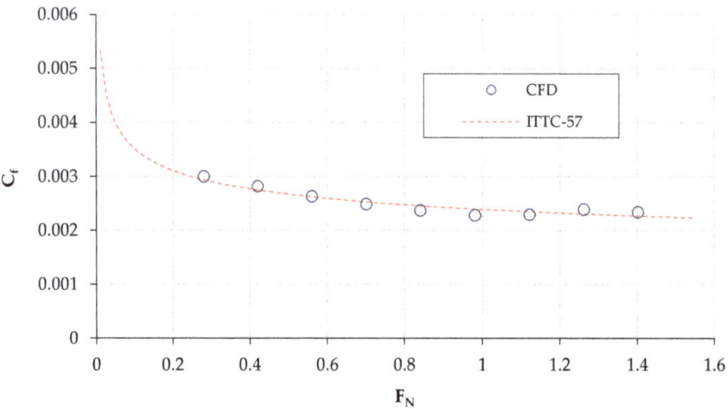

Figure 5. Skin friction coefficient of the reference demihull.

A significant simplification was chosen to model the residuary drag coefficient C_w. A combination of two quadratic formulations with Froude number, for speeds lower and higher than a critical value and linearly function of the operative weight, was adopted:

$$C_w = \begin{cases} (W_{BO} + w_w)\left(k_{w_1} + \frac{k_{w2}}{F_{Ncr}} + \frac{k_{w3}}{F_{Ncr}^2}\right)F_N^2, & F_N < F_{Ncr} \\ \left(W_{BO} + w_w\right)\left(k_{w_1}F_N^2 + (k_{w_2}F_N + k_{w3})\right), & F_N \geq F_{Ncr} \end{cases} \tag{12}$$

The factor $(W_{BO} + w_w)$ accounts for the dependency from the operative weight and is tuned by the term w_w. The values of w_w, k_{w_1}, k_{w_2} and k_{w_3} are to be tuned against the matrix of the known demihull solutions. If a large database of hull solution is available, the combination of values that best match the data might be identified with a trial-and-error procedure. The values adopted for the reference hull were identified by a numerical optimization procedure that converges toward the combination of values that minimize the absolute difference between the analytical formulation and the computed CFD values. To further best match the data, the boundary Froude number might differ from the theoretical critical value of 0.4 referred to the waterline length. The Figure 6 compares the solutions of the analytical wave drag model with the CFD solutions of the reference hull at two operative weights.

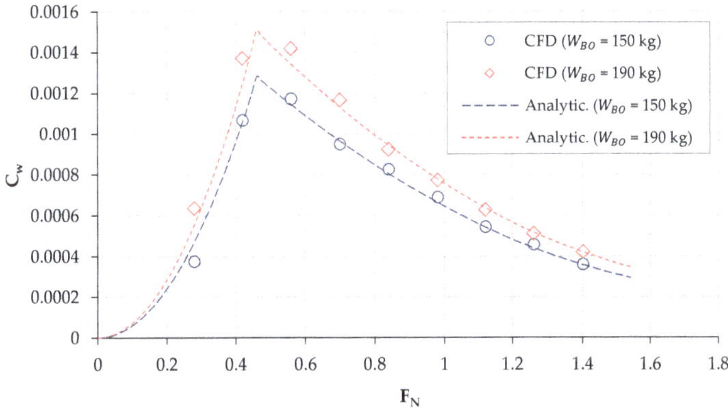

Figure 6. Wave drag coefficient of the reference demihull.

Figure 7 compares the computed (by CFD) and the modelled (by the developed analytical models) viscous and residuary drag of the reference hull. It is evident how the viscous component is dominant in most of typical A-Cat speed range. Therefore, the relative roughness of the wave drag model poorly affect the accuracy of the total hull drag estimation.

An additional factor that accounts for the drag increase due to the leeway angle was also included. Such dependency was assumed to be quadratic with leeway angle. From the CFD computations it was also observed to be linearly dependent to the operative weight and exponentially to velocity. The proposed factor to be included is:

$$1 + k_\beta V^{\tau_\beta} (W_{BO} + w_\beta) \beta^2 \tag{13}$$

The terms k_β, τ_β and w_β are tuned against the known hull solutions. The final analytical drag formulation assumes then the form:

$$D_H = \frac{1}{2} \rho_w V^2 S_{wet} \left[(1+k) C_f + C_w \right] \left[1 + k_\beta V^{\tau_\beta} (W_{BO} + w_\beta) \beta^2 \right] \tag{14}$$

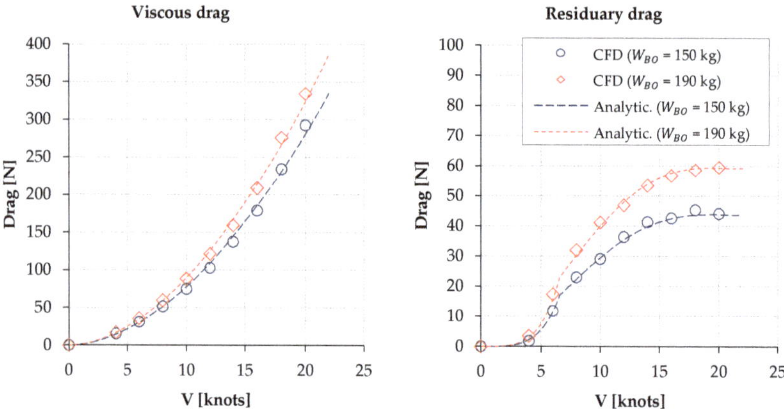

Figure 7. CFD and modelled drag breakdown for the reference demihull.

Figure 8 compares, for the reference hull, the modelled hull drag increase due to the leeway angle with the CFD computations at two velocities.

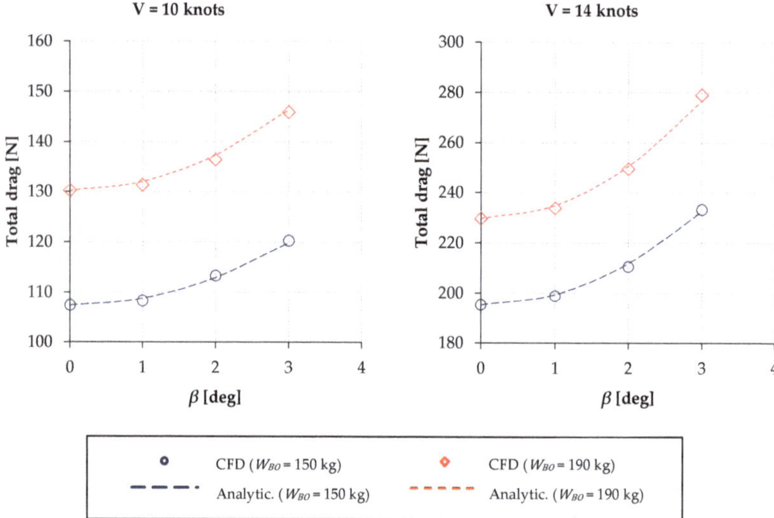

Figure 8. Hull resistance increase due to leeway angle for the reference demihull.

The analytical formulations can be tuned to model new hulls knowing a total of three CAD measurements at three displacements and a minimum of six CFD solutions or experimental forces measurements at two values of velocities, attitudes and leeway angles. The coefficients adopted to model the reference *Flyer S* demihull forces are listed in Table 1.

Table 1. Values of the parameters adopted to model the *Flyer S* demihull forces.

Reference Surfaces	Side Force	Wave Drag	Leeway Drag
$k_{S_{H1}} = 0.00437$	$k_{H1} = 6 \times 10^{-7}$	$w_w = 80$ kg	$k_\beta = 2 \times 10^{-6}$
$k_{S_{H2}} = 0.07$	$k_{H2} = 1.3 \times 10^{-4}$	$k_{w_1} = 2.16 \times 10^{-6}$	$\tau_\beta = 1.5$
$\tau_{S_H} = 0.83$	$\tau_{H_1} = 1.3$	$k_{w_2} = -8.3 \times 10^{-6}$	$w_\beta = 400$ kg
$k_{S_{w1}} = 0.00876$	$\tau_{H_2} = 0.2$	$k_{w_3} = 9 \times 10^{-6}$	
$k_{S_{w2}} = 0.95$			
$\tau_{S_w} = 0.5$		shape factor $k = 0.01$	
$W_{BO_0} = 94$ kg			

2.1.2. Appendages Forces Modelling

Dagger boards and rudders are modelled as wings. Their aerodynamic polars are estimated applying preliminarily design criteria from literature. The formulation for lift is:

$$L = \frac{1}{2} \rho_w V_{eff}^2 S C_L \tag{15}$$

The lift coefficient, in the linear region of the lift curve of a non-symmetric foil, can be expressed as:

$$C_L = \frac{\partial C_L}{\partial \alpha} \alpha_{eff} + C_{L0} \tag{16}$$

where $\frac{\partial C_L}{\partial \alpha}$ is the slope of the lift curve and C_{L0} is the lift generated by the foil at zero incidence. The effective velocity V_{eff} is the component of the boat velocity vector normal to the foil leading edge (for rectangular planform) and α_{eff} is its angle of incidence. V_{eff} is the only velocity component responsible for the generation of lift (the friction contribution can be neglected). The spanwise component, in fact, does not affect the lift but only causes a shifting of the boundary layer [22]. From geometrical considerations it can be demonstrated that, for moderate values of the leeway angle, the effective velocity can be approximated to the boat speed:

$$V_{eff} \approx V \tag{17}$$

and the effective incidence can be approximated, for instance for the downwind appendage, to:

$$\alpha_{eff} \approx \beta \cos(\varphi + \delta_D) \tag{18}$$

From the above considerations and referring to the Figure 1, the lift coefficients of dagger boards and rudders are then expressed in function of β as follows:

$$\begin{aligned}
C_{L_{D_D}} &= \left(\frac{\partial C_L}{\partial \alpha}\right)_{D_D} [\beta \cos(\varphi + \delta_D) + r] + C_{L0_{D_D}} \\
C_{L_{D_U}} &= \left(\frac{\partial C_L}{\partial \alpha}\right)_{D_U} [\beta \cos(\varphi - \delta_D) - r] - C_{L0_{D_U}} \\
C_{L_{R_D}} &= \left(\frac{\partial C_L}{\partial \alpha}\right)_{R_D} [\beta \cos(\varphi + \delta_R) + \gamma] \\
C_{L_{R_U}} &= \left(\frac{\partial C_L}{\partial \alpha}\right)_{R_U} [\beta \cos(\varphi - \delta_R) + \gamma]
\end{aligned} \tag{19}$$

The 3D lift curve slopes are estimated by empirical formulations used in aeronautics in the preliminary design phase [23]. Assuming a linear twist and constant airfoil section along the full span, it is modelled (with dimension $\frac{1}{deg}$) as:

$$\frac{\partial C_L}{\partial \alpha} = f \frac{\left(\frac{\partial C_l}{\partial \alpha}\right)_{2D} \frac{b}{2p}}{1 + 57.3 \frac{\left(\frac{\partial C_l}{\partial \alpha}\right)_{2D} \frac{b}{2p}}{\pi \lambda}} \tag{20}$$

The terms p is the foil planform perimeter subtracting the root chord length, b is the appendage draft and λ the aspect ratio of the mirrored full span geometry. The term $\left(\frac{\partial C_l}{\partial \alpha}\right)_{2D}$ is the 2D lift curve slope of the adopted airfoil.

The dagger boards lift coefficients at zero incidence C_{L0_D} are obtained solving the lift curve equations for $C_{L_D} = 0$ (first two expressions of Equation (19)) substituting to the factor between squared brackets the 3D angle of incidence at which the foil generates zero lift that, for the downwind dagger board, for instance, is given by:

$$\alpha_{C_{L0}} = \alpha_{C_{L0_{2D}}} + J\varepsilon - r \tag{21}$$

$\alpha_{C_{L0_{2D}}}$ is the zero-lift incidence of the airfoil, ε is the foil twist and r the root stagger angle.

The appendages drag formulation is:

$$D = \frac{1}{2}\rho_w V^2 S C_D \tag{22}$$

where the drag coefficient C_D is expressed in function of the lift coefficient.

$$C_D = C_{D0} + \frac{C_L{}^2}{\pi\lambda e} + C_L\varepsilon\left(\frac{\partial C_l}{\partial \alpha}\right)_{2D} v + \left[\varepsilon\left(\frac{\partial C_l}{\partial \alpha}\right)_{2D}\right]^2 w \tag{23}$$

The 2D lift curve slope of the airfoil $\left(\frac{\partial C_l}{\partial \alpha}\right)_{2D}$ required in Equations (20) and (23), the zero lift incidence $\alpha_{C_{L0_{2D}}}$ of Equation (21), the drag at zero lift C_{D0} in Equation (23) refer to the characteristics of the selected airfoil and can be provided in several ways. In the method described in this paper three options were implemented: they can be provided as an external experimental database, in a form of coefficients of analytical 2D polars or can be computed "on the fly" by a coupled panel/boundary layer code [24] in which the airfoil is parameterized by a NURBS control polygon or provided in formatted coordinates of points. The values of f in Equation (20), J in Equation (21), e, v and w in Equation (23) are reported in the literature as a function of aspect and taper ratio [25].

The analytical formulations above described are valid for isolated wings. The effect on rudders of the daggerboard downwash was considered moderate and neglected at this stage. From the downwash chart reported in [26], it was estimated that this approximation introduces uncertainness on the total drag in the order of fractions of percentage. Other phenomena such as wall interference, ventilation and the effects of the moderate curvature of Classic A-Cat daggerboards were also not considered. An activity to refine the models is on progress by fine tuning additional factors against an extended database of CFD solutions on hulls with appendages (Figure 9). For sail optimization purpose, nevertheless, moderate uncertainness in the accuracy of the VMG is not expected to invalidate the search direction of the shape that maximize the sail thrust.

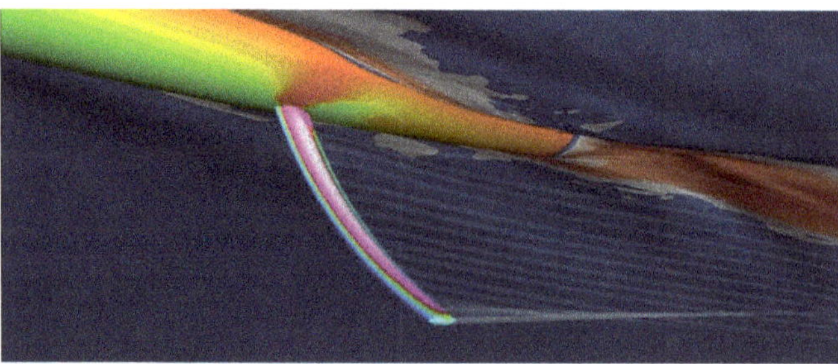

Figure 9. Example of a CFD solution of the hull with appendage used to fine tune the foils analytical models.

Substituting the foils forces formulation in the equilibrium system of equations—Equations (1)–(4)—and including the hull side force model—Equation (5)—we obtain, assuming the boat velocity V and the height of sail aerodynamic centre of effort h_h to be given as input, a system of five equations and five unknowns (D_{TOT}, F_H, W_{BO}, L_H and β). The solution of the equations system is implemented as a script function (written in *Scilab*) that produces as output the boat total resistance D_{TOT} and the sail heeling force F_H (which are the parameters to be compared with the CFD sail solutions) at a given speed V, centre of effort height h_h and set of parameters characterising the boat configuration (Figure 10).

$$
\begin{bmatrix} \text{Boat total resistance } (D_{TOT}) \\ \text{Sail heeling force } (F_h) \end{bmatrix} = Fn
\begin{pmatrix}
\text{Velocity } (V) \\
\textbf{Boat:} \\
\text{Distance between hulls centerline } (d) \\
\text{Empty weight } (W_{BE}) \\
\text{Crew weight } (W_M) \\
\text{Height of center of gravity } (h_B) \\
\text{Parameters of the hull forces model} \\
\text{Exposed boat drag coefficient} \\
\text{Crew drag coefficient} \\
\textbf{Sail:} \\
\text{Height of sail aerodynamic center } (h_h) \\
\textbf{Dagger board(s):} \\
\text{Reference surfaces } (S_D) \\
\text{Aspect ratio } (\lambda_D) \\
\text{Dihedral angle } (\delta_D) \\
\text{Root stagger angle } (r) \\
\text{Twist } (\varepsilon) \\
\text{Airfoil aerodynamic database} \\
\textbf{Rudders:} \\
\text{Reference surfaces } (S_R) \\
\text{Aspect ratio } (\lambda_R) \\
\text{Dihedral angle } (\delta_R) \\
\text{Setting } (\gamma) \\
\text{Airfoil aerodynamic database}
\end{pmatrix}
$$

Figure 10. Scheme of the equilibrium equations function.

2.2. Closure of the Performance Solution Problem

No sail aerodynamic model is included in the function modelling the boat performance. As anticipated, it requires an input of two unknown parameters that are not related to the boat geometry or setting: the velocity of the boat V and the height of the sail centre of effort h_h. The closure of the problem is provided, iterating with the CFD aerodynamic solution of the sail at fixed sailing conditions.

Figure 11 describes the workflow to estimate the VMG for a given combination of parameters characterising the boat configuration. The procedure begins guessing an initial sailing speed V and course TWA. A CFD analysis, with the selected sail plan, shape and trim, is then run at these conditions. Sail forces and centre of effort are extracted and used to verify the equilibrium system. The verification consists in checking if the boat total resistance and the sail heeling force, computed by the analytical model for the given hull and appendage configuration, are equal, respectively, to the sail thrust force and the heeling force deriving from the CFD computation:

$$
\begin{cases} D_{TOT} = F_{tCFD} \\ F_h = F_{hCFD} \end{cases} \tag{24}
$$

If the two solutions are different, new values of boat speed and true wind angle are selected. The CFD computation is restarted at the new conditions and the procedure is repeated until the equilibrium equations criteria are verified (within a prescribed tolerance). The VPP problem is completed with the production, as output, of the "Velocity Made Good" ($VMG = V\cos TWA$), which represents the velocity of the boat toward the wind direction [27] with the selected sail geometry (considered rigid) and appendages.

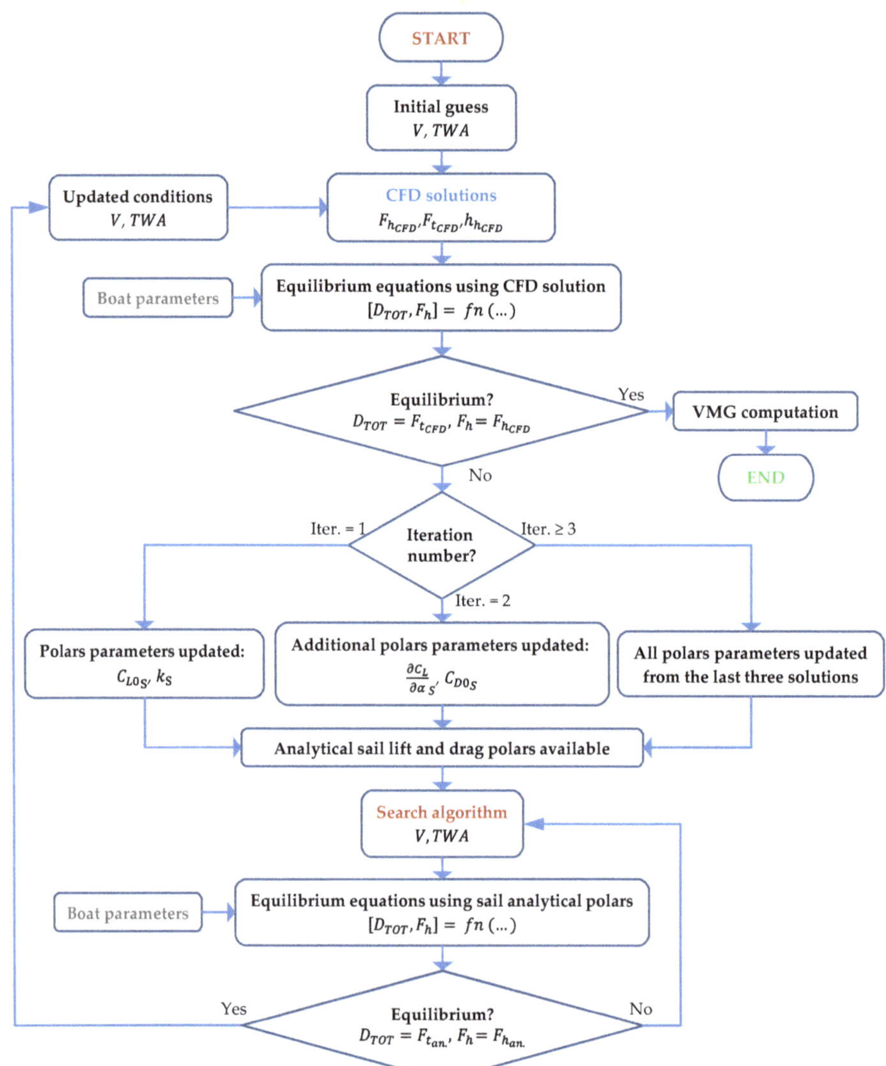

Figure 11. Flow chart of the VPP module.

To speed up the convergence of the VPP solution, the procedure of sailing conditions exploration was split into two nested cycles. The principle is to use an external loop, which involves the RANS computation, to model analytical polars of the sail aerodynamics to indicate the inner search algorithm the direction where to find the sailing conditions that verify the equilibrium. The estimated sail aerodynamic model is then refined every external cycle until the equilibrium is verified in both loops. The analytical polars formulations used to model the sail aerodynamics are similar to the one adopted to model the appendages:

$$C_{LS} = \left(\frac{\partial C_L}{\partial \alpha}\right)_S AWA + C_{L0S} \tag{25}$$

$$C_{DS} = C_{D0S} + k_S C_{LS} + \frac{C_{LS}^2}{\pi \lambda_S e_S} \tag{26}$$

The term e_S is the sail induced drag factor (always lower than 1) or "Oswald efficiency factor" and is related to the shape of the spanwise load distribution. For preliminary design purpose it is reported as function of aspect and taper ratio. λ_S is the sail aspect ratio. The coefficient k_S in related to the sail camber. C_{L0S} and C_{D0S} are, respectively, the lift and drag coefficients the sail rig would exhibits at zero angle of incidence if it was rigid with the current shape.

The thrust and heeling forces are expressed in function of sail lift and drag coefficients by the equations system [28]:

$$\begin{cases} F_{tCFD} = \frac{1}{2}\rho_a AWS^2 S_S(C_{LS}\sin AWA - C_{DS}\cos AWA) \\ F_{hCFD} = \frac{1}{2}\rho_a AWS^2 S_S(C_{LS}\cos AWA + C_{DS}\sin AWA)\cos\varphi \end{cases} \tag{27}$$

where ρ_a is the air density and S_S is the sail reference surface. The apparent wind speed AWS and angle AWA, which are the sail reference freestream velocity and angle of incidence, are obtained as function of the true wind speed TWS (for convention measured at 10 m from the sea level) and its angle TWA by the relations:

$$AWA = \tan^{-1}\left[\frac{TWS\sin TWA\cos\varphi}{TWS\cos TWA + V}\right] \tag{28}$$

$$AWS = \sqrt{(TWS\sin TWA\cos\varphi)^2 + (TWS\cos TWA + V)^2} \tag{29}$$

Substituting the Equations (25), (26), (28) and (29) into the system (27) we obtain the formulation of F_{tCFD} and F_{hCFD}, in function of V and TWA, that will be used in the verification criteria of the equilibrium equations system (24).

The drag polar is a quadratic formulation with the lift coefficient. The sail aerodynamics requires then at least three iterations to be completely modelled. Its progressive update follows different criteria during the first three iterations of the external cycle of the flow chart in Figure 11:

- In the first iteration the sail lift curve slope $\left(\frac{\partial C_L}{\partial\alpha}\right)_S$ and the induced drag factor e_S are estimated from literature as function of sail aspect and taper ratio. The value of zero-lift drag coefficient C_{D0S} is roughly guessed. The sail lift and drag coefficients C_{LS} and C_{DS}, obtained from the CFD analysis, are used to estimate C_{L0S} from Equation (25) and k_S from Equation (26).
- In the second iteration the additional CFD solution is used to complete the analytical lift curve formulation adjusting the values of the lift curve slope $\left(\frac{\partial C_L}{\partial\alpha}\right)_S$ and zero-incidence lift coefficient C_{L0S}. The parameters updated in the polar curve are C_{D0S} and k_S while the value of e_S is still guessed.
- In the third iteration the analytical drag polar formulation is completed with the computation of the induced drag factor e_S which is last unknown parameter. The lift curve is updated connecting a quadratic formulation to the previous computed linear part.
- In all the following iterations the sailing condition estimation are performed modelling the polars regions under investigation updating both curves by a generic quadratic formulation using the closest three solutions.

Figure 12 reports, for a typical A-Class sail plan, an example of the evolution of the sail polars computation during the progress of the first three iterations and the estimation of the values to be used for the computation of the sailing conditions in the fourth iteration (green circles). If the sail aerodynamic conditions fall in the linear region of the lift curve three iterations are in general sufficient to converge. If not, the reported analytical polars formulation are no more valid. The quadratic formulations, with which the non-linear sail aerodynamics is modelled in the following iterations, simply constitute interpolating curves whose coefficients have no particular physical meaning.

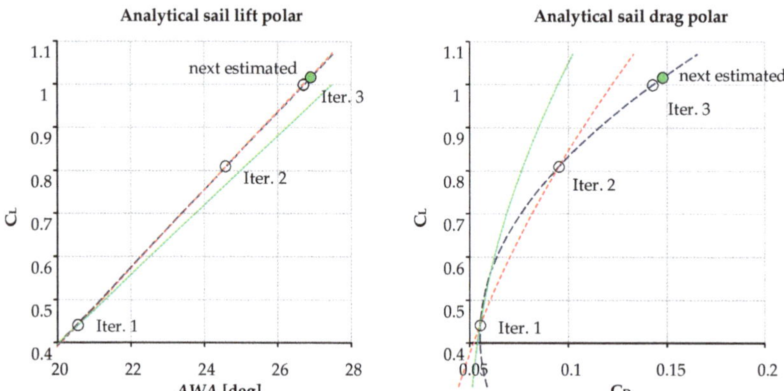

Figure 12. Example of sail analytical polars computation progress.

The searching criterion of the aerodynamic coefficients in the inner cycle is driven by an optimization procedure, based on the Nelder–Mead Simplex algorithm [29], whose objective function is the minimisation of the differences between the forces derived from the boat analytical model and from the CFD computation:

$$Obj.Func. = |D_{TOT} - F_{tCFD}| + |F_h - F_{hCFD}| \tag{30}$$

The values of V and TWA that satisfy the equilibrium are used in the next external loop where the sail polars are updated. The iterations continue up to the satisfaction of the equilibrium system in both loops. When the convergence is reached, the boat VMG is computed and produced as output.

It was experienced that the number of RANS computations required to reach a convergence rarely was higher than four or five (if sail is not stalled or, in general, if separations are not too large). Furthermore, a restarting procedure from the previous solution and a progressive reduction of the CFD iterations, was implemented. This strategy showed to be very efficient in boosting the convergence, but its robustness is related to the capability to select starting sailing conditions as realistic as possible. The procedure fails in case of sudden sail separation. To reject such solutions a check if complete stall occurs was implemented.

3. Optimization Environment

The above-described performance prediction procedure was integrated in an optimization environment in which the optimal sail plan, trim and appendage configuration is researched. Several approaches are possible to parametrize the geometry. A very efficient method consists in adopting mesh morphing techniques [30]. Such approach has the advantage to operate directly on the numerical domain avoiding the remeshing requirement. The adoption of structured meshes or efficient remeshing algorithms, however, might allow to develop procedures with computational efforts comparable to mesh morphing strategies. The method here presented integrates, in an automatic process, the sail parametric CAD model, the computational domain generation, the RANS analysis and the VPP model.

3.1. Sail Parametric Geometric Module

The selected strategy to parametrise the computational domain is based on the update of a parametric CAD model and in the regeneration of the CFD mesh. The software used is the Open-Source *FreeCAD* (www.freecadweb.org accessed on 1 April 2021) tool, a general-purpose parametric 3D CAD modeller [31]. The software can also be used as a library by other programs. The geometry generation and its exportation are managed by a *Python* script (Figure 13).

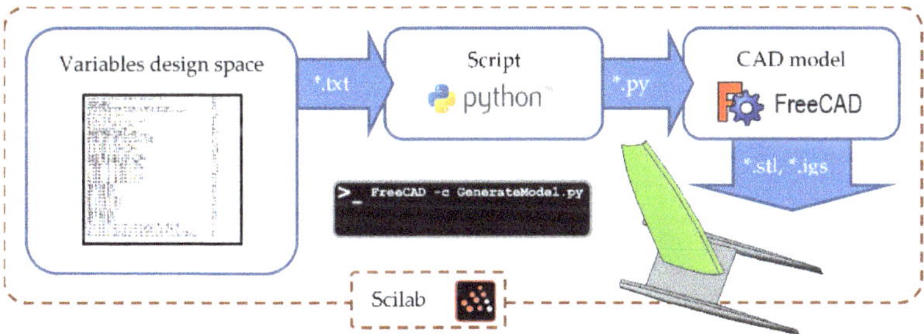

Figure 13. Scheme of the geometry generation module.

The topology of the sail plan consists in a single mast/mainsail configuration. The CAD parameters were selected with the aim to investigate the largest possible range of geometries. Traditional sail plans, wing masts or wing sails with a small portion of flexible sail can be generated. The model is built by a loft surface through a foot, an intermediate arbitrarily positioned and a head curve that are used as control sections. The luff curve is used as guide. In a similar manner, the mast is generated from three geometries at the same stations. The planform is controlled by reference surface, aspect ratio, taper ratio and by other parameters that give the possibility to investigate a wide range of shape. The examples in Figure 14 give the sense on the flexibility of the parametric model.

Figure 14. Examples of sail planforms that can be generated by the parametric CAD module.

The sail sections are generated by cubic Bezier curves. The first point of the control polygon is connected to the mast luff, the last one coincides with the leech of the sail. The four coordinates of the two intermediate control points are parameters of the geometry (red polylines in Figure 15). The mast sections are generated by spline curves controlled by three parameters. The spanner angle (angle between the mast chord and the boat symmetry plane) is a setting parameter. The input reference surface area is kept unchanged. After the geometry creation, the final sail area is measured and the loft surface cut in order to restore the required value. The procedure is linked to the sail CFD analysis module. Two versions of the fluid dynamic computation were setup: one based on commercial software and one based on an Open-Source solver.

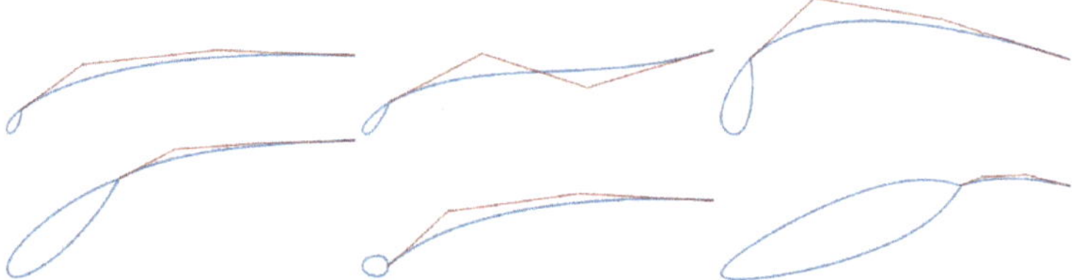

Figure 15. Examples of mast/sail sections that can be generated by the parametric CAD module.

3.2. Sail CFD Analysis Module Implemented Adopting Commercial Software

A well consolidated mesh update procedure [32], widely applied to several aerodynamic optimization problems, was implemented. It is based on an automatic generation of a structured hexahedral multiblock mesh using *ANSYS ICEM/CFD* and in the run of the analysis using the CFD *ANSYS Fluent* solver. Due to the complexity of the geometry (all of the boat, included the helmsman, is modelled) a mixed strategy, in the mesh generation, was adopted. The domain was divided in several regions with common boundaries and each region was meshed applying the more appropriate strategy. A structured CH grid topology was created in a limited volume around the sail and dimensioned to envelope the full range of possible geometries. An unstructured hybrid prism/tetra mesh was generated around the boat in the volume between the sail structured mesh and the water plane. The sail/boat mesh assembly is contained in a box, four boat lengths large and tall, in which another hybrid prism/tetra mesh portion was generated. The remaining volume was meshed with hexahedral cells growing toward the top with a progression aimed to better model the inflow air boundary layer. The full domain is 10 boat lengths wide and is extended 10 boat lengths upstream and downstream the model. The several elements of the mesh are connected by zonal interface boundaries in which a simple "flow-through" condition between the non-conformal zones is imposed. The first image of Figure 16 shows the assembly of the parts with the interfaces in evidence. The other two images detail the structured grid around the sail which is the only part of the mesh subjected to update every optimization iteration.

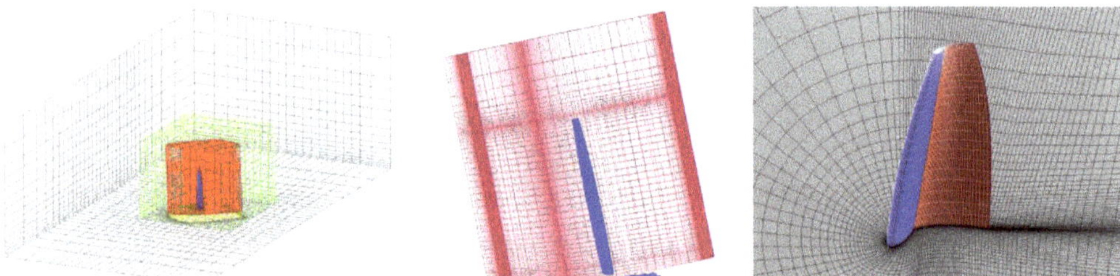

Figure 16. Detail of the parts of the computational domain.

Figure 17 reports the final assembly of the mesh. The total dimension is around half millions of cells. This value was selected after a mesh sensitivity analysis. It was evaluated as a reasonable compromise between accuracy and computational costs being the mesh to be adopted for an optimization procedure in which is more important the difference between candidates than the absolute accuracy of the analysis.

Figure 17. Final assembly of the *Ansys Fluent* computational domain.

The numerical analysis consists in a steady fully turbulent RANS computation. The extracted solutions are the sail heeling and thrust forces, in upwind sailing conditions, together with the resultant aerodynamic centre of pressure. The two-equation $k - \omega$ SST (Shear Stress Transport) turbulence model of Menter [33] was used. Wall Functions were applied to model the wall boundary layer. The boat is moving on a local reference frame at the given speed V and direction TWA with respect to the true wind. These values are updated iterating with the analytical boat model in a process that constitutes the VPP module that provides the performance of the boat with the selected geometric configuration. At far field the wind boundary layer velocity profile [34] in the absolute reference frame is imposed. A pressure outlet is imposed at the outlet boundary behind the boat. Figure 18 reports the solution on a typical geometry with 10 knots of true wind speed at 10 m from the water plane. The streamlines evidence the structures of the sail tip and root vortices. The wind boundary layer velocity magnitude is reported by a contour plot on a plane behind the boat.

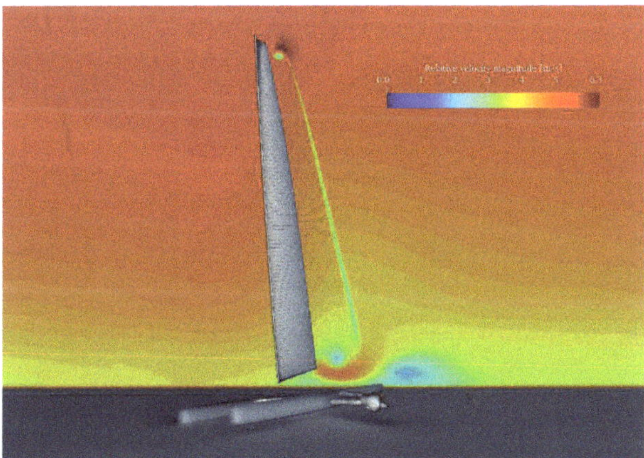

Figure 18. Typical CFD solution on an A-Class catamaran.

Sails often exhibits separations in the region of the mast (Figure 19 reports an example of the typical evolution of this phenomenon behind a traditional A-Class rig). Furthermore, they usually perform in high lift conditions when trailing edge separations might also occur. The solver convergence histories could then be affected by unsteadiness or, in general, by irregularities. The simple extraction of the value of the last iteration might provide misleading information. A routine able to filter and to evaluate the quality of the solutions was then developed. It consists in extracting a linearized interpolation of the last portion of the convergence history, in evaluating its slope, the maximum deviation of the solution from it and in extracting a value applying opportune constraints in order to reject unacceptable solutions. Figure 20 reports an example of how this filter works.

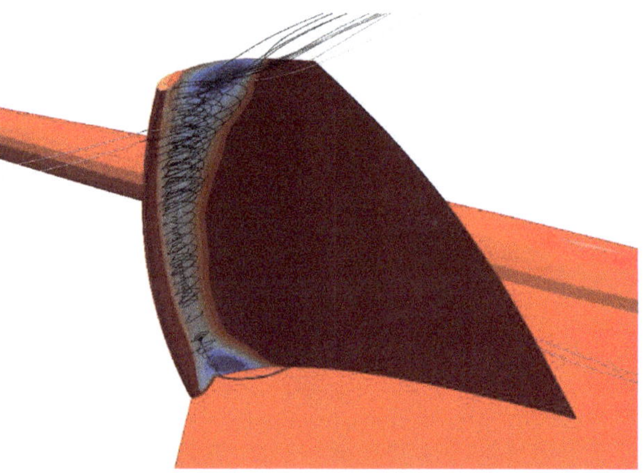

Figure 19. Evolution of separations behind the mast on a traditional A-Cat rig.

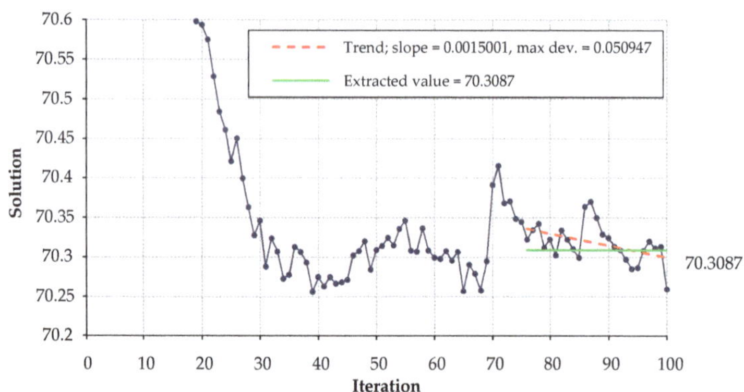

Figure 20. Filter used to extract the solutions from the convergence histories.

3.3. CFD Analysis Module Based on Open-Source Tools

The CFD analysis module was replicated adopting Open-Source codes. The objective is to make available a tool completely unlinked from commercial software. The workflow of the developed procedure is sketched in Figure 21. The steps of the process are summarized below:

- CAD import and pre-processing;
- Geometry meshing;
- Flow field solving;
- Data visualisation and post-processing.

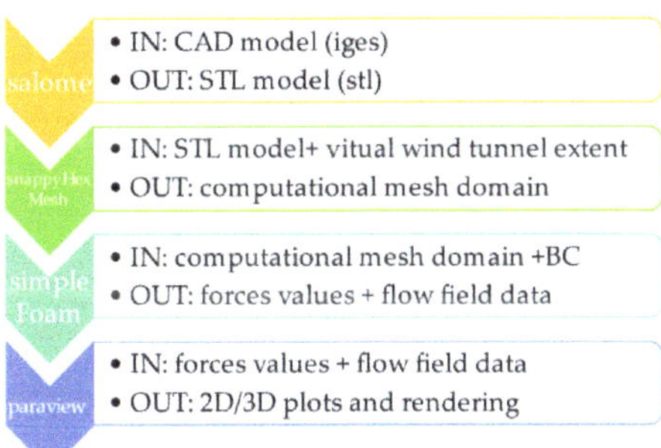

Figure 21. Schematic chart of the Open-Source based CFD analysis workflow.

The baseline CAD model contains the boat (hull, platform, mast and sail) and the body of the sailor. The flow solver used is *OpenFOAM*. The mesh was generated by the mesher *SnappyHexMesh* with the support of *Salome-Mesh*, which was used to generate the triangularisation of the surfaces. The domain is built by defining a background mesh that is iteratively refined accordingly to geometry CAD surface intersection and projection (top-down approach). This technique is very flexible and very low demanding with respect to the quality of the geometry CAD surface definition. The typical resulting computational domain and the mesh are showed in Figure 22.

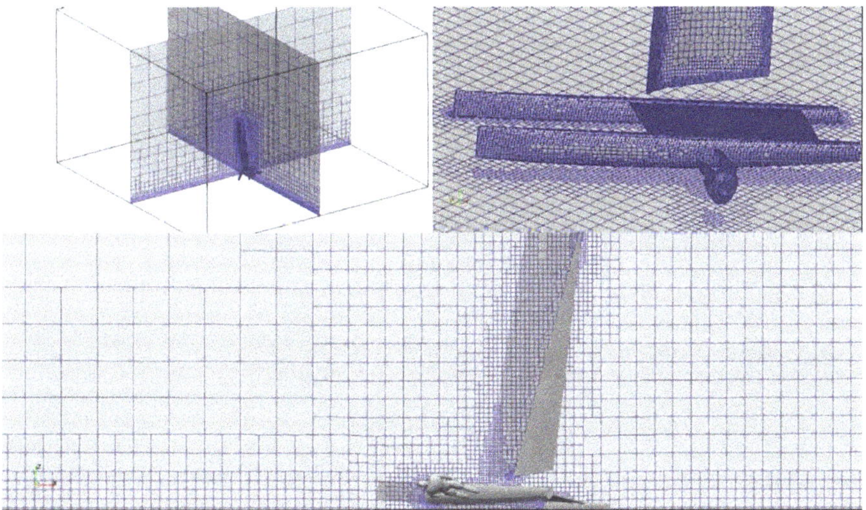

Figure 22. Computational domain and mesh details of the Open-Source based CFD configuration.

The simpleFoam solver with the $k - \omega$ SST turbulence model was used. The simulations have been conducted in the relative reference frame where the boat is stationary and the wind swirling boundary profile is generated at far field by its components tabulated in a formatted file. Wall functions were adopted to model the boundary layer. From preliminary analyses it was observed the Open-Source CFD configuration to provide a solution that differ in the order of 3–5% with respect to the solutions obtained with the commercial solver Figure 23 reports a typical convergence history of the solutions.

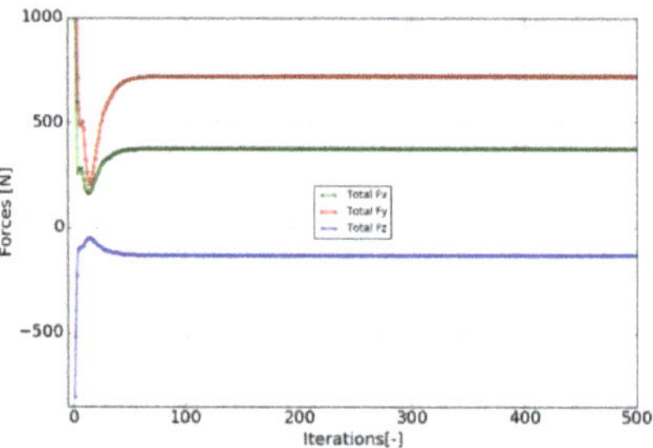

Figure 23. Typical solution convergence obtained by the Open-Source CFD configuration.

To include the Open-Source based analysis in the optimization environment, the following steps were implemented in an automatic procedure by a set of batch scripts and *Python* routines:

- Conversion from CAD to STL;
- Mesh generation and CFD configuration update;
- CFD run and solutions export;
- Post-processing and results extraction.

The obtained procedure is ready to substitute the commercial based one in the optimization environment. Figure 24 lists the software adopted to implement the two CFD analysis modules.

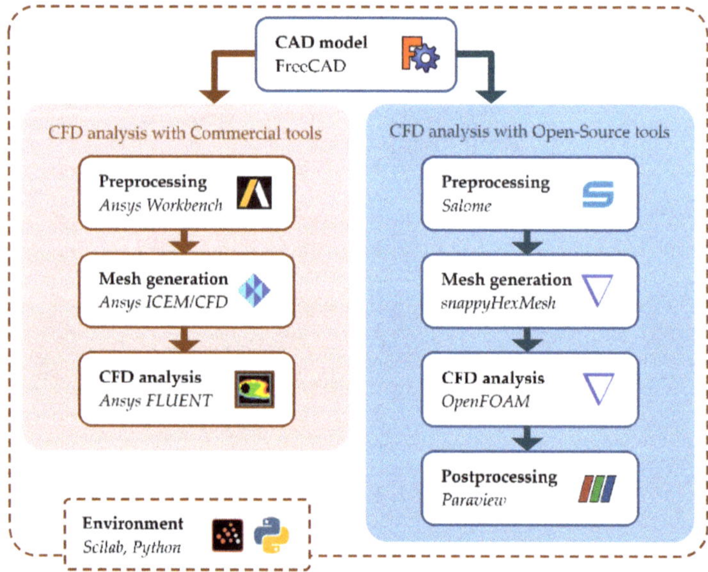

Figure 24. Software used to implement the two analysis modules.

3.4. Implementation of the Optimization Environment

Figure 25 sketches the workflow of the implemented optimization procedure. After the CAD model update the process is guided by a script that loads the new geometry, recomputes the mesh and exports the new grid in the solver format. The CFD configuration is then updated and the VPP module, described in the previous section, executed to provide the performance of the selected configuration. According to the solutions obtained, the optimization algorithm selects a new combination of parameters and the cycle progress until the "Optimum" configuration is identified. All the modules are managed by *Scilab* routines.

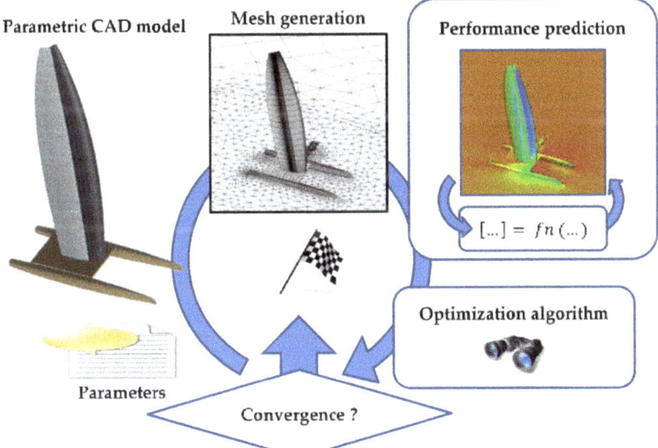

Figure 25. Scheme of the optimization procedure.

More than 40 parameters can be selected to characterise the boat configuration (airfoil, planform, positions and attitude of appendages as well as the parameters controlling the shape and the trim of the sail rig). Such a wide range of potential design variables gives great flexibility in exploring innovative solutions. The optimization environment, furthermore, provides the designer a powerful tool that supports the exploration of their limits.

4. Test of the Analysis Modules

The analysis modules were tested on a simplified optimization problem. It consisted in defining a DOE matrix, using two design variables, and in the definition of a response surface on which to search the optimum. The test has the double objective to find out potential criticisms of the CFD workflows and to compare the performance of the commercial and the Open-Source based analysis procedures. The optimization problem consisted in finding the optimum sail setting of an A-Class traditional rig at a fixed boat speed $V = 10$ knots, a true wind angle $TWA = 45$ deg and a true wind speed $TWS = 10$ knots. The variables were the mast spanner angle and the sail setting intended as the angle between the sail chord at the base and the symmetry plane of the boat. It was decided a range of variation for the spanner from 35 to 50 degree with a step of 5 deg. The range of variation of the sail setting angle was from 1 to 7 degree with a step of 2 deg. The DOE table was then populated with 16 solutions. Figure 26 reports the geometries of two design candidates belonging to two extremes of the variables design space.

The target of the optimization was the maximisation of the sail thrust force. The objective function to maximize was defined as follows:

$$Obj.Func. = F_t = F_Y \sin TWA - F_X \cos TWA \tag{31}$$

where the input forces are referred to a frame aligned to the true wind speed direction.

The optimum sailing condition of classic catamarans is with the upwind hull in flying condition just outside the water. This is also the conditions at which almost the maximum righting moment is generated. In equilibrium conditions, the heeling moment must be equal to the righting moment. Assuming the helmsman is positioned at the trapeze, having a weight of 90 kg and an arm slightly higher than 3 m, the maximum allowable righting moment, including the contribution of the boat weight (M_{hmax}), is around 3500 Nm. This value was implemented as an optimization constraint:

$$M_Y \sin TWA - M_X \cos TWA \leq M_{hmax} \tag{32}$$

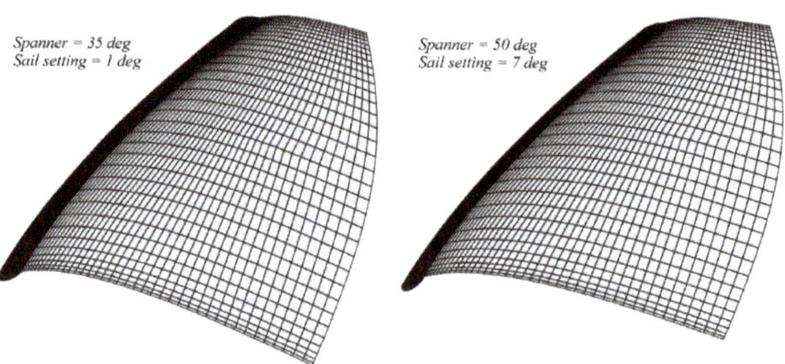

Figure 26. Comparison between two extremes of the variables design space.

Figure 27 compares the response surfaces computed with the solutions obtained by the two solvers. A second-degree polynomial formulation was sufficient to approximate the computed CFD solutions generating a residual error always below 0.3%.

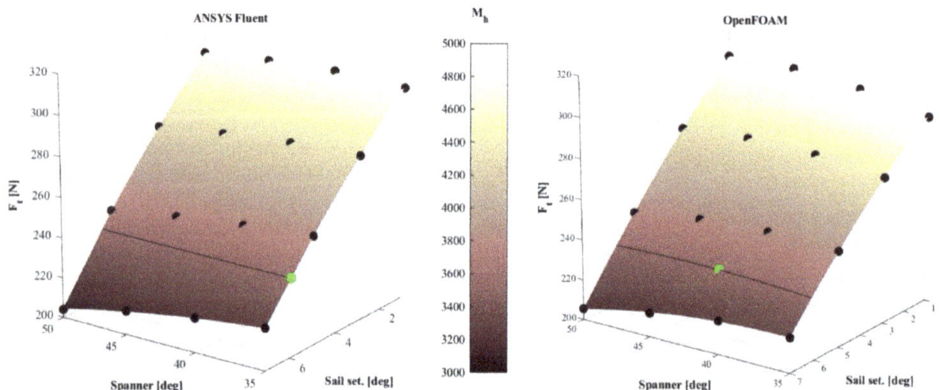

Figure 27. Comparison of the response surfaces obtained with the solutions of the two solvers.

The colours of the response surfaces are associated to the value of the heeling moment. The black curves on the surfaces are the M_{hmax} isolines at 3500 Nm. The optimum solution has then to be searched along these curves. The green points indicate the positions of the optima solutions found with the two methods. The optimum found using *ANSYS Fluent* is located on the boundary of the variables space since the isocurve do not have a maximum within his domain. The maximum found with the *OpenFOAM* based analysis procedure is close to the middle of the spanner range. The Table 2 reports the two optima solutions.

The thrust force estimated by the Open-Source based analysis procedures has a difference in the order of 2.4% with respect to the solution obtained by the commercial based analysis procedure. Both estimated almost the same optimum sail setting while larger differences are observed concerning the spanner setting. This variable, in fact, was shown to cause the generation of an isoline on which to search for the optimum almost horizontal in a wide range of the variable values. As a consequence, small differences in the solutions of the two solvers led to the identification of very different optima that, nevertheless, generated very similar performances. In other words, the spanner angle showed to have a moderate impact on the objective function in a wide range of the variable space.

Table 2. Optima rig setting solutions.

	ANSYS Fluent	OpenFOAM
Spanner	35 deg	41.7 deg
Sail setting	5.9 deg	6 deg
Thrust force	238.2 N	232.5 N

Figure 28 compares the pressure distribution obtained by the two solvers for the configuration with the spanner angle equal to 50 degrees and the sail setting at five degrees. The two solutions are very similar, confirming the quality of the Open-Source solution in comparison to the commercial one.

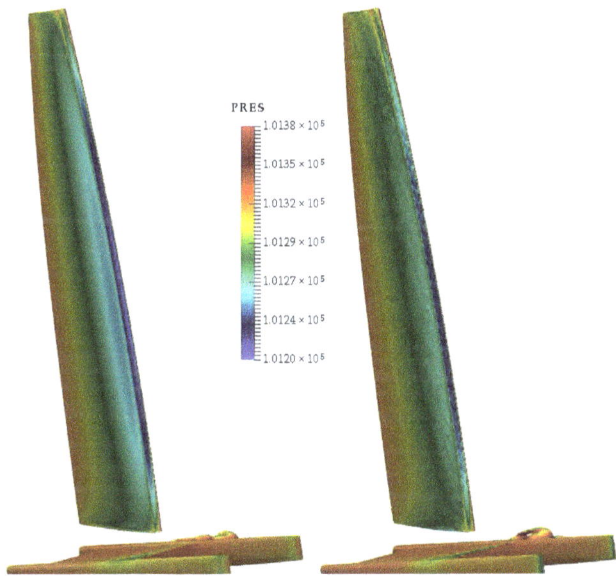

Figure 28. Comparison of pressure distribution obtained with the two solvers.

The *ANSYS Fluent* based analysis procedure ran on a workstation equipped with 20 cores (2 processors Intel Xeon E5-2680 2.8 GHz with 10 cores each). The complete convergence was reached, for each design points, with 200 iterations in less than ten minutes. The *OpenFOAM* based procedure ran on HPC nodes equipped with Intel Xeon 2670 v2 2.5 GHz. A single run required 15 min with 20 cores to complete the CAD setting, mesh generation, computation and solution extraction process. The computational costs of the two methods can then be in general considered comparable. As concluding remarks, it can be stated that the Open-Source based analysis module can be considered a valid candidate to replace the commercial based procedure.

5. Conclusions

A numerical optimization environment for a catamaran's sail plan and appendages, that couples a VPP based on analytical models and on a sail RANS computation, was developed. Two versions of the CFD analysis module were implemented and compared: one based on commercial software and one fully based on Open-Source tools. The procedures that manage the several modules were written using the *Scilab* computing environment. The geometric parametric module was implemented by *Python* scripts used to drive the Open-Source *FreeCAD* software to generate the CAD model. The analytical formulations, used to model the hull and the appendages forces, were implemented in a form of independent functions and coupled to the CFD solutions of the sail to solve the equilibrium system of equations of the boat in an iterative procedure. This procedure constitutes the VPP module that estimates the performance of the selected configuration in terms of boat VMG in upwind sailing conditions. The analysis is performed, in the procedure based on commercial software, using *ANSYS ICEM/CFD* to generate the mesh and *ANSYS Fluent* to provide the aerodynamic solution. In the Open-Source based analysis, the tools adopted were *Salome-Mesh* for pre-processing, *SnappyHexMesh* for the mesh generation, *OpenFOAM* for the CFD solution and *Pareview* for the post-processing. The performance of the two analysis modules were compared applying them on a simple test case: the optimization of the setting of the sail rig of an A-Class catamaran defined by two variables. A DOE approach was adopted, and a response surface generated on the solutions obtained with the two procedures. The objective function was the maximization of the sail thrust force with the constraint of a fixed heeling moment.

The two methods generated relatively similar optima solutions (with a difference in the objective function in the order of 2%). Except for cases where significant separation was present (close to the maximum lift), *OpenFOAM* provided, in general, solutions whose differences were lower than 5% with respect to the forces obtained with *ANSYS Fluent*. The difference of the thrust force of the two optima solutions was in the order of 2.4%. Considering that an optimum solution is expected to have no separations (or at least limited separated regions), it thus can be stated that *OpenFOAM* is a valid candidate to replace *ANSYS Fluent* in the optimization procedure. This assumption is also valid evaluating the computational requirements of the two solvers. Both codes completed an analysis of a case with attached flow in less than 15 min using 20 cpu.

Author Contributions: Conceptualization, U.C.; methodology, U.C., F.S. and R.P.; software, M.E.B.; analysis, U.C., F.S. and R.P.; resources, M.E.B.; writing—original draft preparation, U.C.; writing—review and editing, U.C., F.S., R.P. and M.E.B.; visualization, U.C., F.S. and R.P.; supervision, U.C. All authors have read and agreed to the published version of the manuscript.

Funding: This research was partially funded by the EU's Horizon 2020 research and innovation programme (2014–2020) under grant agreement 653838, by *RBF Morph* (www.rbf-morph.com accessed on 1 April 2021) and by *Design Methods—Aerospace Engineering* (www.designmethods.aero accessed on 1 April 2021).

Institutional Review Board Statement: Not applicable.

Informed Consent Statement: Not applicable.

Data Availability Statement: Not applicable. Deeper information can be provided upon request.

Acknowledgments: The authors wish to thank Filippo Cucinotta and Felice Sfravara from the Engineering department of the University of Messina. They actively contributed to the development of the parametric geometry module. A special thanks to the staff of the ADAG Aerospace Engineering section of the university of Naples "Federico II" and in particular to Agostino De Marco who supported the generation of the CFD solutions database used for the hull analytical model development.

Conflicts of Interest: The authors declare no conflict of interest.

Nomenclature

α	Angle of incidence
β	Leeway angle
γ	Rudder angle
δ	Appendage dihedral angle
λ	Aspect ratio
φ	Heeling angle
ρ_w	Sea water density
AWA	Apparent wind angle
AWS	Apparent wind speed
b	Draft of appendage
d	Distance between hulls centrelines
C_D	Drag coefficient
C_{D0}	Drag coefficient at zero incidence
C_f	Friction drag coefficient
C_L	Lift coefficient
C_{L0}	Lift coefficient at zero incidence
C_w	Wave drag coefficient
D	Drag
D_{B_x}	X component of the boat aerodynamic drag
D_{B_y}	Y component of the boat aerodynamic drag
D_H	Hull drag
D_{M_x}	X component of the crew aerodynamic drag
D_{M_y}	Y component of the crew aerodynamic drag
e	Oswald efficiency factor
F_h	Sail heeling force
F_t	Sail thrust force
h	Appendage aerodynamic centre
h_B	Height of boat centre of gravity
h_h	Height of sail centre of effort
h_g	Height of the boat centre of gravity
L	Lift
L_H	Hull side force (parallel to the sea plane)
l_M	Arm of crew righting moment
p	perimeter of the appendage (excluded root)
r	Daggerboard stagger angle
R_N	Reynolds number
S	Reference surface
TWA	True wind angle
TWS	True wind speed
V	Boat speed
W_{BE}	Boat empty weight
W_{BO}	Boat operative weight
W_M	Crew weight

References

1. Fossati, F. *Aero-Hydrodynamics and the Performance of Sailing Yachts: The Science behind Sailing Yachts and Their Design*; A&C Black: London, UK, 2009.
2. Lasher, W. The determination of aerodynamic forces on sails—Challenges and status. In *Atmospheric Turbulence, Meteorological Modelingand Aerodynamics*; Nova Science Publishers: Hauppauge, NY, USA, 2011; pp. 487–504.
3. Day, A.H.; Cameron, P.; Dai, S. Hydrodynamic Testing of a High Performance Skiff at Model and Full Scale. *J. Sail. Technol.* **2019**, *4*, 17–44. [CrossRef]
4. Graf, K.; Boehm, C. Coupling of ranse-CFD with VPP methods: From the numerical tank to virtual boat testing. In Proceedings of the 2nd International Conference on Innovation in High Performance Sailing Yachts, Lorient, France, 30 June–1 July 2010.
5. Lasher, W.C.; Richards, P.J. Validation of Reynolds-Averaged Navier-Stokes Simulations for International America's Cup Class Spinnaker Force Coefficients in an Atmospheric Boundary Layer. *J. Ship Res.* **2007**, *51*, 22–38. [CrossRef]

6. Peart, T.; Aubin, N.; Nava, S.; Cater, J.; Norris, S. Multi-Fidelity Surrogate Models for VPP Aerodynamic Input Data. *J. Sail. Technol.* **2021**, *6*, 21–43. [CrossRef]
7. Fossati, F.; Muggiasca, S.; Viola, I.M. An investigation of aerodynamic force modelling for IMS rule using wind tunnel techniques. In Proceedings of the 19th International HISWA Symposium on Yacht Design and Yacht Construction, Amsterdam, The Netherlands, 13–14 November 2006.
8. Graf, K.; Freiheit, O.; Schlockermann, P.; Mense, J.C. VPP-Driven Sail and Foil Trim Optimization for the Olympic NACRA 17 Foiling Catamaran. *J. Sail. Technol.* **2020**, *5*, 61–81. [CrossRef]
9. Findlay, M.W.; Turnock, S.R. Investigation of the effects of hydrofoil set-up on the performance of an international moth dinghy using a dynamic VPP. In Proceedings of the Innovation in High Performance Sailing Yachts, Lorient, France, 29–30 May 2008.
10. Bagué, A.; Degroote, J.; Demeester, T.; Lataire, E. Dynamic Stability Analysis of a Hydrofoiling Sailing Boat using CFD. *J. Sail. Technol.* **2021**, *6*, 58–72. [CrossRef]
11. Cella, U.; Salvadore, F.; Ponzini, R. *Coupled Sail and Appendage Design Method for Multihull Based on Numerical Optimisation*; EU SHAPE Project Final Report; PRACE: Bruxelles, Belgium, 2016.
12. Gomez, C.; Bunks, C.; Chancelier, J.P.; Delebecque, F.; Goursat, M.; Nikoukhah, R.; Steer, S. *Engineering and Scientific Computing with Scilab*; Gomez, C., Ed.; Birkhäuser: Basel, Switzerland, 1999.
13. Gerritsma, J.; Onnink, R.; Versluis, A. Geometry, resistance and stability of the Delft systematic yacht hull series. *Int. Shipbuild. Prog.* **1981**, *28*, 276–297. [CrossRef]
14. Keuning, L.J.; Katgert, M. A bare hull resistance prediction method derived from the results of The Delft Systematic Yacht Hull Series to higher speeds. In Proceedings of the Innovation in High Performance Sailing Yachts, Lorient, France, 29–30 May 2008.
15. Kerdraon, P.; Horel, B.; Bot, P.; Letourneur, A.; le Touzé, D. High Froude Number Experimental Investigation of the 2 DOF Behavior of a Multihull Float in Head Waves. *J. Sail. Technol.* **2021**, *6*, 1–20. [CrossRef]
16. Molland, A.F.; Turnock, S.R.; Hudson, D.A. *Ship Resistance and Propulsion*; Cambridge University Press: New York, NY, USA, 2011.
17. De Luca, F.; Mancini, S.; Miranda, S.; Pensa, C. An Extended Verification and Validation Study of CFD Simulations for Planing Hulls. *J. Ship Res.* **2016**, *60*, 101–118. [CrossRef]
18. Zou, L.; Larsson, L. CFD Verification and Validation in Practice—A Study Based on Resistance Submissions to the Gothenburg 2010 Workshop on Numerical Ship Hydrodynamics. In Proceedings of the 30th Symposium on Naval Hydrodynamics, Hobart, Australia, 2–7 November 2014.
19. Insel, M.; Molland, A. *An Investigation into the Resistance Components of High Speed Displacement Catamarans*; RINA: Genoa, Italy, 1992.
20. Hoerner, S.F. *Fluid-Dynamic Drag*; Hoerner Fluid Dynamics: Brick Town, NJ, USA, 1965.
21. ITTC. Resistance Uncertainty Analysis, Example for Resistance Test. In Proceedings of the ITTC International Towing Tank Conference, Recommended Procedures, Venice, Italy, 14–16 January 2002.
22. Losito, V. *Fondamenti di Aeronautica Generale*; Accademia Aeronautica Pozzuoli: Pozzuoli, Italy, 1991.
23. Roskam, J. *Airplane Aerodynamics and Performance*; DARcorporation: Lawrence, KS, USA, 1997.
24. Drela, M. XFoil—An Analysis and Design System for Low Reynolds Number Airfoils. In *Low Reynolds Number Aerodynamics. Lecture Notes in Engineering*; Springer: Berlin/Heidelberg, Germany, 1989; Volume 54, pp. 1–12.
25. Abbott, A.H.; von Doenhoff, A.E. *Theory of Wing Sections*; Dover Publications: Mineola, NY, USA, 1959.
26. Perkins, C.D.; Hage, R.E. *Airplane Performance Stability and Control*; John Wiley & Sons: New York, NY, USA, 1949.
27. Larsson, L.; Eliasson, R. *Principle of Sailing Yacht Design*; Adlard Coles Nautical: London, UK, 1997.
28. Claughton, A.; Wellicome, J.; Shenoi, A. *Sailing Yacht Design: Theory*; Longman: London, UK, 1998.
29. Nelder, J.A.; Mead, R. A Simplex Method for Function Minimization. *Comput. J.* **1965**, *7*, 308–313. [CrossRef]
30. Biancolini, M.E.; Viola, I.M.; Riotte, M. Sails trim optimisation using CFD and RBF mesh morphing. *Comput. Fluids* **2014**, *93*, 46–60. [CrossRef]
31. Cella, U.; Cucinotta, F.; Sfravara, F. Sail Plan Parametric CAD Model for an A–Class Catamaran Numerical Optimization Procedure Using Open Source Tools. In *Advances on Mechanics, Design Engineering and Manufacturing*; Lecture Notes Series in Mechanical Engineering; Springer: Cham, Switzerland, 2016.
32. Cella, U.; Romano, D.G. Winglets design and assessment of optimization algorithms. In Proceedings of the CAE Technologies International Conference, Bergamo, Italy, 1–2 October 2009.
33. Menter, F.R. Two-equation eddy-viscosity turbulence models for engineering applications. *AIAA J.* **1994**, *32*, 1598–1605. [CrossRef]
34. Musker, A.J. Explicit Expression for the Smooth Wall Velocity Distribution in a Turbulent Boundary Layer. *AIAA J.* **1979**, *17*, 655–657. [CrossRef]

Journal of
Marine Science and Engineering

Article

Improving the Downwind Sail Design Process by Means of a Novel FSI Approach

Antonino Cirello [1], Tommaso Ingrassia [1,*], Antonio Mancuso [1], Vincenzo Nigrelli [1] and Davide Tumino [2]

[1] Dipartimento di Ingegneria, Università degli Studi di Palermo, Viale delle Scienze, 90127 Palermo, Italy; antonino.cirello@unipa.it (A.C.); antonio.mancuso@unipa.it (A.M.); vincenzo.nigrelli@unipa.it (V.N.)
[2] Facoltà di Ingegneria e Architettura, Università degli Studi di Enna Kore, Cittadella Universitaria, 94100 Enna, Italy; davide.tumino@unikore.it
* Correspondence: tommaso.ingrassia@unipa.it

Abstract: The process of designing a sail can be a challenging task because of the difficulties in predicting the real aerodynamic performance. This is especially true in the case of downwind sails, where the evaluation of the real shapes and aerodynamic forces can be very complex because of turbulent and detached flows and the high-deformable behavior of structures. Of course, numerical methods are very useful and reliable tools to investigate sail performances, and their use, also as a result of the exponential growth of computational resources at a very low cost, is spreading more and more, even in not highly competitive fields. This paper presents a new methodology to support sail designers in evaluating and optimizing downwind sail performance and manufacturing. A new weakly coupled fluid–structure interaction (FSI) procedure has been developed to study downwind sails. The proposed method is parametric and automated and allows for investigating multiple kinds of sails under different sailing conditions. The study of a gennaker of a small sailing yacht is presented as a case study. Based on the numerical results obtained, an analytical formulation for calculating the sail corner loads has been also proposed. The novel proposed methodology could represent a promising approach to allow for the widespread and effective use of numerical methods in the design and manufacturing of yacht sails.

Keywords: finite element method; computational fluid dynamics; FSI; sail design; gennaker; sail loads

Citation: Cirello, A.; Ingrassia, T.; Mancuso, A.; Nigrelli, V.; Tumino, D. Improving the Downwind Sail Design Process by Means of a Novel FSI Approach. *J. Mar. Sci. Eng.* **2021**, *9*, 624. https://doi.org/10.3390/jmse9060624

Academic Editor: Apostolos Papanikolaou

Received: 29 April 2021
Accepted: 31 May 2021
Published: 4 June 2021

Publisher's Note: MDPI stays neutral with regard to jurisdictional claims in published maps and institutional affiliations.

1. Introduction

In recent years, the exponential growth of computational resources at a very low cost has allowed for the widespread use of numerical methods in the nautical field as well as at an industrial level, not only for scientific research or highly competitive purposes. Nevertheless, to date, advanced numerical methods are commonly used in the nautical field only for competitive applications, but it is conceivable that in the future these methods could also be used in other application fields. Recent racing yachts, as demonstrated by the last edition of the America's Cup, have achieved very high performances thanks to the massive improvements in yacht and sail design, materials, and fabrication. The design of the hulls, sails, and rigging of competitive racing yachts needs more and more detailed research and development that can be obtained by combining advanced computational resources and experimental studies. In particular, with regard to the design of sails, based on traditional and empirical manufacturing processes, nowadays, the best sail designers aim to use more new research and development tools [1].

To improve sail design and to better understand sail aerodynamics, numerical simulations and experimental testing are the most used and reliable methods [2–5]. These methods have led to a better comprehension of the complex phenomena and the physics underlying sailing and, consequently, through them, a remarkable increase of the performances of yacht sails has been obtained in recent years [6,7]. Experimental full-scale testing is the most accurate and reliable method, but, because of cost constraints, its use is only recommended for a few exceptional cases [7,8] or to validate numerical methods [2,9,10]. Instead, numerical simulations are cheaper than experimental methods, but in order to obtain reliable and accurate results, they need long setup times by highly qualified and experienced users and they require high-performing computational resources.

Among the numerical methods, computational fluid dynamics (CFD) simulations are the most widely used in the design process of high-performance sailing yachts [8]. CFD analyses can be effectively used to design high-performance hulls [11] and to simulate both upwind and downwind sail configurations [12–17]. Usually, the flow in upwind sailing is mostly attached and the turbulence effect is quite limited; the flow regime around a downwind sail can be highly turbulent [7] and, consequently, simulating real sailing conditions can be a very difficult task. For this reason, in the past, many studies have concentrated on upwind sails [6], and in recent years, many researchers have focused on simulating the complex downwind sailing conditions [6,18,19]. The first numerical studies on downwind sails frequently treated the topic in a simplified numerical way, considering only the flow effects around a rigid structure [17]. However, the physics of downwind sails is by far more complex than the upwind ones, not only because of the highly turbulent flow, but also because these kinds of sails have an inherent unsteadiness, even in quite stable conditions [7]. This happens mainly because downwind sails are made of very light and flexible cloth, and are attached to the yacht's structure at three points, namely, the head, the clew, and the tack [20]. The shape of a downwind sail is formed by self-generated aerodynamic forces that are strongly affected by the sail shape itself [21]. Moreover, the shape can change remarkably when sailing, depending on the trim settings and wind conditions [7]. All of these aspects must be taken into account when simulating the real sailing conditions of these kinds of sails, and the sail shape must be accurately measured in the flying condition [10,21]. For this purpose, nowadays, several authors focus on the issue of predicting the flying shape a sail develops under the impact of flow forces. In this context, it is well known that the fluid–structure interaction (FSI) is a key feature to study these kinds of problems [11,16,22,23]. Today, sail designers use specific software to define the constructed (molded) sail shape based on their experience, but they do not have any certainty about the real flying shape [6]. The possibility of predicting the real flying shape of a downwind sail could allow designers to improve the sailing performances of yachts [5,7,24,25], for instance, by maximizing the driving forces. However, knowing the real flying shape could also allow for optimizing the mechanical characteristics or the manufacturing process of the sails, in order to improve their stability, to reduce their weight, and to optimize their mechanical behavior [18,22,26,27]. Many papers have focused on evaluating the real shape of sails only to predict sailing performances [20,22,25,28]. However, to the best of our knowledge, no relevant papers have tried to measure the flying shape and the loads of a sail, taking into account both the point of view of sailors, whose primary interest is the sail performance, and of the sailmakers, who are also responsible for suitably dimensioning and manufacturing the sail.

In this work, a new approach based on the perspective of both sailors and sailmakers is proposed. In particular, the authors propose calculating the flying shape and the loads on downwind sails, not only to estimate propulsive forces [2,10], but also to help sailmakers in dimensioning and defining the best arrangement of the sail reinforcements. For this purpose, a weakly coupled FSI procedure [16] is used to find the real flying shape and to estimate the loads on the head, the clew, and the tack of the sail when the sailing conditions change. On the basis of the numerical results obtained, an analytical formulation for the calculation of the forces on the sail is also proposed.

2. Materials and Methods

In this work, to determine the flying shape and to evaluate the loads on downwind sails, a numerical FSI approach was used. Starting from a previous work of the authors [16], a new weakly coupled method was developed and tested. The new procedure is parametric and fully automated, and it only needs some input data to calculate the flying shape and the loads on the sail. As it is known, to perform a numerical FSI analysis, two embedded problems, aerodynamic and structural, need to be solved together [29]. In this study, to solve the aerodynamic and structural problems, the commercial CFD solver Ansys CFX and the Ansys Static Structural solver were used, respectively. JavaScript and a python-based programming language were used to develop the algorithm that managed the whole procedure, including the data exchange between the workbench and spreadsheet, which was developed ad hoc.

2.1. FSI Analysis: Setup of CFD Environment

The proposed procedure was developed to solve the FSI problem of a downwind sail but, with the aim to simulate the flow as accurately as possible, and so a mainsail was also considered in the CFD simulations. As mentioned, the procedure is parametric, in order to study different sailing configurations in a very simple and fast way. The main CFD parameters that must be first set to run the FSI analysis are reported in Table 1.

Table 1. Computational fluid dynamics (CFD) analysis parameters.

Parameter	Unit
Apparent wind speed (AWS)	m/s
Apparent wind angle (AWA)	deg
Mainsail boom angle (αm)	deg
Downwind sail trim angle ($\Delta\alpha_g$)	deg

The downwind sail trim angle, $\Delta\alpha_g$, is a parameter that has been introduced to simulate different sail trims. In particular, the sheet angle of the downwind sail, that is the angle between the tack–clew line and the centerline of the boat (α_g in Figure 1), is defined as $\alpha_g = \alpha_{g\_design} + \Delta\alpha_g$; where α_{g_design} is the design sheet angle of the downwind sail. In this way, once the specific sail type and its design sheet angle have been chosen, different sail trims can be simulated by simply changing the value of parameter $\Delta\alpha_g$.

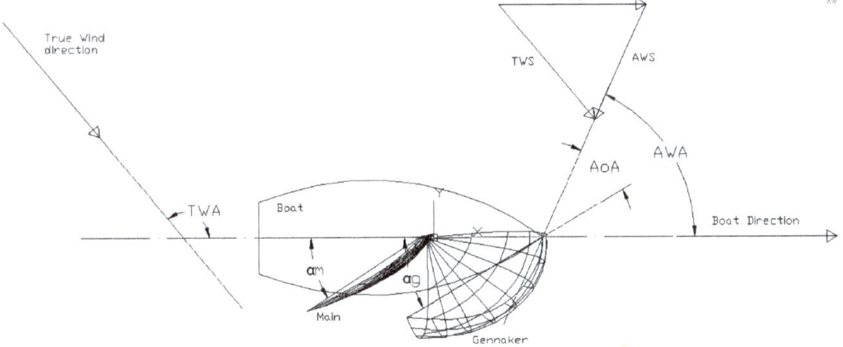

Figure 1. Main CFD parameters and sailing geometric variables.

The domain used for the CFD simulation was made parametric by setting the positions of the main boundary surfaces (inlet, outlet, and top) as a function of the maximum sail chord, C_m, as defined in Table 2.

Table 2. Distance of the boundary surfaces in function of C_m [m].

Inlet	Outlet	Top
$5 \times C_m$	$10 \times C_m$	$1 \times C_m$

Following a largely used approach, the boundary conditions at the inlet surfaces (Figure 2) were set following the formulation proposed by Richards and Hoxey [30]. Concerning the other boundary surfaces, the outlet was set as "opening condition" [16], the bottom surface was set as a rough wall, and the top surface was set as the inlet. The flow velocity at the top surface was calculated using the same formulation of the velocity profile proposed in [30]. The k-ω SST formulation [31] was chosen as the turbulence model.

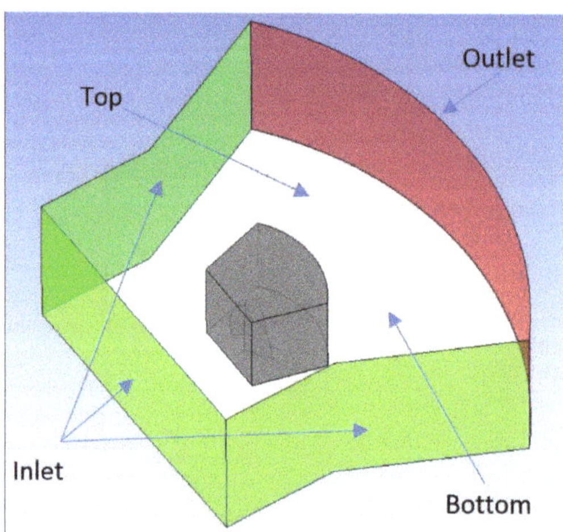

Figure 2. CFD domain: boundary surfaces (top is fully transparent, bottom is white), and inner (grey) and outer regions.

Regarding the shapes of the sails, the geometry of the mainsail was fixed while the downwind sail was updated at each simulation according to the deformed shape calculated using the FEM analysis. To optimize the computational costs, the domain was divided into an inner region (grey in Figure 2), meshed using a mixed hexa-tetrahedral mesh with prismatic layers on the surfaces of the sails, and an outer region, where hexahedral elements were used. Mesh refinements were applied to the cells adjacent and close to the sail surfaces. The height of the first prismatic element attached to the surfaces of the sails was set to achieve the condition of y + ≈ 1.

A convergence study was performed to quantify the influence of the grid resolution on the CFD results. The ratio between the distances of the grid nodes is equal to $\sqrt[3]{2}$. Four similar grids with about 1.5 M (base), 0.75 M, 0.38 M, and 0.18 M cells were investigated. The relative step ratio of the i-th grid is $h_i = \sqrt[3]{\frac{N_{base}}{N_i}}$, with N being the number of cells, resulting in $h = 1.00$ (base), 1.26, 1.59, and 2.00, respectively. The convergence trend is approximated with the following equation [32]:

$$S(h_i) = S_0 + C \times h_i{}^p \tag{1}$$

The values of coefficients S_0, C, and p are computed with the least-squares method. Figure 3 and Table 3 show the drive force computed for the different analyzed grids and the results of the uncertainty estimation. A monotonic convergence can be observed in Figure 3. The value of the estimated uncertainty, U_g, for $h = 1$ (corresponding to 1.5 M cells grid) is 2.1%; this value can be considered acceptable.

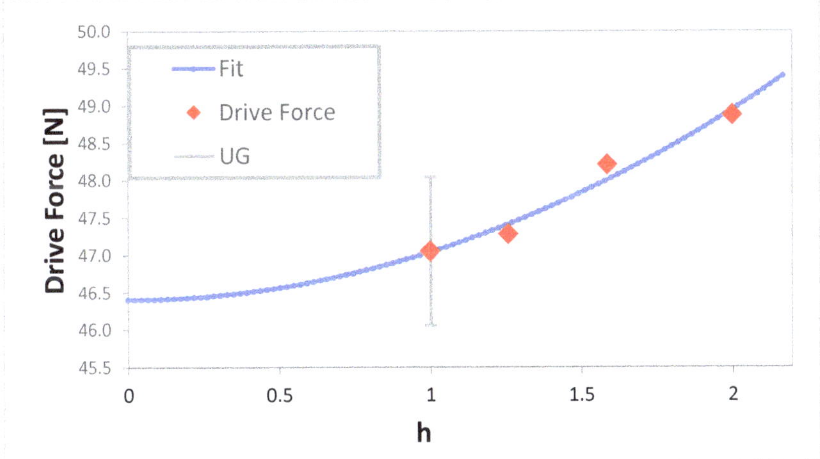

Figure 3. Drive force vs. relative grid size.

Table 3. Estimated uncertainty for different grids.

Number of Cells	Drive Force [N]	h	U_G [N]	U_G%
1,500,000	47.05	1.00	0.99	2.1%
750,000	47.28	1.26	1.44	3.1%
375,000	48.21	1.59	2.54	5.5%
187,500	48.87	2.00	3.25	7.0%

Basing on the grid verification results and the good performance hardware available, we used the 1.5 M cells grid. As a result of the unavailability of experimental data, it is not possible, to date, to present a complete validation of the numerical model based on a comparison of specific CFD results with corresponding experimental data. However, in Section 4, the results obtained with the proposed method are compared with others in the literature. A good level of agreement between the numerical and the experimental results has been noted; this demonstrates that the proposed method is able to give sufficiently reliable results. Of course, a complete validation will be carried out in future work.

The meshed domain and the details of mesh around the sail are shown in Figure 4.

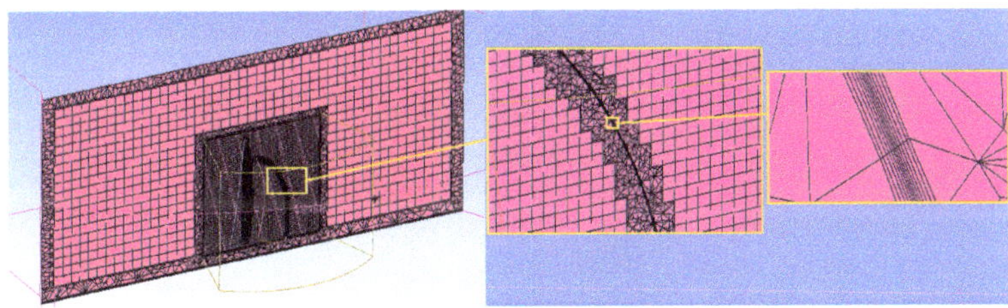

Figure 4. CFD meshed domain and enlarged views of the mesh around the sail.

The CFD analyses have been set with these stop conditions:

- values of the residuals of the continuity and momentum equations lower than 1.0×10^{-5};
- maximum number of iterations equal to 200.

For all of the analyzed configurations, it was observed that the residuals rapidly decreased under the threshold value (1.0×10^{-5}), and the aerodynamic quantities (forces, pressure, etc.) converged towards stable values after a reduced number of iterations. As an example, for AWA = 90° and Δαg = 0°, after 45 iterations, the values of the driving force remained stable around a value of about 47 N; moreover, the standard deviation of the values of the driving force calculated in the last 30 iterations was lower than 0.5%. Figures 5–10 show some results obtained for AWA = 90° and AWA = 115°. In particular, Figures 5 and 6 show the velocity streamlines that develop from the inlet to the outflow boundary surface. For AWA = 90°, the flow was quite clean around the body of the gennaker, even if some divergences occurred near the head and the leech, indicating flow separation in these areas. For AWA = 115°, the flow was slightly less regular than AWA = 90°, and a larger recirculation region could be evidenced at the downstream.

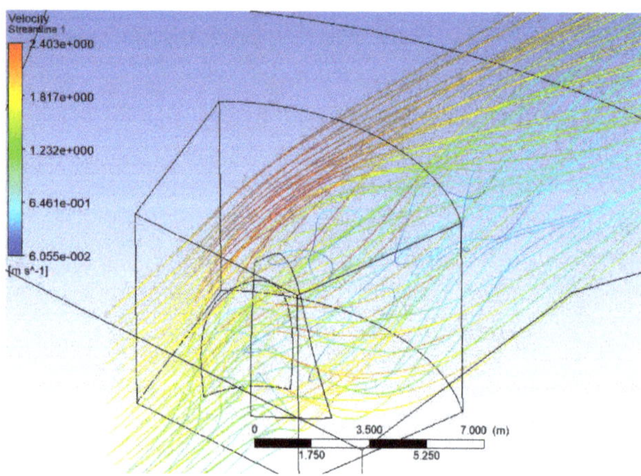

Figure 5. Velocity streamlines for AWA = 90°.

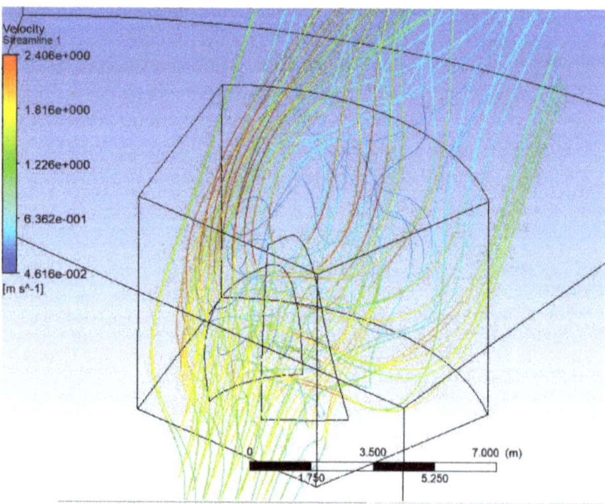

Figure 6. Velocity streamlines for AWA = 115°.

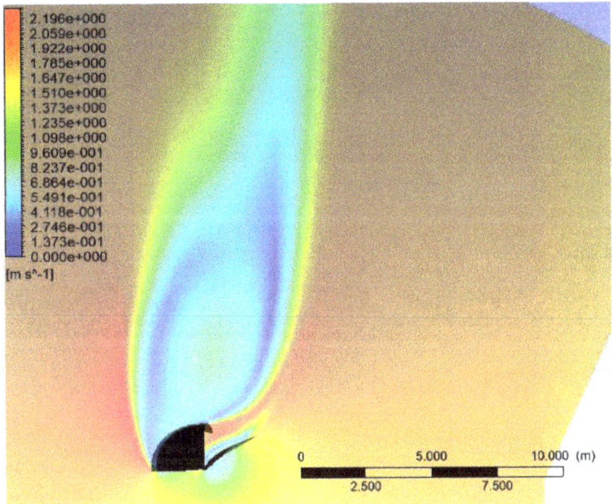

Figure 7. Velocity contour plots at the mid-height section for AWA = 90°.

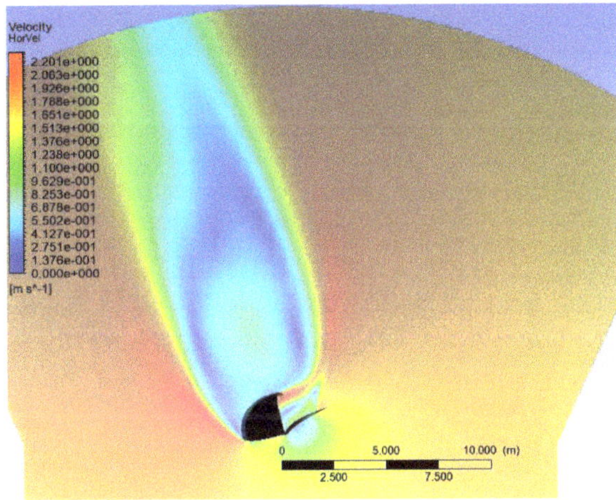

Figure 8. Velocity contour plots at the mid-height section for AWA = 115°.

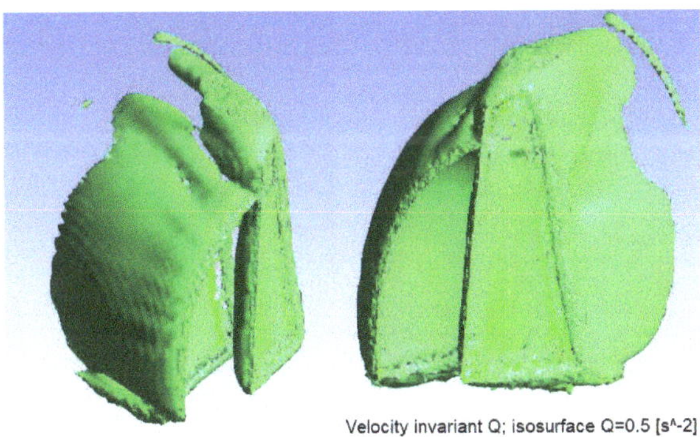

Figure 9. Iso-surfaces of Q-criterion for AWA = 90°.

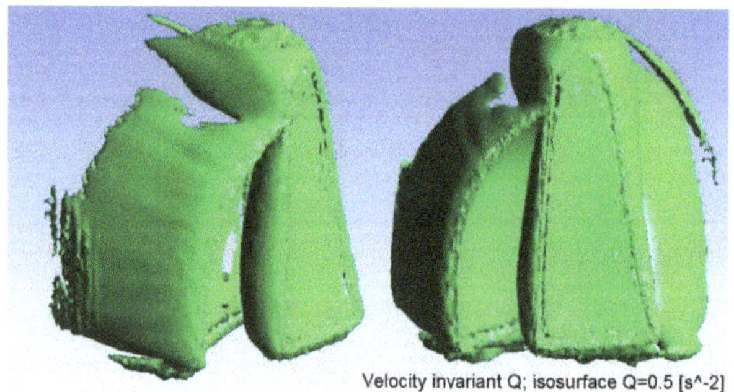

Figure 10. Iso-surfaces of Q-criterion for AWA = 115°.

This phenomenon is also supported by the horizontal velocity contour plots (Figures 7 and 8) at the half downwind sail height. In fact, for AWA = 115° (Figure 8), a larger area with a low flow velocity at the downstream was observed compared with AWA = 90° (Figure 7).

To better visualize the flow structures, the second invariant of the velocity gradient [4], Q, was used. Figures 9 and 10 show the iso-surfaces of Q = 500 s^{-2}. In both cases, it can a strongly turbulent fluid at the downstream can be observed; for AWA = 115° (Figure 10), a large flow vorticity can also be seen at the upper part of the sail.

2.2. FSI Analysis: Setup of FEM Environment

As a result of the assumption of a rigid mainsail, FEM analysis was performed only on the downwind sail. The FEM model is parametric; the main parameters are the sail geometry, the Young's modulus, the Poisson ratio, and the sail thickness. The geometry of the sail was automatically meshed using eight-node shell elements (Shell 281). This kind of shell element has six degrees of freedom at each node, and it is characterized by a mathematical formulation suitable for simulating thin structures (like sails). Suitable mesh refinements were applied on the head, the clew, and the tack.

A convergence study was performed with five structural meshes, with the following average sizes for the shell elements: 200, 100, 5, 30, and 20 mm. The different analyzed meshes had about 1540, 5700, 9200, 12,870, and 18,574 elements, respectively. The result of the convergence study is shown in Figure 11, where the value of the maximum displacement is plotted as a function of the average element size.

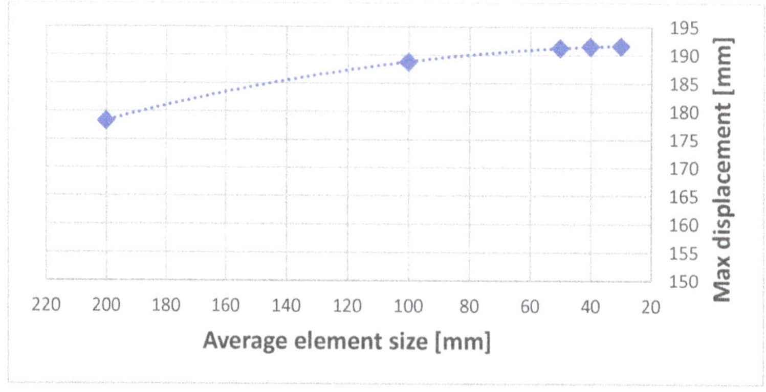

Figure 11. Plot of the maximum displacement vs. average element size.

From Figure 11, it can be observed that convergence is obtained (relative differences between two consecutive meshes lower than 5%) for values of the element size lower than 100 mm. Given the result of this convergence study, the mesh with an average size of 50 mm was chosen. This represents a good compromise between the quality of the results and the computing time.

The chosen mesh, composed of about 9200 elements, is shown in Figure 12.

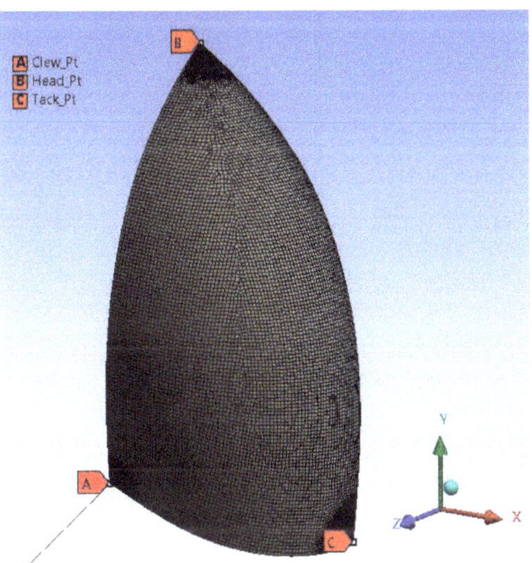

Figure 12. FEM mesh of the sail.

To calculate the loads on the rig, the fluid-dynamic pressure distribution calculated at each step by the CFD module was applied over both sail surfaces. The following boundary conditions were applied on the head and the tack (corners "B" and "C", respectively, in Figure 12): the displacements along the x, y, and z directions were fully constrained and only the rotations along the x, y, and z axes were allowed. The clew (corner "A" in Figure 12) was joined by a spherical joint to a link element, represented by a black dashed line in Figure 12, which simulates a rope. The end of the rope connected to the sail can move freely along the x, y, and z directions; the other end, instead, can only rotate, but no displacement is allowed. A multipoint constraint approach [16] was used to distribute the constraints over many nodes of the head, tack, and clew regions so to avoid peaks of stress due to the application of a load in a single node.

2.3. FSI Procedure

The flow chart of the developed procedure is shown in Figure 13. To launch the procedure, in addition to the CFD and FEM parameters, the following input parameters must be preliminarily defined:

- the sail design shape, S_d;
- the number of steps, N_s, that the FSI analysis is subdivided into;
- the displacement threshold of the current deformed shape, T_d.

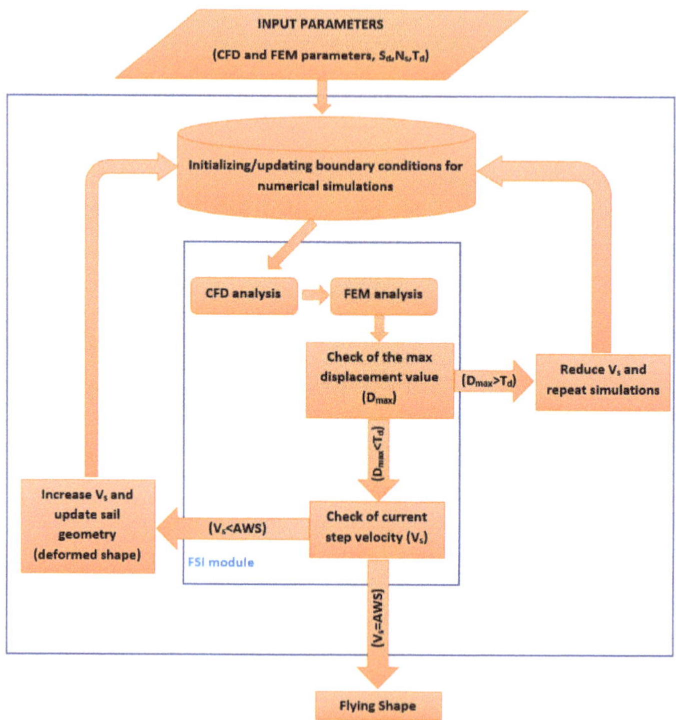

Figure 13. Flow chart of the developed fluid–structure interaction (FSI) procedure.

S_d represents the shape of the sail as defined by the designer. The geometry (shape) file can be loaded in different file formats (iges, sat, dwg, etc.). As "classic" weak coupled procedures could be very unstable because of the large deformations of the sail under the wind load, in the new proposed procedure, the wind velocity was gradually increased in different subsequent steps, until the design value was reached. In this way, the sail shape gradually changed as the wind speed increased and, consequently, no convergence problems of the structural simulations arose. The parameter N_s is used to define the preliminary number of steps and influences the wind velocity, V_s, imposed at each step. The value of the velocity at the *i*-th step, V_s, is calculated as follows:

$$V_s = \text{AWS} \left\{ 1 - 0.5[(N_s - i)/(N_s - 1)]^2 \right\}. \tag{2}$$

Moreover, to better simulate the progressive deformation of the sail, thus avoiding any abrupt change in geometry between two subsequent steps, a check on the maximum value of the displacement was also introduced at the end of each step. For this purpose, a threshold parameter, T_d, has been defined. T_d represents the admissible maximum value of the displacement of the deformed shape measured at each step.

A simple user interface developed in MS Excel (Figure 14) allows for setting the numerical parameters and launching the FSI analysis.

As soon as it is started, the implemented procedure works iteratively in the following way: the sail geometry, which during the first iteration is the designed one (S_d), subjected to V_s < AWS, is analyzed through the CFD module. The calculated fluid-dynamic pressure distribution is used as the boundary condition for the subsequent structural simulation performed by the FEM module. When the FEM simulation is completed, the maximum value of the total displacement of the sail, D_{max}, is calculated and compared with the threshold value, T_d. If D_{max} is higher than T_d, it means that the sail is deformed excessively, so the current velocity is decreased and the procedure restarts from the CFD simulation. If D_{max} is lower than T_d, the deformed shape of the sail is extracted from the FEM solver and is used as the updated input geometry for a new CFD analysis. Before starting the new CFD simulation, the current wind velocity, V_s, is increased by an amount that depends on the value of N_s. When V_s = AWS, the procedure stops and the flying shape of the sail is obtained.

	A	B	C	D	E	F
1	AWS [m/s]	Ns	Td (m)	Young Modulus [Mpa]	Sail Tickness [mm]	
2	3.0	5	0.035	5500	0.7	Start Analisys
3	AWA [°]	Downwind sail trim angle [°]	Mainsail boom angle [°]	Poisson ratio	Status	
4	90	0	35	0.25	Iteration finished	

Figure 14. MS Excel user interface.

2.4. Case Study

A common all-purpose gennaker typically used to equip small size sailing boats, like dinghies, was studied. The sail area was 14.2 m^2, the luff edge length was 5.95 m, and the maximum chord (C_m) length was 3.20 m. A nylon typically used for this kind of applications, characterized by a Young modulus of 5500 MPa, was chosen for this sail. As mentioned in Section 1, the aim of this paper is not only to present a new FSI approach to measure the flying shape of a downwind sail, but also to test the proposed method to evaluate the structural loads. In particular, it was investigated how the loads on the clew, the head, and the tack vary depending of different sailing conditions and, on the basis of the numerical results obtained, an attempt was made to find a possible analytical formulation that would allow for estimating loads as the wind speed varied. Based on the typical sailing conditions of these types of sails, we chose to study the gennaker at different values of AWA ranging between 90° and 130°, with 5° increments. Moreover, for each analyzed AWA value, we decided to investigate different configurations of the gennaker by varying its trim angle ($\Delta\alpha_g$) between 0° and 20° (with increments of 5°), in order to find the best trim depending on the wind angle. This resulted in about 60 different configurations to simulate. For this reason, to limit the computational time, a reduced wind velocity was chosen (AWS = 3 m/s). This choice allowed for investigating all of the sailing configurations in a reduced time using a Dual Intel Xeon E5-2670 processor (12 cores) workstation with 128 GB RAM. The mainsail boom angle (α_m) was set to be constant and equal to 35°. The values of all of the parameters are presented in Table 4.

Table 4. Values of the parameters used for the case study.

Parameter	Value
Displacement threshold value (T_d)	0.035 m
Number of steps (N_s)	5
Sail thickness	0.7 mm
Young's modulus	5500 MPa
Poisson ratio	0.25
Apparent wind speed (AWS)	3 m/s
Apparent wind angle (AWA)	90° ÷ 120°
Mainsail boom angle (α_m)	35°
Downwind sail trim angle ($\Delta\alpha_g$)	0° ÷ 20°

3. Results

3.1. Flying Shapes

Figures 15 and 16 show the flying shapes of the gennaker obtained for two different sailing configurations. In particular, the presented results are the ones obtained for AWA = 90° ($\Delta\alpha_g$ = 0°) and AWA = 110° ($\Delta\alpha_g$ = 15°), respectively. The design shapes are colored in green and the flying ones are blue. The maps of the 3D deviations of the flying shapes from the design ones are also presented in Figures 15 and 16. Different point of views, two from the leech and the luff sides and one frontal, are presented in order to better detect the differences between the real and the flying shape.

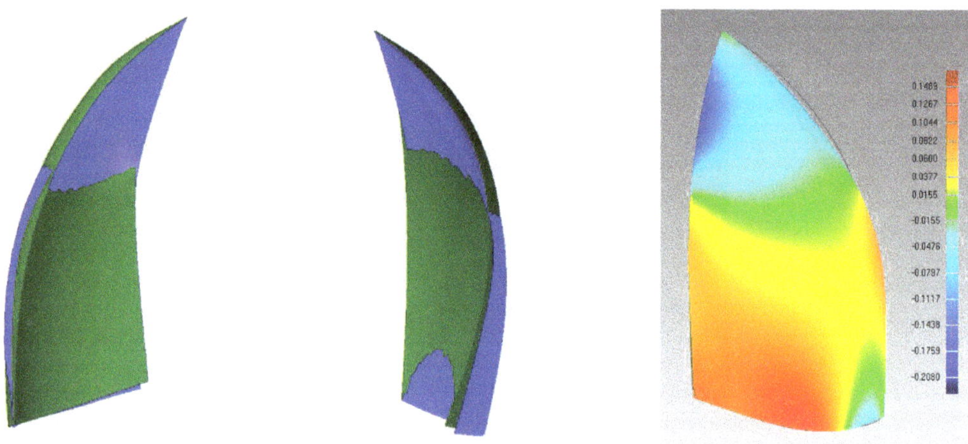

Figure 15. Design shape (green), flying shape (blue), and map of the 3D deviations [m] for AWA = 90° and $\Delta\alpha_g = 0°$.

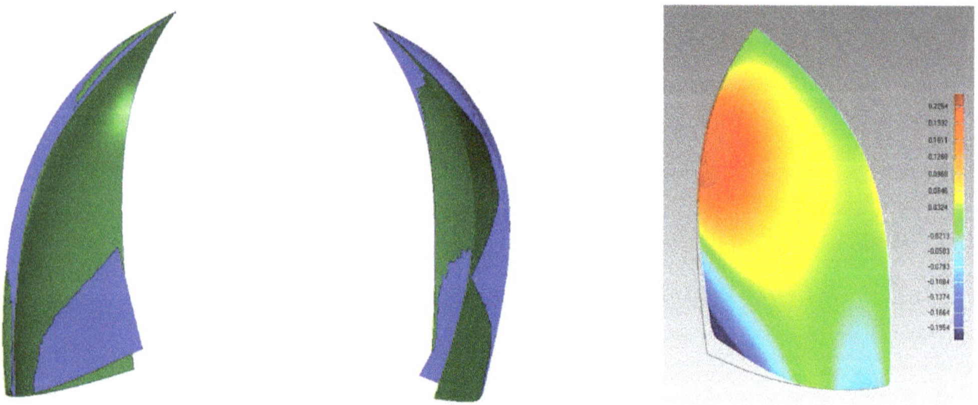

Figure 16. Design shape (green), flying shape (blue), and map of the 3D deviations [m] for AWA = 110° and $\Delta\alpha_g = 15°$.

It can be noticed that the largest positive displacements occur for AWA = 110° and mainly occur along the leech edge. Moreover, it can be observed that for AWA = 90°, the largest positive displacements mainly involve the upper part, while for AWA = 110°, the lower part of the gennaker is the most deformed.

3.2. Loads on the Corners of the Sail: An Analytical Formulation

For all of the analyzed configurations, the driving force and the loads on the corners of the sail, which are the head, the tack, and the clew, were calculated during all of the steps of the FSI simulation. In this way, it was possible to know how these forces varied as the wind speed increased until the chosen value of AWS. To better investigate the structural loads on the head, the tack, and the clew of the sail, for each of the analyzed configurations, the polynomial trendlines of the corner loads and their analytical formulations were extracted by a specific tool of MS Excel. For example, Figure 17 shows the plots of the corner loads over the wind speed for AWA = 120° and $\Delta\alpha_g$ = 20°. The polynomial trendlines and their analytical formulations are also reported.

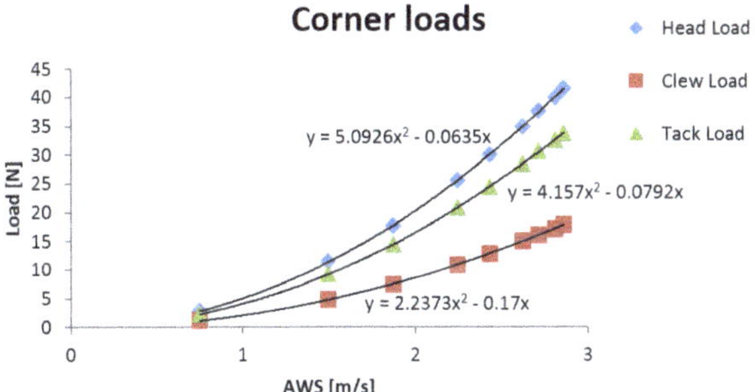

Figure 17. Plots of the corners loads vs. AWS for AWA = 120° and $\Delta\alpha_g$ = 20°.

As soon as the analytical formulations of the trendlines were calculated for all of the configurations, a careful analysis of the values of the first and second order coefficients was performed. From this analysis, it emerged that, in all of the analyzed configurations, the first order coefficient was much smaller than the second order one. In particular, it was calculated that the first order coefficient was, on average, 3.7% of the second order coefficient. By taking this into account and in order to find a simplified analytical formulation for the calculation of the loads on the corners, we neglected the first order coefficient. For this reason, the following equation has been preliminarily proposed for the calculation of the load F on the corners:

$$F = C_C \cdot AWS^2 \tag{3}$$

where C_C [kg/m] is the load coefficient that has to be defined.

To determine the most suitable value of C_C to make the Equation (3) usable in different sailing conditions, the second order coefficients were analyzed for all of the configurations. Figure 18 shows the values of the second order coefficients calculated for the clew, the tack, and the head in all of the analyzed configurations.

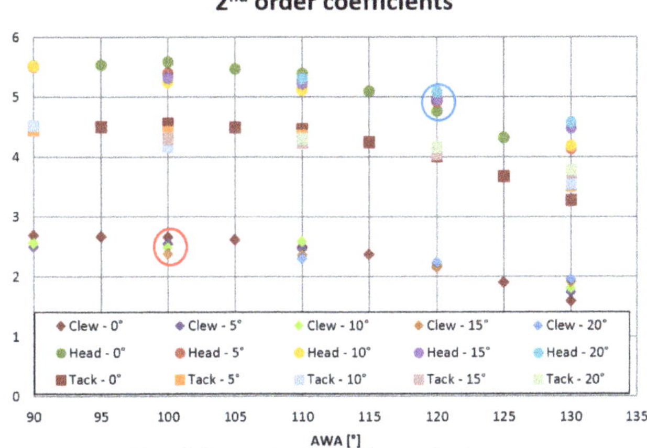

Figure 18. Plots of second order coefficients vs. AWA at different values of $\Delta\alpha_g$.

It can be seen that for each corner (clew, tack, and head), the coefficient values change as AWA and $\Delta\alpha_g$ vary. For example, it can be noted that for the clew corner at AWA = 100°

(red circle in Figure 18), the highest value of the coefficient was calculated when $\Delta\alpha_g = 0°$, while for the head corner at AWA = 120° (blue circle in Figure 18), the highest value was obtained when $\Delta\alpha_g = 20°$. To simplify the data analysis and to determine the most suitable values of C_C to be used in Equation (2), we considered only the maximum values of the second order coefficients at different values of AWA. Figure 19 shows the plots of the maximum values of the second order coefficients for the clew, the head, and the tack at different values of AWA.

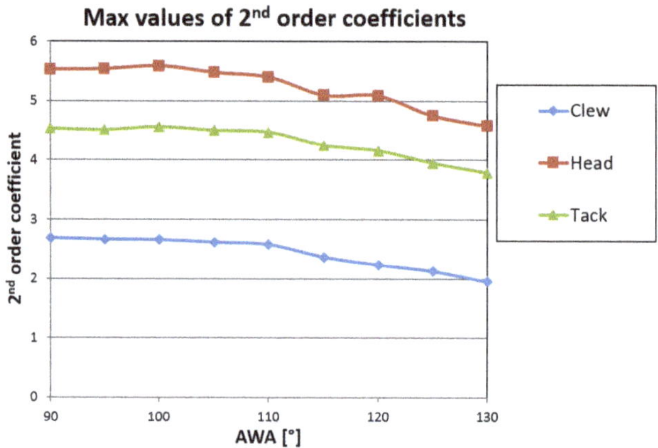

Figure 19. Plots of maximum value of second order coefficients vs. AWA.

From Figure 19, it can be observed that the trends of the maximum values of the second order coefficients are substantially constant up to the value of AWA≈110°; beyond this value, a constant decrease can be. Based on this data, adopting a conservative approach, it was decided to use the average value of the second order coefficients calculated between AWA = 90° and AWA = 110° as the load coefficient C_C for Equation (3). Table 5 reports the values of the second order coefficients and their average values calculated for the clew, the head, and the tack.

Table 5. Second order coefficients: single and average values.

AWA [°]	Clew	Tack	Head
90	2.69	4.53	5.53
95	2.67	4.51	5.54
100	2.66	4.56	5.59
105	2.62	4.50	5.48
110	2.58	4.47	5.40
Average values (C_C)	2.64	4.51	5.51

Following the proposed approach, the loads on the head, the tack, and the clew could be estimated using the following analytical formulations:

$$F_{clew} = C_{C\_clew} \cdot AWS^2 = 2.64 \cdot AWS^2 \tag{4}$$

$$F_{tack} = C_{C\_tack} \cdot AWS^2 = 4.51 \cdot AWS^2 \tag{5}$$

$$F_{head} = C_{C\_head} \cdot AWS^2 = 5.51 \cdot AWS^2 \tag{6}$$

The values of the loads evaluated through CFD simulations and by Equations (4)–(6) are plotted in Figure 20.

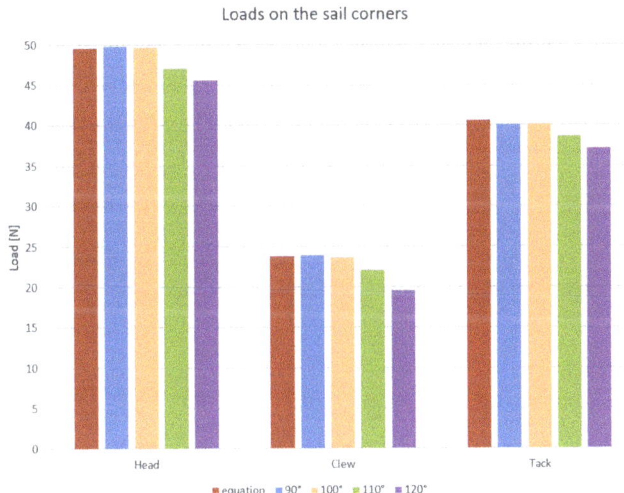

Figure 20. Plots of the loads on the sail corners calculated by the CFD simulations and by Equations (4)–(6) for AWS = 3 m/s and different values of AWA.

It can be seen that for AWA < 110°, the values of the loads evaluated through the proposed analytical formulations are quite similar to the corresponding ones calculated using the CFD simulations. For values of AWA > 110°, as expected, the proposed equations returned slightly overestimated values for the loads, but, considering that a correct design approach should take into account the highest values of the loads on a structure, this could be acceptable.

4. Discussion and Conclusions

The study of the fluid–structure interaction is still one of the most interesting topics in order to predict the performances of sails. A sail, in fact, is usually made of thin plastic material or fabric-based composite, and it forms a three-dimensional camber shape with wind pressure. The shape of the sail camber can remarkably change as the wind conditions vary, and this could result in a varied lift and drag performance by the sail [26]. These phenomena are relevant above all in the case of downwind sails, which are only attached to the yacht's structure at three points—the head, the clew, and the tack.

In this work, the issue of the FSI problem of downwind sails has been addressed by developing and testing a new weakly coupled procedure. Regarding the presented procedure, one of the main innovative features is related to the possibility of simulating the progressive deformation of the sail, through the gradual application of aerodynamic loads. This feature allows for reducing convergence problems and to avoid solution instabilities, in order to overcome the typical drawbacks of classic weak coupled procedures [19,33]. Another key feature of the new procedure is that it is parametric and fully automated. In fact, both CFD and FEM modules have been setup in such a way as to automatically update the models and the boundary conditions when simulation parameters change. Of course, the procedure is parametric within certain limits; if the type of sail to be analyzed changes remarkably, it is necessary to manually create the preliminary setup of the numerical models. Concerning the automation of the procedure, thanks to a specifically developed algorithm that fully manages the process and the data exchange between the CFD and the FEM modules, different sailing conditions can be simulated by simply introducing some parameters in a MS Excel user-interface.

The new procedure has been used to study a gennaker of a small sailing boat to evaluate its flying shape and to calculate the forces on the corners of the sail. The FSI study has been completed with no interruption or convergence problems within the considered sailing conditions. This demonstrates that, for the specific analyzed test case, the new procedure is effective and stable.

An extensive literature search was carried out to find experimental data to validate the numerical proposed methodology. Unfortunately, to the best of our knowledge, in the literature, there are no experimental studies on sails identical or very similar to the one studied here. However, some experimental studies performed on downwind sails show that the results obtained with the proposed methodology can be considered sufficiently reliable. As evidence of this, the values of the drive force coefficient, CFx, have been calculated at different AWAs using the numerical method proposed here (Figure 21) and compared with the values measured experimentally by Motta et al. [34]. Some similarities can be evidenced.

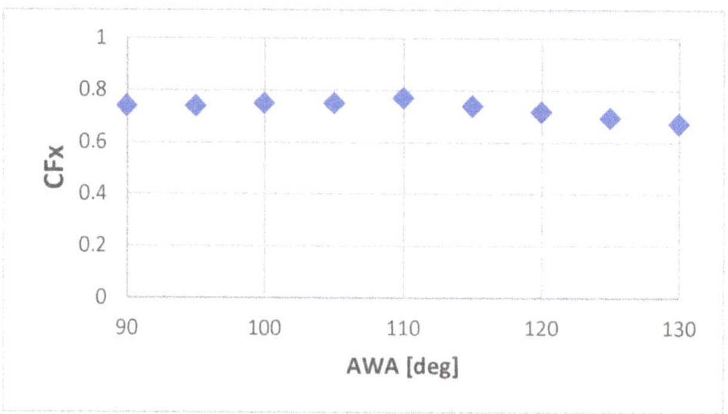

Figure 21. Drive force coefficients calculated with the numerical proposed method.

The trends of the coefficient CF_X as the AWA varies are very similar; in fact, they are almost constant with a slight increase around 100°–110°. Moreover, the order of magnitude of the coefficients is comparable, with 0.73 being the average value calculated with the numerical methodology proposed here and about 0.58 being the average value measured experimentally. Of course, as the compared sails are different, it is not possible to have identical or very similar values of CF_X, but considering that the order of magnitude is identical, it could be deduced that the numerical results are sufficiently reliable. With reference to the loads calculated numerically with the proposed methodology, good levels of agreement were found with other case studies in the literature. In particular, it can be observed that the ratios of the loads on the sail corners and the trends of the forces, when AWA varies, are very similar to those found in other studies [2,20,34]. For example, the force ratios F_{clew}/F_{head} and F_{tack}/F_{head} calculated at AWA = 90° using the new procedure are about 48% and 81%, respectively; the same ratios have been experimentally measured by Deparday et al. [2] for a similar case study and are equal to about 51% and 85%, respectively. Regarding the trends of the loads, it can be observed that, similarly to what was found with the numerical procedure proposed here, as well as in other experimental studies on similar kinds of sails [2,34], a decrease in the forces was detected when AWA > 110°. This good level of agreement between the numerical data presented and other experimental data, even if it cannot be considered a direct validation of the numerical results, demonstrates the proposed method could give enough reliable results and can be used as a useful tool for sail design.

With regard to the analytical formulation for calculating the sail corner loads discussed in Section 3.2, the proposed Equation (3) of course cannot be used to calculate the corner loads for all types of sails. It applies only for the specific kind of gennaker; however, in our opinion, the proposed approach could be interesting because it could be effectively repeated to find other formulations for different sail types. Still, nowadays, in fact, most sailmakers design the reinforcements to apply on the head, the tack, and the clew, based on their experience. Analytical formulations, grouped by similar types of sails, could be an inexpensive and easy to use method to suitably dimension the reinforcements on the sail corners. Moreover, knowledge of the loads of the sail corners could allow for a better comprehension on wind/rig/sail interaction [2,10]. Such an approach could allow for remarkably improving the design and manufacturing process of sails, not just for competitive sailing yachts, but also for cruising ones.

For future work, the authors believe that further developments could be addressed to the setup and validation of a database of analytical formulations in order to evaluate the loads on the sail corners and to the improvements of the numerical models. A database of equations for an initial evaluation of the sail loads could be developed by analyzing numerous sails different in shape and size; in this way, sailmakers could use a simple tool to design the most suitable reinforcements for the sails, based not only on their experience, but also on numerical data. CFD simulations could also be improved by investigating unsteady sailing conditions, while FEM models could be enhanced by introducing additional information on different types of sail materials regarding structure non linearities, and by evaluating the rig deformation. In this way, very accurate and complete information could be obtained and the sail performance could be optimized. In fact, as the developed procedure is based on a weak coupled FSI method, the FEM module could be effectively interfaced with an optimization module to find the best structural parameters for the sails (e.g., optimal thickness of different regions of the sail and most suitable layout of composite materials).

In conclusion, in our opinion, the novel proposed methodology is surely promising, and further enhancements could allow for getting more information on the dynamics of downwind sails and could lead to the widespread use of this procedure in the design and manufacturing of yacht sails.

Author Contributions: Conceptualization, A.C., T.I., A.M., V.N. and D.T.; data curation, A.C.; methodology, A.C., T.I., A.M., V.N. and D.T.; software, A.C. and T.I.; validation, A.C.; writing—original draft, A.C. and T.I.; writing—review and editing, A.C., T.I., A.M., V.N. and D.T. All authors have read and agreed to the published version of the manuscript.

Funding: This research received no external funding.

Institutional Review Board Statement: Not applicable.

Informed Consent Statement: Not applicable.

Data Availability Statement: Not applicable.

Acknowledgments: The authors gratefully acknowledge ANSYS Inc. for the support and the issue of academic licenses.

Conflicts of Interest: The authors declare no conflict of interest.

Abbreviations

The following abbreviations are used in this manuscript:

CFD	Computational fluid dynamics
FSI	Fluid–structure interaction
AoA	Angle of attack
AWA	Apparent wind angle
AWS	Apparent wind speed
TWA	True wind angle
TWS	True wind speed
C_m	Maximum sail chord
α_m	Mainsail boom angle
$\Delta\alpha_g$	Downwind sail trim angle
α_{g_design}	Design sheet angle of the downwind sail
CF_X	Drive force coefficient

References

1. Deparday, J.; Bot, P.; Hauville, F.; Augier, B.; Rabaud, M. Full-scale flying shape measurement of offwind yacht sails with photogrammetry. *Ocean Eng.* **2016**, *127*, 135–143. [CrossRef]
2. Deparday, J.; Bot, P.; Hauville, F.; Motta, D.; Le Pelley, D.J.; Flay, R.G. Dynamic measurements of pressures, sail shape and forces on a full-scale spinnaker. In Proceedings of the 23rd HISWA Symposium on Yacht Design and Yacht Construction, Amsterdam, The Netherlands, 17–18 November 2014; pp. 61–73.
3. Campbell, I.M.C. A comparison of downwind sail coefficients from tests in different wind tunnels. *Ocean Eng.* **2014**, *90*, 62–71. [CrossRef]
4. Viola, I.M.; Bartesaghi, S.; Van-Renterghem, T.; Ponzini, R. Detached Eddy Simulation of a sailing yacht. *Ocean Eng.* **2014**, *90*, 93–103. [CrossRef]
5. Ranzenbach, R.; Armitage, D.; Carrau, A. Mainsail Planform Optimization for IRC 52 Using Fluid Structure Interaction. In Proceedings of the 21st Chesapeake Sailing Yacht Symposium, Annapolis, MD, USA, 15–16 March 2013.
6. Durand, M.; Leroyer, A.; Lothodé, C.; Hauville, F.; Visonneau, M.; Floch, R.; Guillaume, L. FSI investigation on stability of downwind sails with an automatic dynamic trimming. *Ocean Eng.* **2014**, *90*, 129–139. [CrossRef]
7. Ramolini, A. Implementation of a Fluid-Structure Interaction Solver for a Spinnaker Sail. *J. Sail. Technol.* **2019**, *4*, 1–16. [CrossRef]
8. Le Pelley, D.; Morris, D.; Richards, P.; Motta, D. Aerodynamic force deduction on yacht sails using pressure and shape measurements in real time. *Int. J. Small Craft Technol.* **2016**, *158*, 123–124. [CrossRef]
9. Renzsch, H.; Graf, K. An experimental validation case for fluid-structure interaction simulations of downwind sails. In Proceedings of the 21th Chesapeake Sailing Yacht Symposium, Annapolis, MD, USA, 15–16 March 2013.
10. Augier, B.; Bot, P.; Hauville, F.; Durand, M. Experimental validation of unsteady models for fluid structure interaction: Application to yacht sails and rigs. *J. Wind Eng. Ind. Aerodyn.* **2012**, *101*, 53–66. [CrossRef]
11. Miyata, H.; Akimoto, H.; Hiroshima, F. CFD performance prediction simulation for hull-form design of sailing boats. *J. Mar. Sci. Technol.* **1997**, *2*, 257–267. [CrossRef]
12. Augier, B.; Bot, P.; Hauville, F.; Durand, M. Dynamic behaviour of a flexible yacht sail plan. *Ocean Eng.* **2013**, *66*, 32–43. [CrossRef]
13. Viola, I.M.; Bot, P.; Riotte, M. Upwind sail aerodynamics: A RANS numerical investigation validated with wind tunnel pressure measurements. *Int. J. Heat Fluid Flow* **2013**, *39*, 90–101. [CrossRef]
14. Ciortan, C.; Soares, C.G. Computational study of sail performance in upwind condition. *Ocean Eng.* **2007**, *34*, 2198–2206. [CrossRef]
15. Richter, H.J.; Horrigan, K.C.; Braun, J.B. Computational Fluid Dynamics for Downwind Sails. In Proceedings of the SNAME 16th Chesapeake Sailing Yacht Symposium, Annapolis, MD, USA, 1 March 2003.
16. Cirello, A.; Cucinotta, F.; Ingrassia, T.; Nigrelli, V.; Sfravara, F. Fluid–structure interaction of downwind sails: A new computational method. *J. Mar. Sci. Technol.* **2019**, *24*, 86–97. [CrossRef]
17. Viola, I.M. Downwind sail aerodynamics: A CFD investigation with high grid resolution. *Ocean Eng.* **2009**, *36*, 974–984. [CrossRef]
18. Graf, K.; Renzsch, H. Ranse investigations of downwind sails and integration into sailing yacht design processes. In Proceedings of the 2nd High Performance Yacht Design Conference, Auckland, New Zealand, 14–16 February 2006; pp. 1–12.
19. Lombardi, M.; Cremonesi, M.; Giampieri, A.; Parolini, N.; Quarteroni, A. A strongly coupled fluid-structure interaction model for wind-sail simulation. In Proceedings of the 4th High Performance Yacht Design Conference, Auckland, New Zealand, 12–14 March 2012; HPYD 2012, pp. 212–221.
20. Le Pelley, D.J.L.; Richards, P.J.; Berthier, A. Development of a directional load cell to measure flying sail aerodynamic loads. In Proceedings of the 5th High Performance Yacht Design Conference, Auckland, New Zealand, 8–12 March 2015; HPYD 2015, pp. 66–75.
21. Tahara, Y.; Masuyama, Y.; Fukasawa, T.; Katori, M. Sail Performance Analysis of Sailing Yachts by Numerical Calculations and Experiments. *Fluid Dyn. Comput. Model. Appl.* **2012**, 91–118. [CrossRef]
22. Deparday, J.; Augier, B.; Bot, P. Experimental analysis of a strong fluid–structure interaction on a soft membrane—Application to the flapping of a yacht downwind sail. *J. Fluids Struct.* **2018**, *81*, 547–564. [CrossRef]
23. Renzsch, H.; Müller, O.; Graf, K. Flexsail—A fluid structure interaction program for the investigation of spinnakers. In Proceedings of the RINA—International Conference—Innovation in High Performance Sailing Yachts, Southampton, UK, 28–29 October 2015; pp. 65–78.
24. Bergsma, F.M.J.; Moerke, N.; Zaaijer, K.S.; Hoeijmakers, H.W.M. Development of Computational Fluid-Structure Interaction Method for Yacht Sails. In Proceedings of the Third International Conference on Innovation in High Performance Sailing Yachts (INNOVSAIL), Lorient, France, 26–28 June 2013.
25. Abel, A.G.; Viola, I.M. Force generation mechanisms of downwind sails. In Proceedings of the 7th High Performance Yacht Design Conference, Auckland, New Zealand, 11–12 March 2021.
26. Bak, S.; Yoo, J. FSI analysis on the sail performance of a yacht with rig deformation. *Int. J. Nav. Arch. Ocean Eng.* **2019**, *11*, 648–661. [CrossRef]
27. Lombardi, M. Numerical Simulation of a Sailing Boat: Free Surface, Fluid Structure Interaction and Shape Optimization. Ph.D. Thesis, Ecole Polytechnique Fédérale de Lausanne, Lausanne, Switzerland, 2012.

28. Motta, D.; Flay, R.; Richards, P.; Pelley, D.L.; Bot, P.; Deparday, J. An investigation of the dynamic behaviour of asymmetric spinnakers at full-scale. In Proceedings of the 5th High Performance Yacht Design Conference, Auckland, New Zealand, 8–12 March 2015; HPYD 2015, pp. 76–85.
29. Peri, D.; Parolini, N.; Fossati, F. Multidisciplinary design optimization of a sailplan. In Proceedings of the MARINE 2015—Computational Methods in Marine Engineering VI, Rome, Italy, 15–17 June 2015; pp. 177–187.
30. Richards, P.; Hoxey, R. Appropriate boundary conditions for computational wind engineering models using the k-ε turbulence model. *J. Wind Eng. Ind. Aerodyn.* **1996**, *46–47*, 145–153.
31. Collie, S.J.; Gerritsen, M.; Jackson, P. *A Review of Turbulence Modelling for Use in Sail Flow Analysis*; Report No. 603; Department of Engineering Science, University of Auckland: Auckland, New Zealand, 2001. Available online: https://www.library.auckland.ac.nz/ (accessed on 26 May 2021).
32. Viola, I.M.; Bot, P.; Riotte, M. On the uncertainty of CFD in sail aerodynamics. *Int. J. Numer. Methods Fluids* **2013**, *72*, 1146–1164. [CrossRef]
33. Jackins, C.L.; Tanimoto, S.L. Oct-trees and their use in representing three-dimensional objects. *Comput. Graph. Image Process.* **1980**, *14*, 249–270. [CrossRef]
34. Motta, D.; Flay, R.; Richards, P.; Le Pelley, D.; Deparday, J.; Bot, P. Experimental investigation of asymmetric spinnaker aerodynamics using pressure and sail shape measurements. *Ocean Eng.* **2014**, *90*, 104–118. [CrossRef]

Journal of
*Marine Science
and Engineering*

Article

Laser Powder Bed Fusion of a Topology Optimized and Surface Textured Rudder Bulb with Lightweight and Drag-Reducing Design

Alessandro Scarpellini [1], Valentina Finazzi [1], Paolo Schito [1], Arianna Bionda [2], Andrea Ratti [3] and Ali Gökhan Demir [1,*]

[1] Department of Mechanical Engineering, Politecnico di Milano, Via La Masa 1, 20156 Milan, Italy; alessandro1.scarpellini@mail.polimi.it (A.S.); valentina.finazzi@polimi.it (V.F.); paolo.schito@polimi.it (P.S.)

[2] Department of Management, Economics and Industrial Engineering, Politecnico di Milano, Via Lambruschini, 4/B, 20156 Milan, Italy; arianna.bionda@polimi.it

[3] Department of Design, Politecnico di Milano, Via Giovanni Durando 10, 20158 Milan, Italy; andrea.ratti@polimi.it

* Correspondence: aligokhan.demir@polimi.it

Abstract: This work demonstrates the advantages of using laser powder bed fusion for producing a rudder bulb of a moth class sailing racing boat via laser powder bed fusion (LPBF). The component was designed to reduce weight using an AlSi7Mg0.6 alloy and incorporated a biomimetic surface texture for drag reduction. For the topological optimization, the component was loaded structurally due to foil wing's lift action as well as from the environment due to hydrodynamic resistance. The aim was to minimize core mass while preserving stiffness and the second to benefit from drag reduction capability in terms of passive surface behavior. The external surface texture is inspired by scales of the European sea bass. Both these features were embedded to the component and produced by LPBF in a single run, with the required resolution. Drag reduction was estimated in the order of 1% for free stream velocity of 2.5 m s^{-1}. The production of the final part resulted in limited geometrical error with respect to scales 3D model, with the desired mechanical properties. A reduction in weight of approximately 58% with respect to original full solid model from 452 to 190 g was achieved thanks to core topology optimization. Sandblasting was adopted as finishing technique since it was able to improve surface quality while preserving fish scale geometries. The feasibility of producing the biomimetic surfaces and the weight reduction were validated with the produced full-sized component.

Keywords: biomimetic design; lightweight structure; computer fluid dynamics; design for additive manufacturing

Citation: Scarpellini, A.; Finazzi, V.; Schito, P.; Bionda, A.; Ratti, A.; Demir, A.G. Laser Powder Bed Fusion of a Topology Optimized and Surface Textured Rudder Bulb with Lightweight and Drag-Reducing Design. *J. Mar. Sci. Eng.* **2021**, *9*, 1032. https://doi.org/10.3390/jmse9091032

Academic Editor: Md Jahir Rizvi

Received: 9 July 2021
Accepted: 9 September 2021
Published: 19 September 2021

1. Introduction

Metal additive manufacturing (AM) has imposed itself as a competitive alternative with respect to "conventional" production in multiple industrial fields. Supply chain reduction [1], relative design freedom and material waste reduction are only some of the innovations introduced. In the naval field, however, the technology results to be still in early introduction stages. Small to single-unit batches are what is often required, with sensible degree of customization. Lightweight and fluid dynamic resistance reduction are persecuted, involving light materials and optimized shapes, with aim of minimizing operational costs and obtaining more sustainable solutions.

With metal AM, higher degree of design and customization freedom can unlock new levels of product performance, in terms of shape optimization and single client needs in the naval industry. The localized production preserves from transport issues and allows design centers to provide schemes on multiple distributed manufacturing sites. Standardization is required with the aim to validate products and guarantee reliability of the innovative production technique. Despite the limited presence of AM in naval sector, these advantages are already being explored [2].

Up to medium size vessels, composite materials lead the manufacturing path, while for large, commercial or tourism cruise vessels, welding, riveting and bolted joins among metal plates are the main approaches for production and assembly. AM technologies can deal with both categories, relying on fused deposition modeling (FDM) techniques for composites [2–4], eliminating any mold need [4,5], while metal techniques such as wire arc additive manufacturing (WAAM) can satisfy plates-based constructions [6,7] and laser powder bed fusion (LPBF), direct energy deposition (DED) or even electron beam melting (EBM) can deal with smaller components, allowing fabrication of relatively complex and curved shapes, unfeasible otherwise in a single production run [8,9].

On metal AM, lightweight is often achieved by means of topology optimization of prior existing geometries [10], as well as the utilization of light metal alloys as could be Al-alloys. In this approach, the design of the component is less constrained by fabrication and can be driven by working condition needs. Al-alloys are commonly employed for aerospace and automotive applications requiring lightweight design, where the desired mechanical properties rely on the careful selection of the process parameters [11]. Fluid dynamic behavior can be involved in the design and particularly on surface interaction [10,12–18]. As geometrical features can be manufactured directly during the production run of the component, a new generation of products can be obtained, equipped with small-size surface features. These may allow to work with fluid resistance phenomena, that previously had to be accepted as they were, in a functionalization of the surface approach.

In this innovative environment, Politecnico di Milano's Sailing Team (PoliMi Sailing Team) explores the feasibility of using metal AM for the Sustainable Moth Challenge (SuMoth Challenge) hosted by the Foiling Week. It requires teams to design, manufacture and sail a moth class boat according to a common budget-based ruleset, that promotes sustainability and innovation of materials and manufacturing. The moth class boat to be realized is a one-sailor foiling vessel. The research interest over this type of vessel has been concerning the design of passive and active elements to optimize the fluidodynamic behavior against water and air, while the use of textured surfaces and topologically optimized components for weight reduction appear to be neglected [19–21]. In this context, the project also provides several exploitable points using the AM technologies. Even in a relatively small size sailing boat, where most of the manufacturing process is linked to fiber reinforced materials, metal parts play a vital role, often because of impossibility to obtain a reliable assembly of multiple composite parts without them. However, the introduction of metal components must not negatively affect the system, as a local increase of mass may disrupt center of gravity position and so behavior of the boat, as well as the exposed surface should not promote excessive dynamic resistance. The parts to be produced by metal AM processes should address such difficulties. Despite the great potential and the possibility to work in a complete digital environment from design to production, to the authors' knowledge, no previous work in literature addressed different points concerning design, FEM, CFD, and production of a naval component as a whole.

The rudder bulb for the new generation racing boat of PoliMi Sailing Team was the component under study (see Figure 1). Such component is loaded structurally both from the boat and from the fluid environment. The concept is based on the optimization of the main structure of the component for weight reduction as well as to have biomimetic textured surface in contact with water. The combined features were produced via laser powder bed fusion (LPBF) using an Al-alloy. The biomimetic surface texture was inspired by scales of the European sea bass in order to reproduce the hydrodynamic advantage, which results in drag reduction. Both features were designed and tested in a digital environment and later on produced by the AM method. The described concept allows to reduce material waste as well as providing the speed advantage during the race.

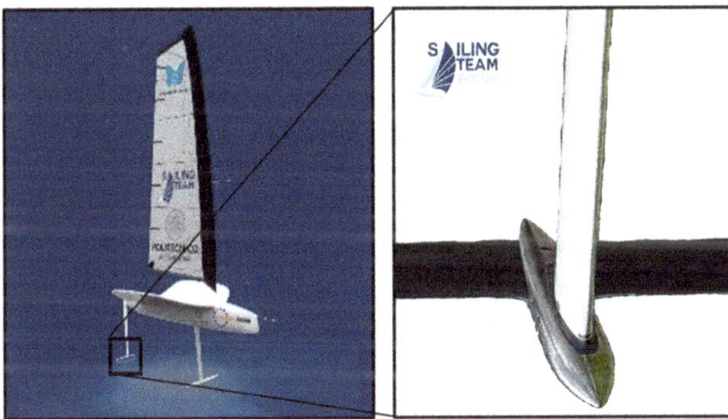

Figure 1. PoliMi Sailing Team rudder bulb concept model, from the team's archive.

Accordingly, the present work aims to demonstrate the potentiality of using metal AM in the naval industry with the selected case study. Hence, this work discusses the design and production stages of a novel lightweight and drag reducing rudder bulb via LPBF. The work shows the finite element modeling (FEM) phases for weight reduction and computer fluid dynamics (CFD) for identifying the advantage of the biomimetic fish scale surfaces. Finally, the component production stages are explained showing high geometrical fidelity with respect to the digital model.

2. Design of the Rudder Bulb

2.1. Design Requirements and Choices

The PoliMi sailing boat is equipped with two foiling wings, namely the main one, in keel position and the rudder one. The rudder bulb has the aim of structural connection between the rudder vertical and the horizontal wing. This latter is the one that generates approximately 30% of the lift force required to the moth in order to "fly" above the waterline. The other 70% is provided by the keel wing. The keel wing has no reason to be displaced with respect to its vertical, as instead is an advantage in the rudder, as allowed by the bulb that in the proposed design, provides regulation in such matter. The resulting spatial translation, along the main boat axis, of the lift force with respect to the rudder vertical below it, allows the sailor to have higher handling, especially while turning, by increasing the span between the keel and the rudder lift force components. The beneficial effect cannot be obtained by displacing all the rudder structure at once because moth class rules impose specific limits in boat body length [22].

The bulb must be realized with a metal alloy, because of the high strength required. The chosen Al alloy was AlSi7Mg0.6, which is already used in aeronautical applications but novel to LPBF for this field. [23]

A reduction of mass down to 10–30% of the original internal bulb core volume (design space) is desired, with no excessive loss of stiffness. For the surface, the objective was to investigate the feasibility of using biomimetic surface textures. The feature was evaluated by CFD for drag reduction and its producibility was assessed by the production of the full-scale bulb.

2.2. Topology Optimization Strategy

The bulb base model embraces mainly low resistance external shape (i.e., elongated bulb) and foiling wing position regulation (longitudinally). The first is obtained starting from the existing applications as well as PoliMi Sailing Team's experience. The overall longitudinal length of the model is 290 mm. An innovative solution was introduced in order to have the possibility to increase and reduce, during foil set up, the distance between keel lift and rudder lift components. This approach helps to study and find an improved cruising stability when boat is in "flying" condition (raised above water level).

At the front side, the bulb body must host the rudder vertically (Figure 2a). The joining was made by means of interference fitting, fixed in position with a central bolt. 3D modeling of bulb base geometry was obtained by means of Dassault Systèmes Solidworks environment while Altair Inspire was used for the topology optimization.

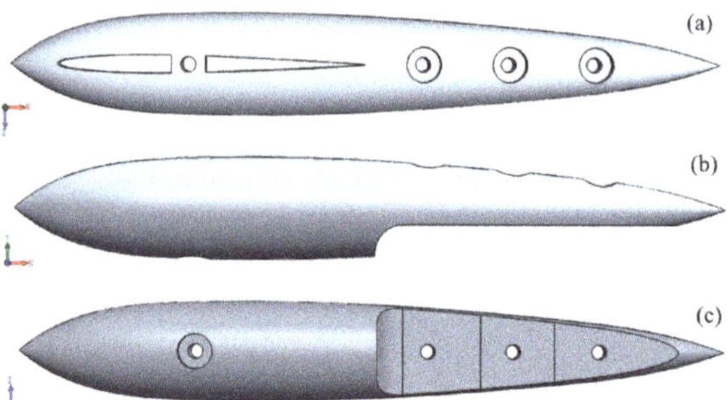

Figure 2. Bulb 3D base model views, (**a**) top, (**b**) side, (**c**) bottom. The overall longitudinal length of the component is 290 mm.

The wing was connected by means of bolts without interference fitting so two bolts at a time are used to constrain the wing in translation and rotation. Hence, two different wing positions (X-wise) could be used. This aspect was considered when defining optimization load cases, as three area subdivisions are sketched in the model to pre-partition the wing connection surface (visible in Figure 2c) for the configuration of the load cases.

Topology optimization was carried out starting from a base model shown in Figure 2a, weighing 452 g with the same Al-alloy used in LPBF. The model was partitioned geometrically in sub-bodies with the aim to identify the core from all the rest. The core was set as the volume that can be modified by optimization iterations (design space), while the rest, the surrounding "case" or "shell", is the fixed portion, not modifiable and onto which loads are applied. In Figure 3, longitudinal sections of the partitioned model are presented with core (Figure 3a) and without core (Figure 3b). Target objective was to reduce the core mass by 10–30% with respect to the initial design while maximizing the stiffness. The weight performance was evaluated at the FEM level only.

Figure 3. (**a**) Bulb 3D model partitioning, longitudinal section view; the main green internal portion is the "core" while its contour (shown in green) is part of the "case", (**b**) case sectioned view, uniform thickness of 1 mm. The overall longitudinal length of the component is 290 mm.

The introduced load cases were obtained considering actions on the bulb structure, mainly transmitted by the down standing wing. The loads were applied contemporarily, and varied according to modeled working condition, while some were common for all load cases. All actions are assumed at maximum criticality of the event (boat speed of about 20 kts or an eventual sailing accident), no extremely rapid impact condition is considered for the optimization. Reasonable impact cases as could be debris impacts or extreme accident were considered to be sufficiently compensated by the minimum safety factor imposed (1.8) with respect to the yield stress (YS). Five optimization runs were conducted as shown in Appendix B.

2.3. Definition of the Load Cases

Prior to the definition of the load cases, all actions on the component structure were identified. Modeling of the actions was made by means of sailing team experience, balance analysis on rudder loads and practical assumptions. The following are the considered actions.

i. Lift force (L) generated on rudder accounts approximately for the 30% of the overall lift required, however, due to the changing attack angle, the wing may transmit a variable load. To account for this variation, it is considered to have L to be equal to 600 N for 0° attack angle, with a direction perpendicular with respect to bulb–wing interface. A magnitude of 400 N was assumed with inclinations of $\pm 10°$ with respect to the interface, to account for magnitude and direction variability load cases. The overall drag force (D_w) due to the relative motion of the wing and the water, transmitted to the bulb is estimated to be approximately 34 N at 0° while increasing up to about 50 N in magnitude for $\pm 10°$ attack angle.

ii. Exceptional circumstances may lead to boat roll up to capsize, during such event, or after, while recovering, the wing opposes resistance to fluid, or it may be used as lever for boat recovery. These situations may be emulated on the bulb by means of a torque action (T). The value considered is 6.25 Nm. This value simulates a force of $F = 12.5$ N applied at 0.5 m, as it is approximately the distance of one wing extremity.

iii. It may happen that the rudder is carried upward due to boat rapid pitch angle change. As a consequence, the wing acts as hydrodynamic resistance, trying to force downward the bulb rear portion with it. The load (P_c) is estimated to be at max 400 N. This load is transmitted to the bulb body by means of wing connecting bolts.

iv. Bulb overall drag force (D_b) at max boat speed condition was estimated to be about 50 N. However, for simplicity no change in magnitude was set when dealing with different attack angle conditions.

v. During the assembly, the compression state imposed by bolts preload should be considered. Rudder vertical bolt preload (P_v) is 60 N while preloads on wing bolts ($P_{w,front}$ and $P_{w, rear}$) are 125 N each.

All the actions are collected in Table 1. Further details are presented in Appendix A.

Table 1. The used load cases in the topological optimization stage.

Load	Description	Symbol	Unit	Magnitude
		$L(0°)$	[N]	600
Lift force	Rudder wing lift action	$L(+10°)$	[N]	400
		$L(-10°)$	[N]	400
		$D_w(0°)$	[N]	34
Wing drag force	Rudder wing motion with respect to water	$D_w(+10°)$	[N]	50
		$D_w(-10°)$	[N]	50
Capsize torque	Torque due to capsizing of boat	T	[Nm]	6.25
Wing vertical resistance	Resistance to vertical translation of the wing	P_c	[N]	400
Bulb drag force	Bulb motion with respect to water	D_b	[N]	50
Rudder vertical bolt preload	Compression due to assembly	P_v	[N]	60
Wing bolts preloads	Compression due to assembly	$P_{w,front}$	[N]	125
		$P_{w,rear}$	[N]	125

Topological optimization was run by means of successive trials. The followed methodology was based on performing an optimization with stiffness maximization and design space mass decrease targets. The obtained model was then analyzed with respect to all load cases and if displacements and safety factors were still acceptable, it was proceeded with a new optimization characterized with a more ambitious core mass reduction with respect to the full core model. All topology optimizations were performed with a symmetry plane requirement with respect to the only symmetry plane of the base model (congruent with section plane of Figure 3), it means that results had to satisfy the symmetry of material spatial distribution with respect to this plane.

Design for laser powder bed fusion rules was also implemented in the optimization work. In order to avoid internal supports, surfaces with angles lower than 45° with respect to the build platform plane of the LPBF machines should be avoided. The overhang control with 45° critical angle was not inserted initially to freely optimize the mass with the required rigidity. The optimized geometries were evaluated for overhang regions first. Later, the overhang control was implemented, setting the printing direction and the overhang critical angle.

Minimum thickness of core part was set to be kept above 3–5 mm for all optimizations. It is known from experience that too thin optimized geometry is likely linked to fitting failure when dealing with surface definition. Set material properties are YS, Young modulus, Poisson's ratio and mass density, respectively, at 200 MPa, 70 GPa, 0.33 and 2.68 g/cm^3 using the LPBF as-built properties of the AlSi7Mg0.6 without considering material anisotropy.

2.4. Biomimetic Texture Design

Muthuramalingam et al. [24] have studied the hydrodynamic behavior of 3D scanned model-based sea bass scales pattern, reporting a decrease in the overall drag with respect to a surface with no scales. The geometrical model was obtained from a CAD reconstruction tool that used microscopy analysis to assess dimensions directly on fish skin. The fish skin can be represented as a pattern of partially superimposed scales. It was chosen to model each scale as a half circle, linearly growing in thickness and triangular in vertical section. Starting from the midsection (Figure 4, section A-A), it is a triangle, with base length l (equal to the circle's radius), height h_s as the maximum scale thickness and the two angles, α as the "scale angle" and β.

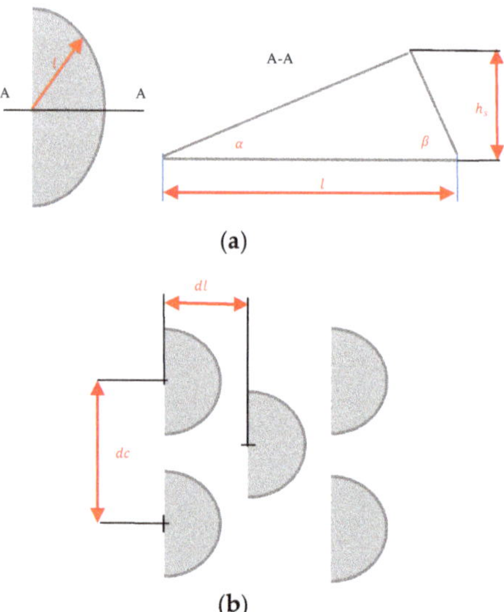

Figure 4. (a) Scale model geometry and (b) pattern parameters.

As seen in Figure 4b, the pattern should allow the scales to be oriented according to the longitudinal axis of the considered carrying body for the growing thickness of the scales themselves and transversally (i.e., circumferentially in case of curvature) with the "scale diameter" (as it corresponds to angle α's origin position). The longitudinal distance among neighbor-row scales is named dl, while among aligned not row-neighboring ones was set to be $2 \times dl$. The transversal spacing among transversally aligned scales is named dc, while among not transversally aligned scales the spacing is modeled as $0.5 \times dc$. To determine the single scale geometry, at least three parameters among all mentioned should be set. The chosen ones were l, h_s and α, while dl and dc are used for the pattern. A surface pattern was determined after a sensitivity analysis, which is not reported here for brevity. The chosen values are listed in Table 2.

Table 2. Chosen scale pattern geometrical parameters.

α	l	dl	dc
[deg]	[mm]	[mm]	[mm]
3	3	0.9	6

2.5. Fluid Dynamic Behavior of the Fish Scale

Reynolds-averaged Navier–Stokes equations were solved with a k-ω SST turbulence model using the open-source framework OpenFOAM. The geometry considered for the fluid-dynamic study consists of a flat surface covered with scales: Rhinoceros environment was used to generate the fish scale patterns used for the CFD analysis. A domain sensitivity study, and a mesh independency study was performed, to achieve a reliable numerical model to assess the fluid-dynamic performance of the fish scales. A plate equipped with scales with a dimension of 50 mm in length and 6 mm in width was placed in the domain reported in Figure 5. The boundary conditions are cyclic in the transversal direction to emulate an infinitely large plate and has an inlet and an outlet. The mesh selected after the mesh independency study for the numerical computations has a number of cells around 11 million cells, and the calculation was performed on the CFDHub HPC infrastructure at Politecnico di Milano.

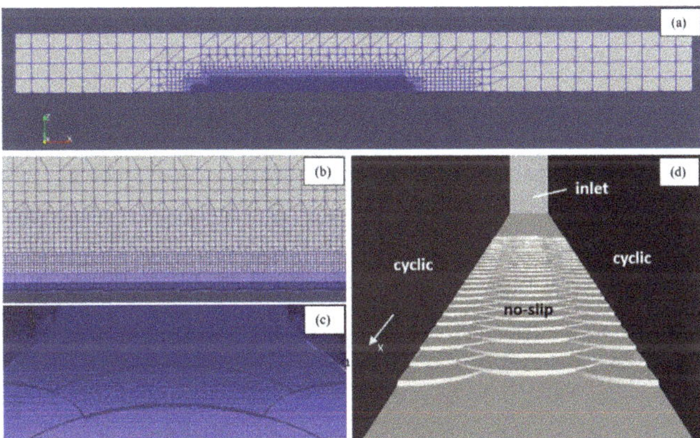

Figure 5. (**a**) Full domain, lateral view; (**b**) detail on scale pattern mesh refinement, (**c**) perspective, (**d**) boundary conditions.

For comparison the initial design was modeled as a flat surface in the design domain with no fish scales. The drag value was compared to the surface with the scale design determined in Table 2.

3. Materials and Experimental Systems

3.1. Laser Powder Bed Fusion System

The full-scale rudder bulb bas produced using a TRUMPF TruPrint 3000 LPBF machine (Ditzingen, Germany). The AlSi7Mg0.6 powder employed had a 20–63 μm grain size (Carpenter Additive, Philadelphia, PA, USA). The powder was produced by means of gas atomization. Process chamber was set to work on Ar inert atmosphere to prevent oxidation-related issues. The main process parameters are reported as in Table 3. The part density was measured at 99.3% ± 0.2% by means of optical microscopy. LPBF built processor software used to prepare the building platform in digital environment was Materialize Magics 19 (12/2019–10/2020).

Table 3. Overall volume LPBF process parameters involved.

Process Parameters		Level
Chamber oxygen concentration	[%]	0.1
Inert gas type	[–]	Ar
Shielding gas flow rate	[m/s]	0.8
Preheating temperature (T_{ph})	[°C]	100
Laser spot diameter (d_s)	[μm]	100
Layer thickness (z)	[μm]	50
Layer scan strategy	[–]	No pattern
Scan direction rot. layer by layer	[deg]	67
Laser power (P)	[W]	345
Scan speed (v)	[mm/s]	1500
Hatch distance (h)	[mm]	0.10

3.2. Surface Finishing

In the study, manual sandblasting was used to improve the surface quality of scale equipped LPBF built surface. The system used to perform the treatment was a Guyson FL600 Blast system equipped with an 8 mm nozzle diameter, ejecting a 6 bar pressurized mixture of air and abrasive grains. Abrasive grains were dried silica-based sand grains of type BACCHI 510 PLUS.

Sandblasting was performed for approximately 30 s on all of the surface to be treated, keeping the nozzle at a distance between 10 and 15 cm from the surface. For the manufactured fish scales, it was observed that small features are preserved and enhanced in quality if the blasting direction is not perpendicular with respect to the surface. An inclination of approximately 45° was kept with respect to the longitudinal axis of the part.

3.3. Tensile Test

Tensile tests were performed on sandblasted tensile specimens to verify the mechanical properties of the build according to ISO 6892. Equipment used to perform the test was an MTS Alliance RT/100 tensile test machine along with an MTS 685.22 Hydraulic Grip Supply (Eden Prairie, MN, USA).

3.4. Surface Analysis

The geometrical fidelity of the produced biomimetic surface was evaluated through focus variation microscope (Infinite Focus, Alicona Imaging GmbH, Graz, Austria). Surface height maps were acquired with vertical and lateral resolutions, respectively, at 1 and 15 μm. The acquisitions were followed by a form removal procedure tool together with a coordinate system adjustment, to center the dataset over the origin of the reference system. To distinguish between surface roughness elements and fish scales geometrical features, a cut off wavelength of 0.8 mm (ISO4287) was used for all roughness measures. This value was smaller than the minimum periodicity length imposed by scale pattern geometries.

Primary profile analysis was used to compare the produced surface profile with the theoretical dimensions. For the roughness measurements, paths of 10–15 mm were selected to comply with the cut-off wavelength. Three paths were selected (Figure 6a) measuring the average roughness (R_a).

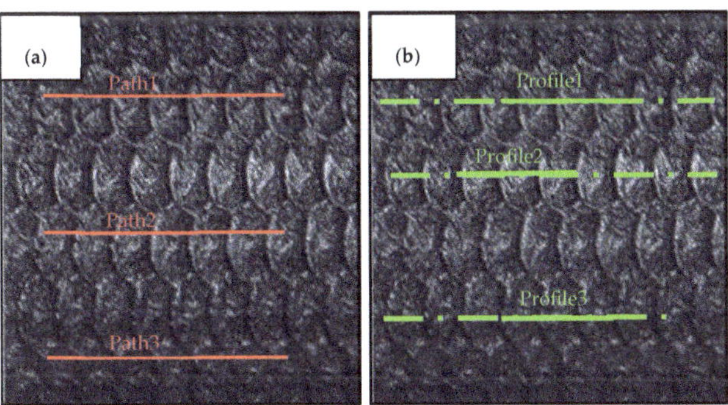

Figure 6. Roughness profile paths example (**a**), primary profile study paths example (**b**).

Along with the linear roughness parameters, the areal average roughness (S_a) was measured. Primary profiles analysis was performed choosing three profiles laying on the mid planes of the scale rows (Figure 6b). For each primary profile, the mean peak to valley distance (Δpv_{meas}) was collected and distance among correspondent scales' points ($dl_{sr,i}$) was sampled three times (Figure 7a). The considered distance among correspondent scales' points ($dl_{sr,\,meas}$) for the analyzed profile was the average between the sampled ones. The final considered values for the analyzed scale pattern were then the average results among the three observed profiles. These distances were used to generate a prior assessment of the manufactured scales' quality, comparing with the theoretical distances imposed by the 3D original model, respectively called Δpv_{theo} and $dl_{sr,\,theo}$ (Figure 7b).

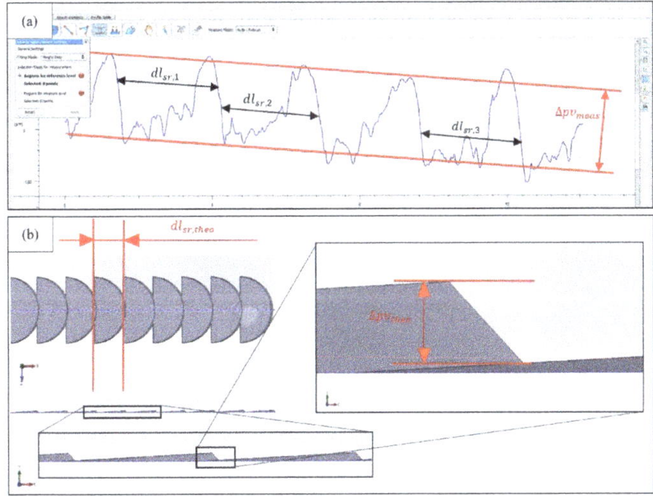

Figure 7. (**a**) Primary profile mean peak to valley distance (red) and scale to scale correspondent point distance (black) examples, (**b**) theoretical measures from scale pattern CAD model.

Each $dl_{sr,i}$ value was taken at a different scale height, preventing eventual nuisance effect in the measure due to a particular sampling position. The "*sr*" subscript stands for "same row" and it was inserted to distinguish these measures from the previously defined dl parameter, which refers to neighboring different rows, with $dl = 0.5 \times dl_{sr,\,theo}$.

4. Results and Discussion

4.1. Topology Optimized Model

The successful and sequential topology optimization runs are summarized in Table 4. The observation of mass variability is an indicator of optimization quality. Limited variation (up to 10%) is considered to suggest good optimization results, therefore it is possible to consider analyzing the model.

Numerical details are presented in Appendix B.

Table 4. Optimization runs.

Run Name	Mass Target	Overhang Control	Core Mass	Core Mass
	[%] wrt. Original	[Yes/No]	[kg]	[%] wrt. Original
A	50	No	0.218	50
B	15	No	0.061	14
C	10	No	0.038	9
D	15	Yes—45° printing orientation	0.061	14
E	15	Yes—90° printing orientation	0.059	14

Run A was the first optimization performed to assess settings and load case behaviors. Target core mass was set to be reduced down to 50% of the original value. The result was satisfying in terms of optimization success, but still far from mass reduction target. The minimum safety factor resulting from stress analysis was 3.22, experienced in Working $(0°, l)$ load case condition. The original full model started with a minimum safety factor of 3.97. The high safety factor suggested the possibility to further remove the core material. The geometry of run A was the starting point in run B. Resulting core relative mass, at target level condition was about 13.9%. The model, once tested on the load cases, resulted in minimum safety factor of 3.29. The result suggested that the previous 50% target was not a limit for structural resistance of the optimized shape. Since B condition was still characterized by a sensible gap from the full model, run C had the aim to explore further mass reduction. Target was set to 10% and successful run results in 1.82 minimum safety factor (Working $(0°, l)$ load case). However, even if the local high stress did not lead to critical safety factor, maximum displacements reached were too high to be considered acceptable. In particular, bulb geometry would displace in X and Y directions of about 2 and 1.6 mm, respectively. Displacements did not decrease sensibly by increasing the optimization index, not even at index 9. This condition was not experienced in run B where maximum displacement in any direction is not greater than 0.2 mm. At such point then, mass reduction condition was set to be kept at 15%.

In the final stage of the topological optimization the overhang control was introduced according to the build direction. Run D and run E embraced this need. Run D considered generating a support free core while printing the bulb inclined by 45° with respect to platform plane with its longitudinal axis and orienting the wing interface surface upward (Figure 8D). Run E employed instead a vertical build of the bulb (Figure 8E).

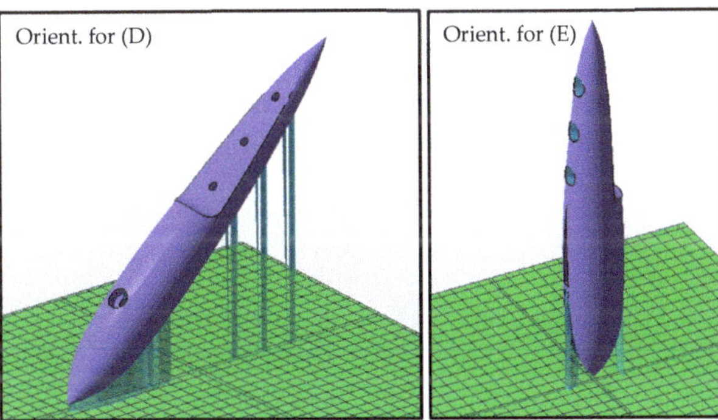

Figure 8. Run (D,E) with printing orientation emulation in order to account for support free core (overhang critical angle 45). In light blue, the support structures are shown. The overall longitudinal length of the component is 290 mm.

Both result in about 14% core mass (at target level) but D minimum safety factor falls to 1.6, while in E is preserved at 3.18. No excessive displacement concerns results. Since the preferred printing orientation is the one used in D, a more massive version of D result was analyzed. After a few trials, optimization level was set to index 9; results satisfy the safety factor requirement, resulting in a minimum of 2.46, while core mass reached 19.7% with respect to the original core. With the achieved significant reduction of the core mass, the final product mass comprised of the core and shell in the optimized condition was 190 g, corresponding to a 58% weight reduction.

Such ratio is considered to be acceptable and optimization run D at index 9, named "D(9)" was set as definitive core geometry. In Figures 9–11, a comparison of D(9) behavior with respect to original model (solid) and "no core" model (null core mass) is reported for the most relevant load cases. No-core condition was tested for comparison to the minimization results achieved with core mass reduction of about 80%. D(9) model achieves similar results as the solid original version while the total lack of core is associated to excessive displacements. Minimum safety factor against yielding results critical only for no core condition, on the majority of load cases. The behavior is related to local stress concentration due to the only presence of the relatively thin case structure bearing loads.

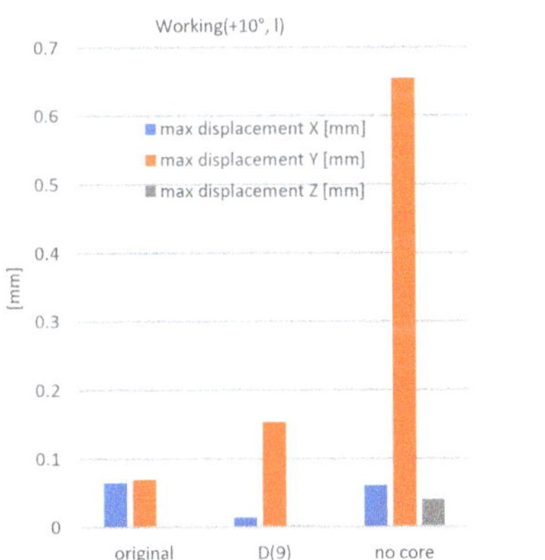

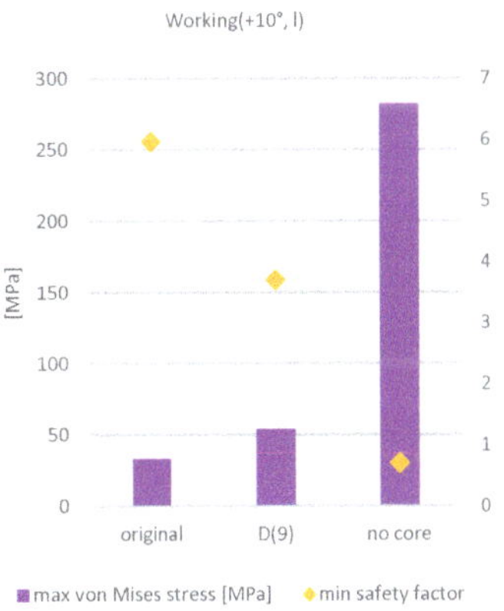

Figure 9. Working (+10°, *l*) load case analysis results.

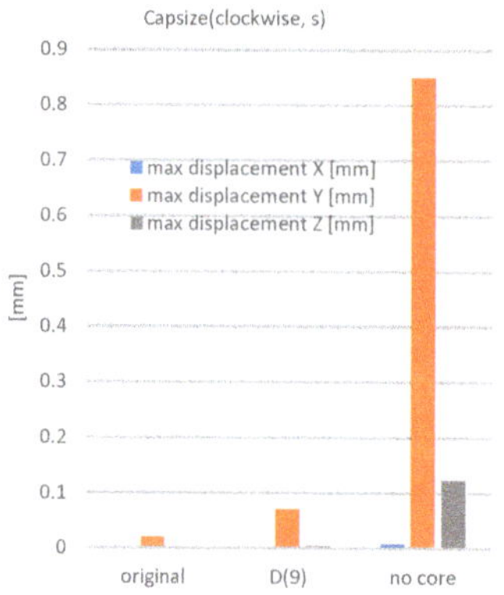

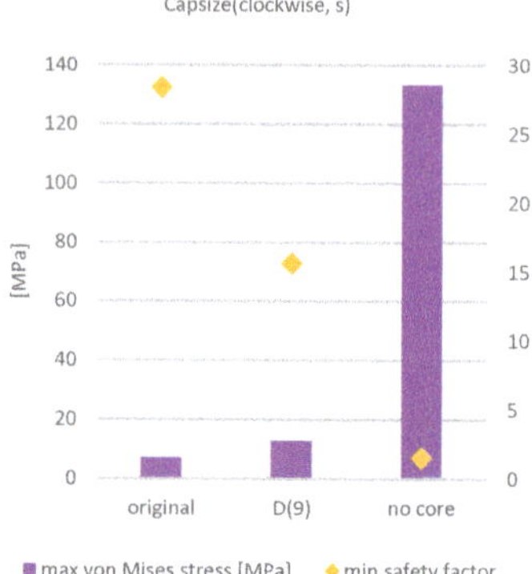

Figure 10. Capsize (clockwise, *s*) load case analysis results.

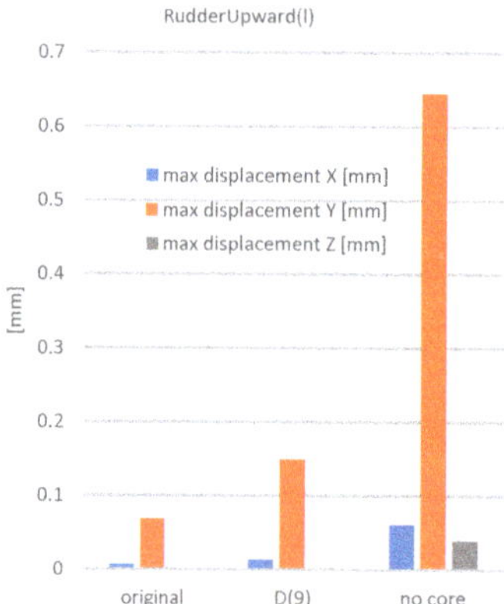

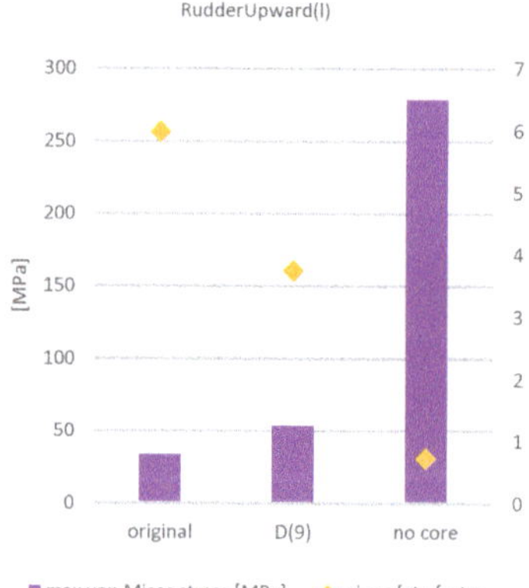

Figure 11. Rudder upward (*l*) load case analysis results.

Optimized bulb interior definitive shape results as in Figure 12a,b and isolated optimized core geometry is observable in Figure 12c,d. It is also visible the evacuation site for internal powder removal. It is located in a low stress region as well as a hidden and sufficiently covered position in terms of isolation with respect to eventual water penetration since it would be surmounted by rudder vertical encastre extremity.

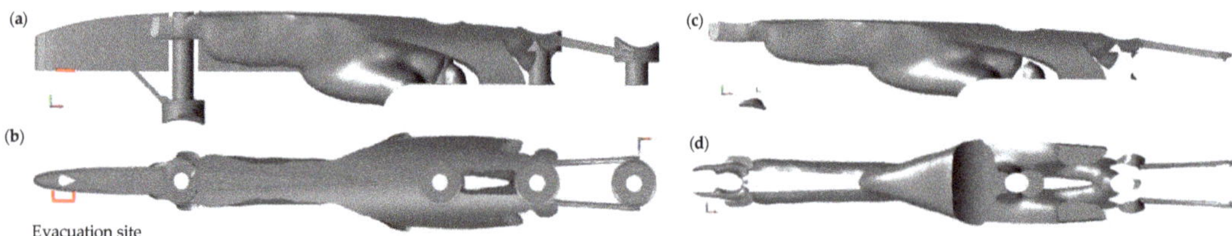

Figure 12. D(9) core together with interior partitions of bulb; (**a**) side horizontal view, (**b**) top horizontal view; while isolated optimized core D(9) shape; (**c**) side, (**d**) bottom. The overall longitudinal length of the component is 290 mm.

Stress analysis, on original and optimized shapes, reports localized stresses in the rear edge region on the rudder vertical socket (Figure 13a). The localized stress concentration is present in all load cases and models and it is considered to not represent a material resistance critical location since stress levels reached comply in the minimum safety factor (Figure 13c) reached and the local stress state is mainly compressive (Figure 13b); these considerations are valid for all load cases.

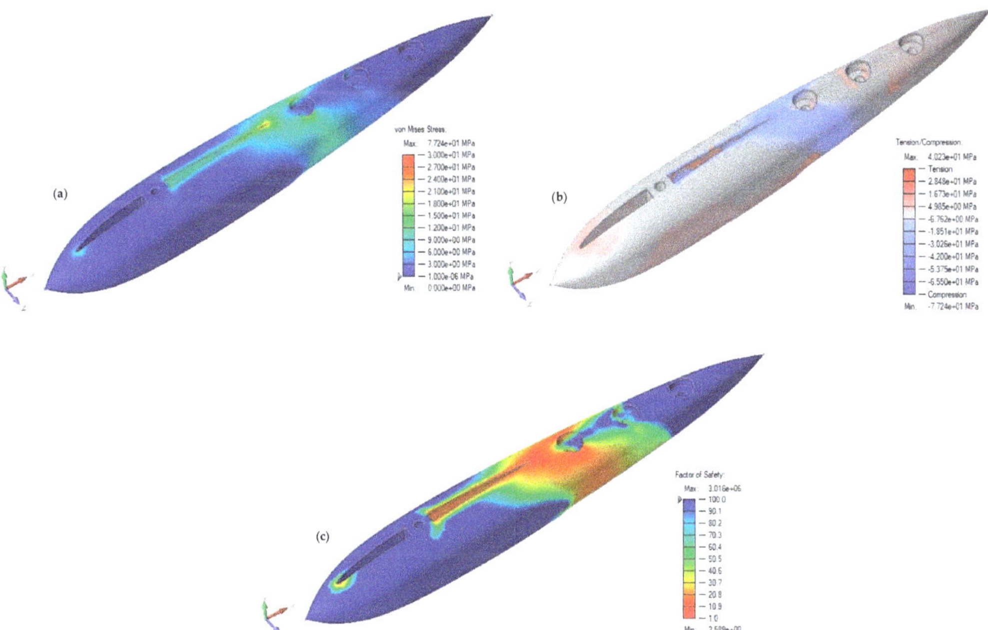

Figure 13. (a) von Mises stress, (b) tension-compression, (c) safety factor (wrt. YS) analyses on Working (0°, s) load case. The overall longitudinal length of the component is 290 mm.

4.2. Drag Reduction

The plate of fixed dimensions is proved to be capable of reaching a reduction of overall drag force of about 1% at 2.5 m/s water speed with respect to a scales-unequipped plate, by means of viscous drag component reduction. Drag force over the test plate is given by the sum of viscous and pressure components along the main stream direction. The viscous component undergoes 19.9% reduction while the pressure drag force component is increased by about three times due to scale behavior as a series of walls. However, the viscous drag reduction is enough to compensate the pressure component increase and slightly overcome it, determining the reduction of the overall drag force. In Figure 14, the upper view of the tested domain and analogy with the natural behavior on European bass skin by means of speed corridors generation (streaks effect) is linked to viscous drag reduction. The results were found to be coherent with previous research that modeled fish skin [24].

The speed of 20 kts is considered as the highest boat speed. In this condition, loads are of relevant entity for the involved metal component. Moth class vessel speeds may lay around 5–10 kts (2.5–5 m/s). In particular, the ones built for the competition by research-university teams are assumed to sail near such value and have 20 kts as sensibly high limit for boat speed. The speed around 2.5 m/s is the most important one, since the boat starts flying and take-off with lower driving force is an advantage when racing. According to the novelty of the work, the presented speed of 2.5 m/s is also coherent in terms of possible swimming speed of the involved species and so of the scale geometries involved. While the drag reduction may appear very limited, for the competition purposes such reduction is expected to provide an advantage. Beyond the demonstrated results, the work shows that biomimetic surface textures can be further exploited to enhance the sailing performance.

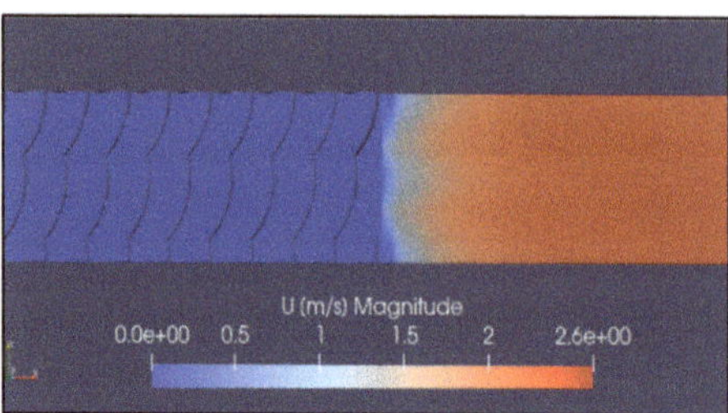

Figure 14. Streaks formation observed on the velocity field at back tested plate.

4.3. Bulb Production

Optimized bulb model was at first textured with the chosen scale type. Details are reported in Appendix C. The final bulb geometry was then oriented as optimization case D, with respect to building platform reference plane. With the aim to assess production run mechanical properties, tensile specimens were inserted. As common good practice, control specimens for destructive testing are frequently manufactured together with the final product. This approach increases reliability of product material properties since they can be assumed to be not sensibly distant from the one of control specimens. The whole platform geometries are to be subjected to the same optimized process parameters. Tensile specimens are oriented vertically, horizontally and with same longitudinal bulb axis inclination. This may allow to assess material strength anisotropy among printing orientations. Three specimens are inserted for each orientation. Tensile specimens are supported with software-generated thin geometries (Materialize Magics 19, accessed on September 2020).

Production results in continuative and uninterrupted process. No issues related to evident deformations nor layer scanning errors are reported, with an exception made for horizontally oriented tensile specimens that gained slight curvature with respect to their main axis. Deformation is likely due to high residual stresses achieved during their larger area scanned layers (approximately in correspondence of their mid height). The internal powder evacuation site allowed to remove all powder during the unpacking step, by means of building platform inclination. The built platform is as in Figure 15.

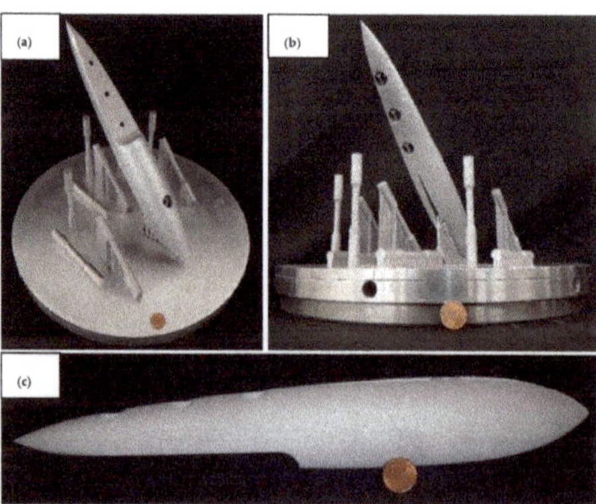

Figure 15. (**a**) Frontal, (**b**) back perspectives on built platform, (**c**) bulb after sandblasting.

4.4. Surface and Material Properties

The analyzed surface was taken in a portion of the bulb with reduced curvature. In Table 5, the resulting roughness and primary profile quality measures are presented. Sandblasting was applied to the bulb external surface. The surface analysis results in improved quality on both roughness and primary profiles. In particular, sandblasting is observed capable of smoothing the highest peaks preserving scale geometrical features.

Table 5. Roughness and primary profiles measures for as built and sandblasted conditions.

Condition	R_a	S_a	$dl_{sr,\ theo}$	$dl_{sr,\ meas}$	Δpv_{theo}	Δpv_{meas}
	[μm]	[μm]	[μm]	[μm]	[μm]	[μm]
As-built	11.9	15.8	1800	1787	89.94	99.11
Sandblasted	7.13	9.09	1800	1888	89.94	84.29

The final product mass is 220 g, slightly higher than the forecasted one by optimization software of 190 g (Altair Inspire 19.3, accessed on May 2020). The discrepancy is related to scale pattern geometrical features addition, which have increased the case thickness. Considering the volume occupied by bulb geometry when immersed in water, approximately 114 cm^3, the overall force acting on bulb due to simultaneous actions of weight and Archimedes' force, leads to a vertical resultant magnitude of 0.24 N (oriented as gravity). This means that the bulb is not provided of buoyancy by itself, which is wanted because it would have acted as a destabilizing force in boat rolling behavior, but at same time does not affect boat buoyancy capabilities significantly.

Tensile specimens result in YS of approximately 250 MPa for all printing directions, while UTS and elongation at break (A%) result smaller for horizontal specimens of about 50 MPa and 3–4% with respect to the means of 45° inclined and vertical specimens which both end up in similar values; UTS of about 400 MPa and A% at 5–6%. Bulb material mechanical properties can be assumed to be similar to the ones of the specimens. Horizontal specimens performed worse in strength and elongation at break. The reason may lay in residual stresses-caused deformations that could have determined an earlier failure of specimens, due to eventual superimposition of stress states. The bulb is oriented according to 45° inclined configuration with its main axis. It may be inferred that by the fact that the most relevant load cases would act as bending moment, they encounter the majority of material disposed as such configuration, which appears to lead to acceptable mechanical properties. The rather low elongation at break values observed of the AlSi7Mg0.6 alloy are related to the inherent porosity and the layered structure of the material. While the achieved density is acceptable for most of the applications, the level of porosity (approximately 0.7%) can be detrimental for fatigue behavior especially. Mechanical properties of the material depend on the process parameters as well as the machine configuration and the applied heat treatments. The measured properties are comparable to what is reported in the literature [25–30].

Concerning the mechanical properties, AlSi7Mg0.6 can be hardened by an aging treatment. In general practice, a stress-relieving treatment is applied to Al-Si-Mg alloys in order to reduce the thermal stresses generated during the LPBF process [30]. Such treatments generally result in a reduction of the mechanical properties and improve elongation at break compared to the as-built material. Successively aging treatments can be done to improve the mechanical properties and have an isotropic behavior. As a matter of fact, maintaining the high YS and UTS values after an ageing treatment may be difficult and require extensive research. Given the rudder bulb's geometry that does not have extensive transitions from bulky to thin regions, part deformations during the build could be assumed negligible. Without the necessity to apply stress-relieving, further hardening can be possibly avoided. However, the tailored mechanical properties can be beneficial for further improving the topological optimization.

In this work, isotropic mechanical properties with lower YS and E values during the topological optimization were considered. The resultant mechanical performance of the as-built material was higher. At a level of demonstrating the capability of the metal AM process, the results were sufficient, and no further iterations were carried out with the actual mechanical properties. This can allow to increase the safety factor. Given the fact that the weight reduction was sufficient, and the aim was to maintain the core thickness higher than a limit value, further mass reduction was not also found to be useful.

5. Conclusions

This work shows an applied case study where metal additive manufacturing was exploited for the design of a novel rudder hub. The geometrical flexibility of the laser powder bed fusion process was combined with a lightweight Al-alloy to design the component with biomimetic surface design. The main outcomes can be summarized as follows.

- The design step satisfied the mass reduction target. The core mass was reduced by 80.3% (corresponding to 19.7% of the original core mass) with acceptable increase in elastic deformations and maximum displacements in the order of 0.1 mm for the most critical load cases. The final mass composed of the shell and core was reduced by 58% from 452 to 190 g.
- The biomimetic fish scale surfaces were modeled to understand their fluid dynamic behavior. The results showed improved behavior in terms of viscous drag force component which sees a reduction of about 19.9%.
- The designed rudder hub was manufactured along with specimens to verify the mechanical behavior of the build. The results showed high geometrical fidelity of the scale details, acceptable surface roughness after sandblasting and the desired mechanical properties.
- The results confirm the great potential of using metal AM processes in the naval industry as it combined great geometrical flexibility as well as design and manufacturing in a digital environment. Along with the obtained results, the work shows a framework for the design, manufacturing and verification of the metal AM products for the naval industry with novel features.

In this work, the produced component had relatively small dimensions for the naval sector, while larger parts can be required for other applications. Larger parts can be produced by welding separate additively manufactured parts, which can be also designed for the assembly purposes. The use of topological optimization methods can also be beneficial for designing the component for successive assembly by welding methods. Laser welding can be a viable option for the welding of parts with complex geometries. While Al-alloys have overall good weldability with lasers despite their high optical reflectivity [31], the weldability of LPBF produced Al-alloys can be difficult due to very large pore formation observed with these materials [32]. Solid state welding processes such as friction stir welding may be another option, but the complexity of the LPBF parts may require complex weld paths not easily applicable to such processes due to the used tools and applied forces. While this work demonstrated the feasibility of producing biomimetic features and topological optimization together, the final verification of the component performance remains an open issue. The work validated the feasibility of producing the biomimetic surfaces and the weight reduction was verified. On the other hand, the drag reduction estimated with the simulations requires further validation.

Author Contributions: Conceptualization, A.S., P.S. and A.G.D.; data curation, A.S.; formal analysis, A.S. and V.F.; investigation, A.S. and V.F.; methodology, P.S. and A.G.D.; resources, A.G.D.; software, A.S. and P.S.; supervision, P.S., A.B., A.R. and A.G.D.; writing—original draft, A.S. and A.G.D.; writing—review and editing, V.F., P.S., A.B. and A.R. All authors have read and agreed to the published version of the manuscript.

Funding: The Italian Ministry of Education, University and Research is acknowledged for the support provided through the Project "Department of Excellence LIS4.0—Lightweight and Smart Structures for Industry 4.0".

Data Availability Statement: The data presented in this study are available on request.

Acknowledgments: The authors are grateful to Trumpf for technical assistance and providing the TruPrint 3000 system.

Conflicts of Interest: The authors declare no conflict of interest.

Appendix A. Load Cases

Load cases were defined according to the load acting simultaneously on the considered condition. However, exceptions existed for $P_{r,v}$, $P_{w,front}$ and $P_{w,\,rear}$ that were inserted uniformly on all load cases, since there is no condition in which they were not present. The capsizing directions were distinguished considering watching the boat from its forward position. Clockwise and counterclockwise directions were considered, and consequently, by being T a resistance action, it will be characterized by opposite direction. Attack angle is named positive if rotation is related to boat forward position to move downward with respect to boat stern height while negative in the opposite case. In Table A1 the distinguished load cases are presented.

Table A1. Considered load cases.

Load Case	Description	Loads
Working (0°)	Rudder cruising with 0° wing attack angle	$P_{r,v}$, $P_{w,front}$, $P_{w,\,rear}$, D_b, $D_w(0°)$, $L(0°)$
Working (+10°)	Rudder cruising with +10° wing attack angle	$P_{r,v}$, $P_{w,front}$, $P_{w,\,rear}$, D_b, $D_w(+10°)$, $L(+10°)$
Working (−10°)	Rudder cruising with −10° wing attack angle	$P_{r,v}$, $P_{w,front}$, $P_{w,\,rear}$, D_b, $D_w(−10°)$, $L(−10°)$
Capsize (clockwise)	Rudder rotation in correspondence of bulb longitudinal axis clockwise	$P_{r,v}$, $P_{w,front}$, $P_{w,\,rear}$, $T(counterclockwise)$
Capsize (counterclockwise)	Rudder rotation in correspondence of bulb longitudinal axis counterclockwise	$P_{r,v}$, $P_{w,\,front}$, $P_{w,\,rear}$, $T(clockwise)$
RudderUpward	Rudder carried upward by boat vertical translation	$P_{r,v}$, $P_{w,front}$, $P_{w,\,rear}$, P_c

Load cases must also account for position of the wing with respect to the bulb. All loads transmitted by the wing to the bulb by means of bolted connection must be applied in different positions according to the usage of the couple of bolted connections involved. Naming (1), (2) and (3), respectively, the front, middle and rear bolt (Figure A1) and the correspondent interface surface region (Figure 2c), load cases were duplicated as in Table A2, where the utilization of (1) and (2) positions is denoted as "*s*" (i.e., short configuration) while the utilization of (2) and (3) is denoted as "*l*" (i.e., long configuration).

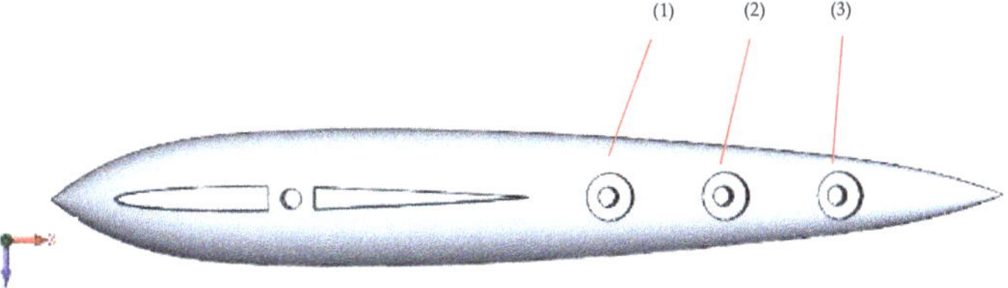

Figure A1. Bolt indexing. (1), (2) and (3), front middle and rear bolt positions. The overall longitudinal length of the component is 290 mm.

Table A2. Definitive load cases list.

Load Case	Description	Loads
Working $(0°, s)$	Rudder cruising with $0°$ wing attack angle and short configuration	$P_{r,v}, P_{w,(1)}, P_{w,(2)}, D_b, D_w(0°, s), L(0°, s)$
Working $(+10°, s)$	Rudder cruising with $+10°$ wing attack angle and short configuration	$P_{r,v}, P_{w,(1)}, P_{w,(2)}, D_b, D_w(+10°, s), L(+10°, s)$
Working $(-10°, s)$	Rudder cruising with $-10°$ wing attack angle and short configuration	$P_{r,v}, P_{w,(1)}, P_{w,(2)}, D_b, D_w(-10°, s), L(-10°, s)$
Capsize (clockwise, s)	Rudder rotation in correspondence of bulb longitudinal axis clockwise and short configuration	$P_{r,v}, P_{w,(1)}, P_{w,(2)}, T(counterclockwise, s)$
Capsize (counterclockwise, s)	Rudder rotation in correspondence of bulb longitudinal axis counterclockwise and short configuration	$P_{r,v}, P_{w,(1)}, P_{w,(2)}, T(clockwise, s)$
RudderUpward (s)	Rudder carried upward by boat vertical translation and short configuration	$P_{r,v}, P_{w,(1)}, P_{w,(2)}, P_c(s)$
Working $(0°, l)$	Rudder cruising with $0°$ wing attack angle and long configuration	$P_{r,v}, P_{w,(2)}, P_{w,(3)}, D_b, D_w(0°, l), L(0°, l)$
Working $(+10°, l)$	Rudder cruising with $+10°$ wing attack angle and long configuration	$P_{r,v}, P_{w,(2)}, P_{w,(3)}, D_b, D_w(+10°, l), L(+10°, l)$
Working $(-10°, l)$	Rudder cruising with $-10°$ wing attack angle and long configuration	$P_{r,v}, P_{w,(2)}, P_{w,(3)}, D_b, D_w(-10°, l), L(-10°, l)$
Capsize (clockwise, l)	Rudder rotation in correspondence of bulb longitudinal axis clockwise and long configuration	$P_{r,v} P_{w,(2)}, P_{w,(3)}, T(counterclockwise, l)$
Capsize (counterclockwise, l)	Rudder rotation in correspondence of bulb longitudinal axis counterclockwise and long configuration	$P_{r,v}, P_{w,(2)}, P_{w,(3)}, T(clockwise, l)$
RudderUpward (l)	Rudder carried upward by boat vertical translation and long configuration	$P_{r,v}, P_{w,(2)}, P_{w,(3)}, P_c(l)$

All loading conditions rely on the same structural constraints. The bulb was joined by interference fitting with rudder vertical, therefore, bolted constraint was placed in correspondence of the bolt hole, while translation constraints were applied on the surface of the socket.

The "common" loads, that shared the same magnitude and direction as the multiple load cases, were the vertical and wing bolts preloads and bulb drag force. All these were applied as uniformly distributed loads over the local surface partition involved. For drag force, all the external surface was considered.

In the following, some of the load cases are detailed while the constraints are hidden in the figures. In Figure A2, "Working $(0°, s)$" load case is presented. It is worth noticing that to comprehend only the first two portions of wing interface surface, both drag and lift, wing caused, actions, are divided equally into two loads each (subscripts (1) and (2) are then added for clarification). Applied loads are not concentrated, they are set as uniformly distributed on the local surface partition of application.

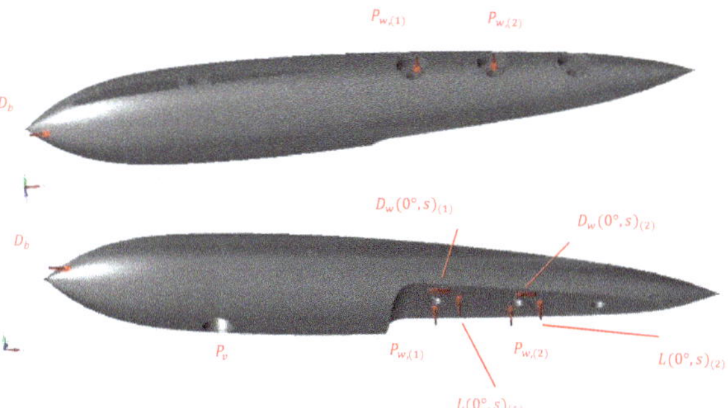

Figure A2. Working ($0°$, s) load case, constraints are hidden for representation purpose. The overall longitudinal length of the component is 290 mm.

In Figure A3 the load case "Capsize (clockwise, *l*)" is presented where the torque is obtained by means of concentrated forces applied at 0.25 m distance with respect to bulb longitudinal axis. Two forces (12.5 N in magnitude each) are used to comply with the "long" configuration. Application points of forces were connected by means of rigid connectors over the correspondent partition of the interface surface.

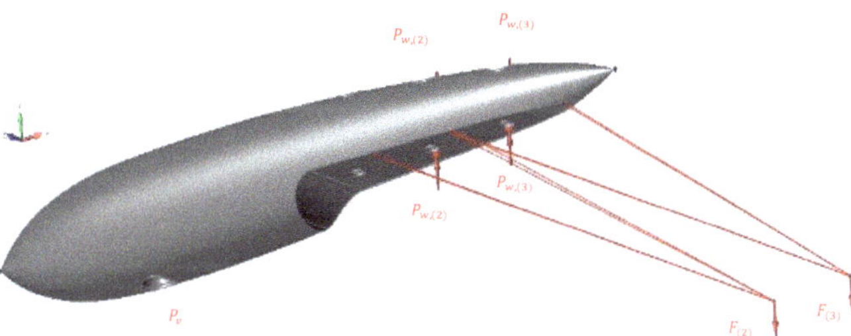

Figure A3. Capsize (clockwise, *l*) load case, constraints are hidden for representation. The overall longitudinal length of the component is 290 mm.

In Figure A4 the "RudderUpward (s)" load case is reported. Since the wing tends to resist the carrying action of the rudder, the bulb ends up subjected to $P_c(s)$. by means of the bolted junction. Forces were then positioned as the head of the bolt is transmitting the load. Two forces (of equal magnitude, 200 N) are used to comply with the "short" configuration.

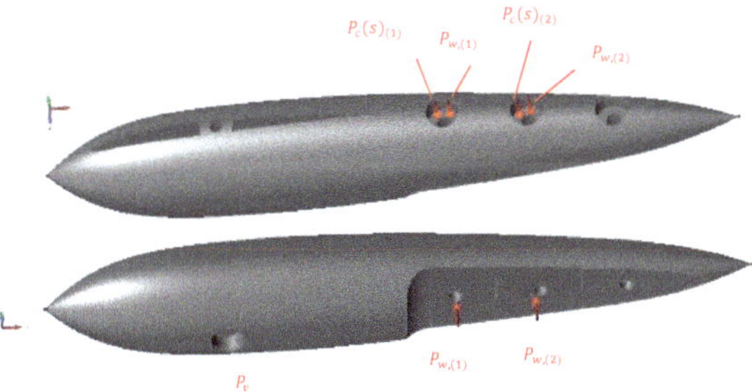

Figure A4. Rudder Upward (s) load case, constraints are hidden for representation purpose.

Appendix B. Topological Optimization Runs

The software allows to observe local variation of optimized mass around the target level, in Figure A5, the variations for all the optimization runs, target level corresponds to index 5 in the topology case slider; from 1 to 9 mass increases.

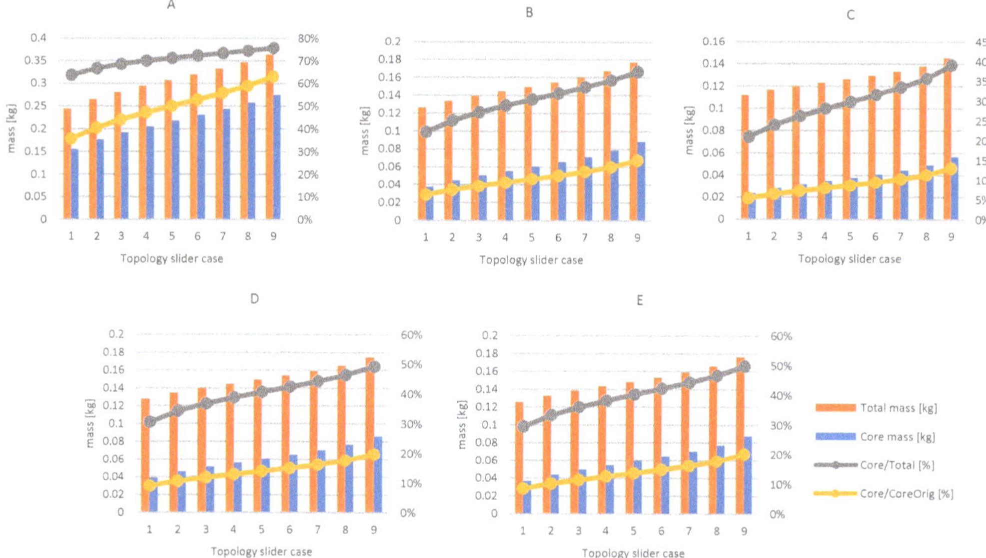

Figure A5. Optimization target neighborhoods. Letters (**A–E**) corresponds to respective optimization run.

Appendix C. Rudder Bulb Model Design and Support Preparation

Scale pattern is applied over the bulb case external surface, increasing its thickness according to the local scale height (Figure A6). Bulb structural behavior is considered to be not sensibly affected by thickness increase and at same time by generation of peaks and valleys surface texture (i.e., due to scale pattern profile) that is considered not capable of determining a sensible localized concentration of stress.

The final bulb geometry is then oriented as optimization case D, with respect to building platform reference plane (Figure A7a). The main support structure is designed manually (Figure A7c), since the area to be sustained embraces part of the rudder vertical encastre socket and it is wanted to minimize influence on the local scale pattern geometry. In order to avoid any excessively high platform-part connecting support structures, localized part to part supports are inserted to preserve bolt holes geometry (Figure A7b), since they experience local limited overhang due to product printing orientation. Building orientation is coherent with recoater wear uniformity (recoater comes from X positive direction), supporting needing area minimization as well as production stability principle in terms of thermal energy to be evacuated per scanned layer. The main bulb support structure is designed to be thicker than what structurally required to satisfy thermal evacuation needs since it remains the only direct connected portion with the baseplate. Contact between bulb and main support is limited to avoid any unnecessary joining with scale pattern, by means of contained anchors dimensions and maximum spacing among neighboring anchors (about 2–3 mm).

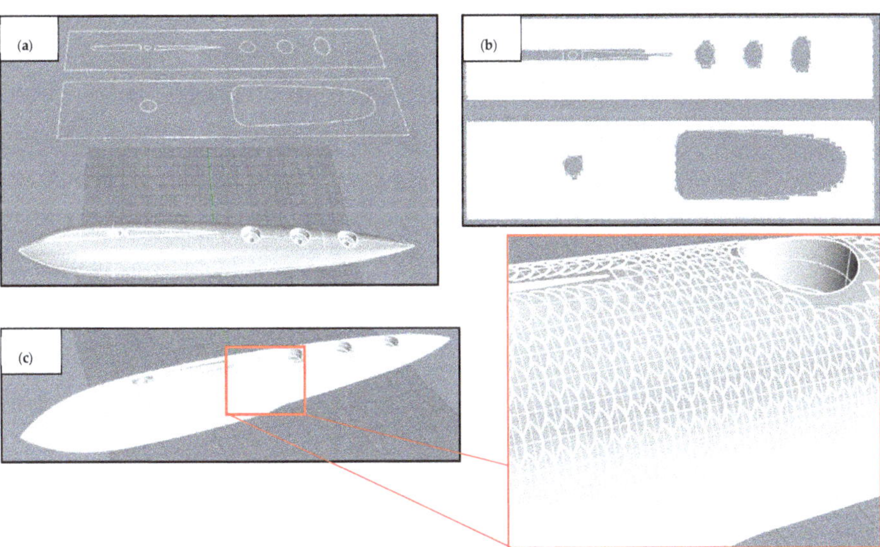

Figure A6. Scale application on bulb external surface; (**a**) reference half-surfaces, (**b**) prepared planar scale pattern for "flow along surface", (**c**) applied scales. The overall longitudinal length of the component is 290 mm.

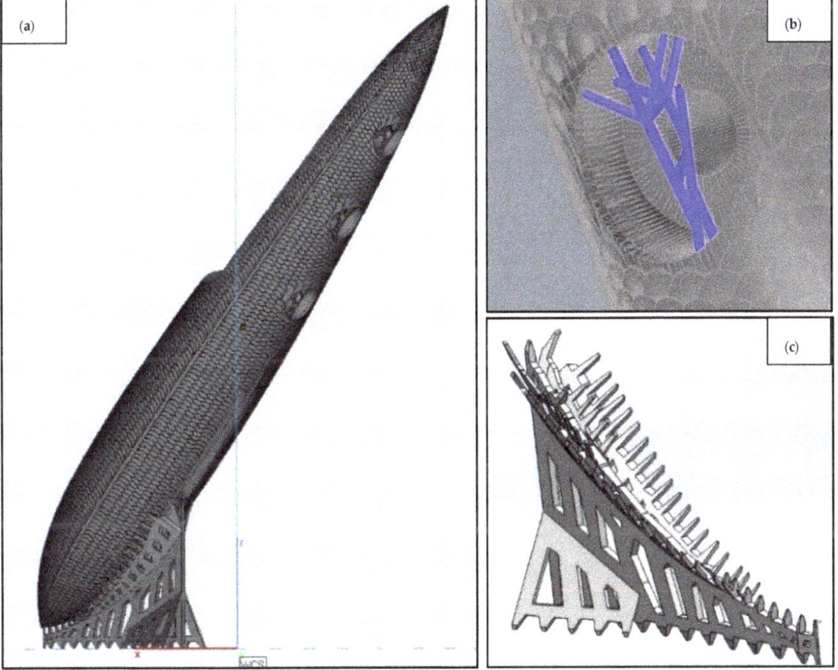

Figure A7. (**a**) Oriented and supported scales-equipped bulb model, (**b**) detail on bolted connection holes support structures and (**c**) isolated main support structure geometry. The overall longitudinal length of the component is 290 mm.

References

1. Ramirez-Peña, M.; Sotano, A.J.S.; Pérez-Fernandez, V.; Abad, F.J.; Batista, M. Achieving a sustainable shipbuilding supply chain under I4.0 perspective. *J. Clean. Prod.* **2020**, *244*, 118789. [CrossRef]
2. University of Maine. Available online: https://composites.umaine.edu/3dirigo-the-worlds-largest-3d-printed-boat/ (accessed on 7 May 2020).
3. Braghin, F. Tecnologie Free-form per la Realizzazione di Componenti Nautici Tramite Fiber Placement. In *NAUTICA +++ | Additive Manufacturing in campo Navale e Nautico | Arianna Bionda e Andrea Ratti*; Edizioni Poli.design: Milano, Italy, 2017; ISBN 978-88-95651-11-8.
4. Tecniche Nuove Spa, Mini 650. Available online: https://www.plastix.it/mini-650-ocore-stampa-3d/ (accessed on 29 April 2020).
5. Cevola, D. Fabbricazione Additiva nel Comparto Nautico: Nuovi Scenari e Prospettive. In *NAUTICA +++. Additive Manufacturing in Campo Navale e Nautico | Arianna Bionda e Andrea Ratti*; Edizioni Poli.design: Milano, Italy, 2017; ISBN 978-88-95651-11-8.
6. Nemani, A.V.; Ghaffari, M.; Nasiri, A. Comparison of microstructural characteristics and mechanical properties of shipbuilding steel plates fabricated by conventional rolling versus wire arc additive manufacturing. *Addit. Manuf.* **2020**, *32*, 101086. [CrossRef]
7. Horgar, A.; Fostervoll, H.; Nyhus, B.; Ren, X.; Eriksson, M.; Akselsen, O. Additive manufacturing using WAAM with AA5183 wire. *J. Mater. Process. Technol.* **2018**, *259*, 68–74. [CrossRef]
8. Torims, T.; Pikurs, G.; Ratkus, A.; Logins, A.; Vilcāns, J.; Sklariks, S. Development of Technological Equipment to Laboratory Test In-situ Laser Cladding for Marine Engine Crankshaft Renovation. *Procedia Eng.* **2015**, *100*, 559–568. [CrossRef]
9. Korsmik, R.S.; Rodionov, A.A.; Korshunov, V.A.; Ponomarev, D.A.; Prosychev, I.S.; Promakhov, V.V. Topological optimization and manufacturing of vessel propeller via LMD-method. *Mater. Today Proc.* **2020**, *30*, 538–544. [CrossRef]
10. Liravi, M.; Pakzad, H.; Moosavi, A.; Nouri-Borujerdi, A. A comprehensive review on recent advances in superhydrophobic surfaces and their applications for drag reduction. *Prog. Org. Coat.* **2020**, *140*, 105537. [CrossRef]
11. Aboulkhair, N.T.; Simonelli, M.; Parry, L.; Ashcroft, I.; Tuck, C.; Hague, R. 3D printing of Aluminium alloys: Additive Manufacturing of Aluminium alloys using selective laser melting. *Prog. Mater. Sci.* **2019**, *106*, 100578. [CrossRef]
12. Park, S.H.; Lee, I. Optimization of drag reduction effect of air lubrication for a tanker model. *Int. J. Nav. Archit. Ocean Eng.* **2018**, *10*, 427–438. [CrossRef]
13. Pujals, G.; Depardon, S.; Cossu, C.; Pujals, G.; Depardon, S.; Cossu, C. Drag reduction of a 3D bluff body using coherent streamwise streaks. *Exp. Fluids* **2010**, *5*, 1085–1094. [CrossRef]
14. Fu, Y.F.; Yuan, C.Q.; Bai, X.Q. Marine drag reduction of shark skin inspired riblet surfaces. *Biosurface Biotribology* **2017**, *3*, 11–24. [CrossRef]
15. Ibrahim, M.D.; Amran, S.N.A.; Yunos, Y.S.; Rahman, M.R.A.; Mohtar, M.Z.; Wong, L.K.; Zulkharnain, A. The Study of Drag Reduction on Ships Inspired by Simplified Shark Skin Imitation. *Appl. Bionics Biomech.* **2018**, *2018*, 1–11. [CrossRef] [PubMed]
16. Ran, W.; Zare, A.; Jovanović, M.R. Model-Based Design of Riblets for Turbulent Drag Reduction. February 2020. Available online: http://arxiv.org/abs/2002.01671 (accessed on 5 June 2020).
17. Song, X.; Zhang, M. Turbulent Drag Reduction Characteristics of Bionic Nonsmooth Surfaces with Jets. *Appl. Sci.* **2019**, *9*, 5070. [CrossRef]
18. Bai, J.; Meng, X.; Ji, C.; Liang, Y. Drag reduction characteristics and flow field analysis of textured surface. *Friction* **2016**, *4*, 165–175. [CrossRef]
19. Hu, H.; Tamai, M.; Murphy, J.T. Flexible-Membrane Airfoils at Low Reynolds Numbers. *J. Aircr.* **2008**, *45*, 1767–1778. [CrossRef]
20. Banks, J.; Giovannetti, L.M.; Taylor, J.; Turnock, S. Assessing Human-Fluid-Structure Interaction for the International Moth. *Procedia Eng.* **2016**, *147*, 311–316. [CrossRef]
21. Giovannetti, L.M.; Banks, J.; Ledri, M.; Turnock, S.; Boyd, S. Toward the development of a hydrofoil tailored to passively reduce its lift response to fluid load. *Ocean Eng.* **2018**, *167*, 1–10. [CrossRef]
22. International Sailing Federation. *International Moth Class Rules.* 1 May 2017. Available online: http://www.moth-sailing.org/history/rules-and-documents/ (accessed on 1 February 2020).
23. Pereira, J.C.; Gil, E.; Solaberrieta, L.; Sebastián, M.S.; Bilbao, Y.; Rodríguez, P.P. Comparison of AlSi7Mg0.6 alloy obtained by selective laser melting and investment casting processes: Microstructure and mechanical properties in as-built/as-cast and heat-treated conditions. *Mater. Sci. Eng. A* **2020**, *778*, 139124. [CrossRef]
24. Muthuramalingam, M.; Villemin, L.S.; Bruecker, C. Streak formation in flow over biomimetic fish scale arrays. *J. Exp. Biol.* **2019**, *222*. [CrossRef]
25. Rao, H.; Giet, S.; Yang, K.; Wu, X.; Davies, C. The influence of processing parameters on aluminium alloy A357 manufactured by Selective Laser Melting. *Mater. Des.* **2016**, *109*, 334–346. [CrossRef]
26. Rao, J.H.; Zhang, Y.; Fang, X.; Chen, Y.; Wu, X.; Davies, C.H. The origins for tensile properties of selective laser melted aluminium alloy A357. *Addit. Manuf.* **2017**, *17*, 113–122. [CrossRef]
27. Trevisan, F.; Calignano, F.; Lorusso, M.; Pakkanen, J.; Ambrosio, E.P.; Lombardi, M.; Pavese, M.; Manfredi, D.; Fino, P. Effects of Heat Treatments on A357 Alloy Produced by Selective Laser Melting. In Proceedings of the World PM 2016 Congress Exhibition: European Powder Metallurgy Association, Hamburg, Germany, 9–13 October 2016; pp. 1–7.
28. Aversa, A.; Lorusso, M.; Trevisan, F.; Ambrosio, E.P.; Calignano, F.; Manfredi, D.G.; Biamino, S.; Fino, P.; Lombardi, M.; Pavese, M. Effect of Process and Post-Process Conditions on the Mechanical Properties of an A357 Alloy Produced via Laser Powder Bed Fusion. *Metals* **2017**, *7*, 68. [CrossRef]

29. Grande, A.M.; Cacace, A.; Demir, A.G.; Sala, G. Fracture and fatigue behaviour of AlSi7Mg0. 6 produced by Selective Laser Melting: Effects of thermal-treatments. In Proceedings of the 25th Conference of the Italian Association of Aeronautics and Astronautics (AIDAA 2019) AIDAA, Rome, Italy, 9–12 September 2019; pp. 1138–1144.
30. Sorci, R.; Tassa, O.; Colaneri, A.; Astri, A.; Mirabile, D.; Iwnicki, S.; Demir, A.G. Design of an Innovative Oxide Dispersion Strengthened Al Alloy for Selective Laser Melting to Produce Lighter Components for the Railway Sector. *J. Mater. Eng. Perform.* **2021**, *30*, 1–11. [CrossRef]
31. Garavaglia, M.; Demir, A.G.; Zarini, S.; Victor, B.M.; Previtali, B. Fiber laser welding of AA 5754 in the double lap-joint configuration: Process development, mechanical characterization, and monitoring. *Int. J. Adv. Manuf. Technol.* **2020**, *111*, 1643–1657. [CrossRef]
32. Mäkikangas, J.; Rautio, T.; Mustakangas, A.; Mäntyjärvi, K. Laser welding of AlSi10Mg aluminium-based alloy produced by Selective Laser Melting (SLM). *Procedia Manuf.* **2019**, *36*, 88–94. [CrossRef]

Journal of
*Marine Science
and Engineering*

Article

AUV Obstacle Avoidance Planning Based on Deep Reinforcement Learning

Jianya Yuan [1], Hongjian Wang [1,*], Honghan Zhang [1], Changjian Lin [2], Dan Yu [1] and Chengfeng Li [1]

1   College of Intelligent Systems Science and Engineering, Harbin Engineering University,
    Harbin 045100, China; yuan040061@hrbeu.edu.cn (J.Y.); oceanzhh@hrbeu.edu.cn (H.Z.);
    cocomomo@hrbeu.edu.cn (D.Y.); lcf1986@hrbeu.edu.cn (C.L.)
2   School of Information and Control Engineering, China University of Mining and Technology,
    Xuzhou 221000, China; LIN_Changjian@cumt.edu.cn
*   Correspondence: wanghongjian@hrbeu.edu.cn

Abstract: In a complex underwater environment, finding a viable, collision-free path for an autonomous underwater vehicle (AUV) is a challenging task. The purpose of this paper is to establish a safe, real-time, and robust method of collision avoidance that improves the autonomy of AUVs. We propose a method based on active sonar, which utilizes a deep reinforcement learning algorithm to learn the processed sonar information to navigate the AUV in an uncertain environment. We compare the performance of double deep Q-network algorithms with that of a genetic algorithm and deep learning. We propose a line-of-sight guidance method to mitigate abrupt changes in the yaw direction and smooth the heading changes when the AUV switches trajectory. The different experimental results show that the double deep Q-network algorithms ensure excellent collision avoidance performance. The effectiveness of the algorithm proposed in this paper was verified in three environments: random static, mixed static, and complex dynamic. The results show that the proposed algorithm has significant advantages over other algorithms in terms of success rate, collision avoidance performance, and generalization ability. The double deep Q-network algorithm proposed in this paper is superior to the genetic algorithm and deep learning in terms of the running time, total path, performance in avoiding collisions with moving obstacles, and planning time for each step. After the algorithm is trained in a simulated environment, it can still perform online learning according to the information of the environment after deployment and adjust the weight of the network in real-time. These results demonstrate that the proposed approach has significant potential for practical applications.

Keywords: autonomous underwater vehicle (AUV); collision avoidance planning; deep reinforcement learning (DRL); double-DQN (D-DQN)

Citation: Yuan, J.; Wang, H.; Zhang, H.; Lin, C.; Yu, D.; Li, C. AUV Obstacle Avoidance Planning Based on Deep Reinforcement Learning. *J. Mar. Sci. Eng.* **2021**, *9*, 1166. https://doi.org/10.3390/jmse9111166

Academic Editors: Davide Tumino and Antonio Mancuso

Received: 30 June 2021
Accepted: 24 September 2021
Published: 23 October 2021

1. Introduction

The ultimate goal of mobile robot research is to process the data measured by the sensing devices carried by the robot and to achieve completely independent decisions through online processing. In a dynamically changing environment, it is extremely important to plan a reasonable, safe path for a given task. The path planning of autonomous underwater vehicles (AUVs) can be divided into two categories: global path planning based on a completely known environment and local path planning based on an uncertain environment, in which the shape and location of the obstacles are unknown. In recent decades, research in this field has produced many achievements. Global path planning methods include the sampling-based A* algorithm [1] and the rapidly-exploring random tree [2,3]. In addition, there are many intelligent optimization algorithms, such as genetic algorithms (GAs) [4], ant colony optimization (ACO) [5], and particle swarm optimization (PSO) [6], which can also realize path planning. Petru et al. [7] proposes a sensor-based neuro-fuzzy controller navigation algorithm. The control system consists of a hierarchical

175

structure of robot behavior. The authors propose the use of segmented 2-D occupancy maps to use ground-based probability grids for application in mobile robot navigation with collision avoidance in [8]. The authors propose an extended Voronoi transform algorithm that imitates the repulsive potential of walls and obstacles and combines the extended Voronoi transform and the fast marching method to realize the robot's navigation in a previously undeveloped dynamic environment [9]. These methods can produce excellent results in simulated environments. Nevertheless, it is difficult to obtain a satisfactory path because information about the real environment is incomplete.

The marine environment has various uncertainties, such as floating objects, fish, and ocean currents. In addition, accurate path planning is challenging due to errors caused by approximations and linearization in the modeling of the system. As AUVs have limited energy, designing an AUV path planner to run in real-time in an uncertain environment while avoiding static and uncertain dynamic obstacles is a critical issue.

In a continuous large-scale space, it is difficult to perform obstacle avoidance using only reinforcement learning (RL), as this requires long-term learning and is relatively inefficient. The continuity of the state and action spaces leads to the so-called dimensionality disaster, and traditional RL cannot be applied in any effective form. Although dynamic programming (DP) can be applied to continuous problems, accurate solutions can only be obtained for special cases. Many scholars combine RL with fuzzy logic, supervised learning, and transfer learning to realize the autonomous planning of robots [10–13]. For example, [10] proposed the use of RL to teach collision avoidance behavior and goal-seeking behavior rules, thus allowing the correct course of action to be determined. The advantage of this approach is that it requires simple evaluation data rather than thousands of input-output training data. In [11], a neural fuzzy system with a hybrid learning algorithm was proposed in which supervised learning is used to determine the input and output membership functions, and an RL algorithm is used to fine-tune the output membership functions. The main advantage of this hybrid approach is that, with the help of supervised learning, the search space for RL is reduced, and the learning process is accelerated, eventually leading to better fuzzy logic rules. Meng et al. [12] described the use of a fuzzy actor-critic learning algorithm that enables a robot to readapt to the new environment without human intervention. The generalized dynamic fuzzy neural network is trained by supervised learning to approximate the fuzzy inference. Supervised learning methods have the advantage of fast convergence, but it can be difficult to obtain sufficient training data.

The authors proposed Q-learning and neural network (NN) planners to solve the obstacle avoidance problem in [14–16]. The robot has the ability to implement collision avoidance planning, but the time taken to reach the target point is too long, and the target point cannot always be identified. In practical applications, the robot's ability to reach the target point is as important as its ability to avoid obstacles.

In the existing literature, there are many research studies in which deep reinforcement learning (DRL) is applied to solve the problem of self-learning. For example, [17] proposed a DRL model that can successfully learn control strategies directly from some high-dimensional sensory input. The authors of [18] developed a DRL approach that adds memory and assisted learning objectives for training agents, enabling them to navigate through large, visually rich environments, including frequent changes to the start and target locations. NNs have been used to perform data fusion from different sensor sources with DRL and then employed for search and pick things [19]. In [20], an improved reward function was developed using a convolutional NN, and the Q-value was fitted to solve the problem of obstacle avoidance in an unknown environment. Long et al. [21] proposed a multi-scenario, multistage training framework for the optimization of a fully decentralized agent-system collision avoidance strategy through a powerful policy gradient. The authors of [22] developed three different deep neural networks to realize the robot's collision avoidance planning in a static environment. The authors of [23] developed a distributed

DRL framework (each subtask passes through the designed LSTM-based DRL network) for the unmanned aerial vehicle (UAV) navigation problem.

Several algorithms take information directly from the original image, without preprocessing, and formulate control strategies through DRL. In [24], the author used CNN and DRL methods to use raw pixels as input to allow the agent to learn navigation capabilities in a 3D environment. For instance, [25] used supervised learning to obtain depth information from a monocular vision image and then used a DRL algorithm to learn a control policy that selects a steering direction as a function of the vision system's output in the virtual scene before finally applying the policy in actual autonomous driving. In [26], a motion planner based on DRL was proposed in which sparse 10-dimensional range findings and the target position are used as the input and the continuous steering commands as output. The DRL approach seems very promising because it does not require deep learning to obtain training samples through additional methods.

A convolutional NN was integrated with a decision-making process in a hierarchical structure, with the original depth image taken as input and control commands generated as the output, thus realizing model-free obstacle avoidance behavior [27]. The convolutions can be replaced with complete connections in standard recurrent NNs, thereby reducing the number of parameters and improving the feature extraction ability to achieve AUV obstacle avoidance planning [28]. Deep learning is an effective method of collision avoidance planning, but there are still some problems to overcome. Prior to learning, it is necessary to use methods such as PSO to generate large amounts of data. Therefore, the effect of deep learning is heavily dependent on the performance of algorithms such as PSO. When encountering dynamic obstacles, the above-mentioned methods cannot guarantee optimal planning.

In this paper, we present a DRL obstacle avoidance algorithm for AUVs under unknown environmental dynamics. The planning performance of the D-DQN-LSTM algorithm is compared with several algorithms. When the GA algorithm encounters dense obstacles and moving obstacles, the real-time planning ability of the algorithm is obviously insufficient. Compared with the GA algorithm, the generalization ability of the DL algorithm is significantly improved. It has achieved certain results, but it still has shortcomings. For example, a large number of training samples need to be generated with the help of the GA algorithm or PSO algorithm. The algorithm is offline, and the weights cannot be updated online in real-time. In [18], the author used the asynchronous advantage actor-critic (A3C) algorithm, which takes the pixels of the maze environment as input and divides the output into eight policies. The network also uses LSTM. Using pixels as input places higher demands on the computer. Inspired by the LSTM network's structure suitability for processing time series sequences and reinforcement learning for online learning, this paper proposes a D-DQN-LSTM algorithm. The input in our paper is only dimensionality-reduced sonar information. The algorithm framework is double-DQN. The network output is also eight policies, and the angle information of the goal point is also used as the output policy. The advantage is that it increases the probability of the AUV reaching the target point and improves training efficiency. We use double-DQN to reduce the overestimation of the Q-value. By learning the reward function to determine the end-to-end mapping relationship between the perception information and the control command, the AUV reactive collision avoidance behavior is realized.

The main contributions of this article can be summarized as follows:

- Increasing the target-oriented policy. Previous article studies have used RL for path planning, resulting in an inefficient problem describing how the AUV should reach the target point. We add the AUV to the target point as a strategy in the algorithm, thus increasing the probability of the algorithm reaching the target point and greatly shortening the time required for the AUV to explore the environment.
- Using the LSTM network instead of NNs. As collision avoidance planning involves decisions based on a series of observation states, and because long short-term memory

(LSTM) is good at processing time series, we use LSTM instead of NNs as part of the Q-network to propose a deep Q-network DQN-LSTM algorithm.

- Double-DQN [29] reduces an overestimated Q-value. As the traditional DQN algorithm has an overoptimistic Q-value, we use a double-DQN (D-DQN) approach that makes the algorithm more stable and reliable.
- The line-of-sight guidance method has a smooth trajectory. The output action of DRL is discrete, so any inconsistency between two actions will result in a drastic change in the heading direction, which is both impractical and unfavorable to the actuator of the AUV. Thus, we introduced a line-of-sight guidance method to make the trajectory switching process smoother.

The rest of this article is organized as follows. Section 2 introduces the perception model of active sonar, the AUV model, and the line-of-sight guidance system. The principle and network structure of the DRL method are introduced in Section 3. Section 4 presents and discusses the simulation results from different planning algorithms. Finally, Section 5 introduces the conclusions of this research.

2. Problem Formulation

2.1. Coordinate System and Perception Model

A moving AUV may encounter static and dynamic obstacles during navigation. In this paper, the state information of the external environment is assumed to be mainly obtained by an active sonar sensor. Consider 256 beams divided into 13 groups $\{g_0, g_1, g_2, \cdots \cdots g_{11}, g_{12}\}$, with $\{g_1, g_2, \cdots \cdots g_{11}\}$ consisting of 20 beams and g_0, g_{12} including 18 beams. These beams are used to judge whether there is an obstacle around the AUV and whether there is an optimal path for the AUV that does collide with the obstacle. Assume that the X-axis of the AUV coordinate system {R} is consistent with the moving direction and that the positive Y-axis points outward from the left side of the AUV. As shown in Figure 1, d_i represents the minimum distance from the obstacle detected by the ith group and ϕ_i is the angle corresponding to the minimum distance. The coordinates of the obstacle in the global coordinate frame {G} can be expressed as:

$$\begin{bmatrix} x \\ y \end{bmatrix} = \begin{bmatrix} \cos\theta & -\sin\theta \\ \sin\theta & \cos\theta \end{bmatrix} \begin{bmatrix} d_i\cos\phi_i + r \\ d_i\sin\phi_i \end{bmatrix} + \begin{bmatrix} x_c \\ y_c \end{bmatrix} \tag{1}$$

where (x, y) are the coordinates of the obstacle in the global coordinate frame {G}, r is the distance from the sonar to the center of mass of the AUV, θ is the angle between the X-axis of the coordinate frames {R}and{G}, θ_g is the angle between the AUV and the target point relative to the X-axis ($\theta_d = \theta_g - \theta$), with positive values running counterclockwise, and (x_c, y_c) is the current position of the AUV in {G}.

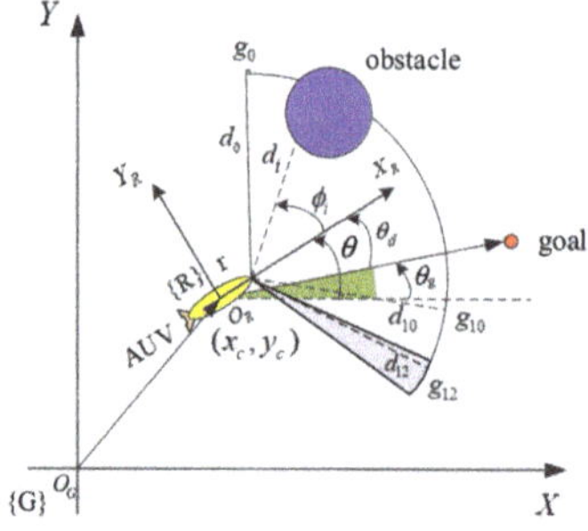

Figure 1. AUV coordinate system and obstacle detection model.

2.2. AUV Model

In this paper, we consider only the three horizontal degrees of freedom (DOFs) when designing the guidance system for the AUV. The 3-DOF kinetics and dynamics can be represented as:

$$\dot{\boldsymbol{\eta}} = \mathbf{R}(\theta)\mathbf{v} \tag{2}$$

$$\mathbf{M}\dot{\mathbf{v}} + \mathbf{C}(\mathbf{v})\mathbf{v} + \mathbf{D}(\mathbf{v})\mathbf{v} = \boldsymbol{\tau} + \widehat{\boldsymbol{\tau}} \tag{3}$$

In this article, the AUV movement process does not consider the interference of wind, waves, and currents. This matrix shows properties of: $\mathbf{R}(\theta)^T\mathbf{R}(\theta) = \mathbf{I}$.

The 3-DOF kinematics can be simplified to:

$$\begin{cases} \dot{x} = u\cos\theta - v\sin\theta \\ \dot{y} = u\sin\theta - v\cos\theta \\ \dot{\theta} = r \end{cases} \tag{4}$$

The 3-DOF dynamics can be represented as:

$$\begin{cases} \dot{u} = (-d_u u + \tau_u)/m_u \\ \dot{v} = (C_1 m_r + C_2 Y_{\dot{r}})/(m_v m_r - Y_{\dot{r}}^2) \\ \dot{r} = (C_2 m_v + C_2 Y_{\dot{r}})/(m_v m_r - Y_{\dot{r}}^2) \end{cases} \tag{5}$$

$C_1 = -d_v v + (Y_{\dot{r}} - m_u u)r$, $C_2 = (N_{\dot{v}} - m_{uv}u)v - d_r r + \tau_r$, $m_u = m - X_{\dot{u}}$, $m_v = m - Y_{\dot{v}}$, $m_r = I_z - N_{\dot{r}}$, $m_{uv} = X_{\dot{u}} - Y_{\dot{v}}$, $d_u = Y_v + Y_{|v|v}|v|$, $d_r = N_r + N_{|r|r}|r|$, $m = 40$ kg, $I_z = 8.0$ N $\cdot$ m^2. The AUV model parameters are listed in Table 1.

Table 1. The parameters of the AUV model.

$X_{\dot{u}}$	X_{u_r}	$X_{|u|u}$	$Y_{\dot{r}}$	$Y_{\dot{v}}$	Y_r
-1.42	0.1	8.2	-2.5	-38.4	5
kg	kg	kg $\cdot$ m^{-1}	N $\cdot$ m^2	kg	kg
Y_{ur}	$Y_{|v|v}$	N_v	$N_{\dot{v}}$	N_r	$N_{|r|r}$
0	200	36	2.2	5	15
kg $\cdot$ m^{-1}	kg $\cdot$ m^{-1}	kg	kg	kg $\cdot$ m	kg

2.3. The Line-of-Sight Guidance System

The planning algorithm in this paper is realized by adjusting the heading of the AUV. When the planning policy is h_0 at time t, then the heading of the AUV at time $t+1$ will increase by 10^0, which is impossible in practical applications. With such a large change in an instant, the AUV heading can only be adjusted gradually through control. To make the trajectory smoother and achieve precise tracking control, a line-of-sight [30] method is used to solve this problem. As seen from Figure 2, assuming that the angle between the velocity $\mathbf{V}$ at time t and the velocity $\mathbf{V}'$ at time $t+1$ is 10^0 to show more clearly, the angle in Figure 2 is not 10^0, we choose a forward-looking vector $\boldsymbol{\Delta}$ as the reference for trajectory tracking to obtain the desired error angle $\theta_d(e)$ (where $\theta_d(e)$ is the achievable target of the propeller and rudder). By continuously calculating $\theta_d(e)$, the AUV is slowly transferred from point A to point B. The following experiments also show that the guidance algorithms can cause the AUV to perfectly track the required trajectory.

A careful inspection of Figure 2 gives the following formulas:

$$\theta(t) = \frac{\pi}{2} + \arctan\left(\frac{x(t) - x_{ac}}{y(t) - y_{ac}}\right) \tag{6}$$

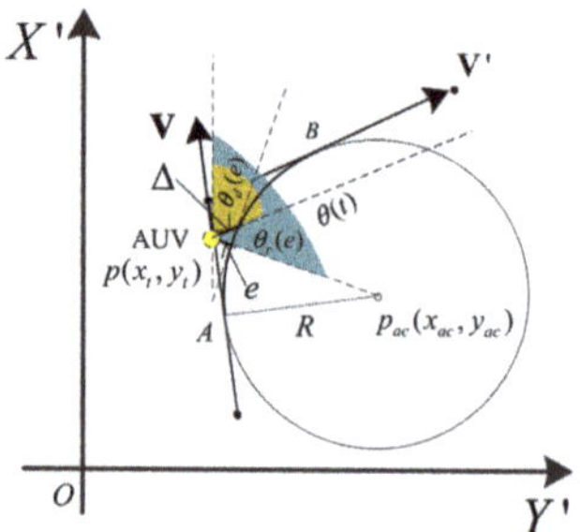

Figure 2. Line-of-sight guidance for AUV trajectory.

The tracking error is:

$$e = \|p_{ac} - p\|_2 - R \tag{7}$$

$$\theta_d(e) = \theta(t) - \theta_r(e) = \begin{cases} \theta(t) - \arccos(\frac{|e|}{\|\Delta\|_2}), \ |e| \leq \|\Delta\|_2 \\ \theta_t(t) - \frac{\pi}{2}, \ \text{else} \end{cases} \tag{8}$$

where $p(x_t, y_t)$ represents the current position of the AUV and $p_{ac}(x_{ac}, y_{ac})$ represents the center position of the transition arc. $\theta_r(e)$ is the angle between the forward-looking vector Δ and the vector $\overrightarrow{pp_c}$, where Δ is the forward view vector parallel to the next desired trajectory. $\theta_d(e)$ represents the desired angle, and $\theta(t)$ is the angle between the vector $\overrightarrow{pp_{ac}}$ and the X'-axis. The coordinate system $X'OY'$ in Figure 2 is not related to the XOY in Figure 1.

Figure 3 verifies the effect of the line-of-sight guidance algorithm. The AUV adopts a new steering strategy at the trajectory transition point and the velocity changes from **V** to **V'**. The heading of the AUV has to be adjusted to a large degree, which is difficult to accomplish if it only relies on the propeller and rudder. The red dotted line represents the trajectory without the guidance algorithm, and the solid blue line represents the trajectory with the guidance added. The red circle represents the radius of the trajectory transition. After introducing this guidance algorithm, the trajectory changes slowly, rather than sharply, near the trajectory switching point, instead of violently changing. The red circles represent the process of track switching.

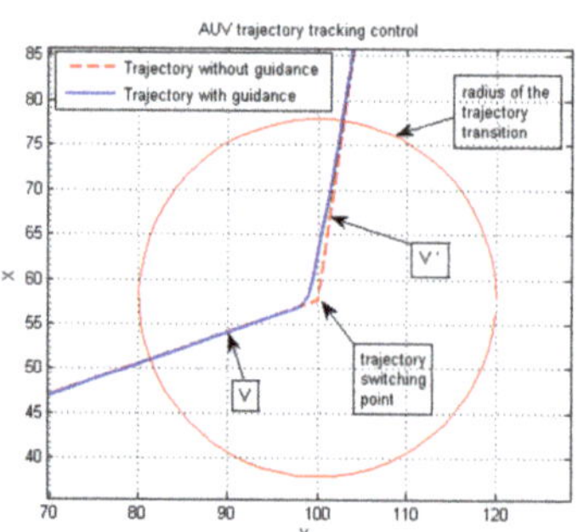

Figure 3. AUV trajectory tracking control based on the Line-of-Sight method.

3. Proposed Deep Reinforcement Learning Algorithm

When using RL to solve robotic control problems, we assume that the robot's environmental model is a finite Markov decision process (MDP). In addition, the whole environment is fully observable. The path planning task considers the results of the AUV's interaction with the environment through a series of actions in discrete time steps.

We assume that the information received by the AUV from the active sonar is $D_t \subseteq \{d_{0t}, d_{1t} \cdots d_{it} \cdots d_{12t}\}$ $(i = 0, 1, 2, \ldots \ldots 12)$ at time t. We assume that the AUV receives observations D_t of the simulation environment E, $D_t \subset \mathbb{R}^{13}$, and then the agent selects a heading adjustment h_t. The AUV receives a numerical reward $r_{t+1} \in \mathbb{R}$ and transfers to the next state according to the probability $p(D_{t+1}, r_{t+1} | D_t, h_t)$, where p is the probability distribution for each state D_t and choice h_t.

The expected reward r_{t+1} for state-action pair (D_t, h_t) is computed as:

$$r_{t+1} \triangleq r(s_t, a_t) = E[R_{t+1} | D_t, h_t] = \sum_{r_{t+1} \in \mathbb{R}} r_{t+1} \sum_{s_t \in S} p(D_{t+1}, h_{t+1} | D_t, h_t) \tag{9}$$

The aim is to select the heading adjustment h_t that maximizes the expected value of the cumulative sum of the received scalar signals. The optimal action-value function $Q^*(D, h)$ defines the maximum expected return for heading adjustment h in state D and the decisions thereafter, following an optimal policy.

$$Q^*(D, h) = \max_\pi E\left[\sum_{t'=t}^{T} \gamma^{t'-t} r^{t'} \middle| D_t = D, h_t = h, \pi \right] \tag{10}$$

where policy $\pi(h|D)$ is the probability of selecting action $h_t = h$ when $D_t = D$, T is the termination time-step of the episode, γ is a discount rate and $0 \leq \gamma \leq 1$.

The optimal action-value function $Q^*(D, h)$ obeys an important identity known as the Bellman equation [31]

$$Q^*(D, h) = E[r_{t+1} + \gamma \max_{a_{t+1}} Q^*(D_{t+1}, h_{t+1}) | D_t = D, h_t = h] \tag{11}$$

Watkins and Dayan proved that $Q^\pi(D_t, h_t) \rightarrow Q^*(D, h)$ with a probability of 1 as $t \rightarrow \infty$ [31].

When actually applying the Q-learning algorithm, we allow t to be sufficiently large to approximate the action value $Q(D, h)$. Sometimes we also use linear or nonlinear NNs with weights w as a Q-network to estimate the action-value $Q(D, h; w) \approx Q^*(D, h)$. Nonlinear NNs diverge in many cases, but they are often successfully used. In this paper, we use an off-policy technique to solve the challenges of exploration and exploitation.

The Q-network can be trained by minimizing the loss function $L_t(w_t)$:

$$L_t(w_t) = E_{D, h \sim \rho}[(y_t - \hat{Q}(D_t, h_t; w_t))^2] \tag{12}$$

where $y_t = E_{D_{t+1} \sim \xi}[r_{t+1} + \gamma \max_{h_{t+1}} Q(D_{t+1}, h_{t+1}; w_{t-1}) | D_t, h_t]$ is the target value for iteration t, and ρ, ξ describes the behavior distributions when the AUV receives observations D_t and makes heading adjustments h_t, and $\hat{Q}(D_t, h_t; w_t)$ is the estimated value of the NN's network output. Differentiating the loss function $L_t(w_t)$ with respect to the weight w_t, the following gradient is obtained:

$$\nabla_{w_t} L_t(w_t) = E[(r_{t+1} + \gamma \max_{h_{t+1}} Q(D_{t+1}, h_{t+1}; w_{t-1}) - \hat{Q}(D_t, h_t; w_t)) \nabla_{w_t} \hat{Q}(D_t, h_t; w_t)] \tag{13}$$

The weight w_t is updated according to:

$$w_{t+1} = w_t + \alpha \nabla_{w_t} L_t(w_t) \tag{14}$$

The stability and convergence of DRL must be considered [32] when using a nonlinear function approximator. Instabilities may result from small updates to $Q(D, h)$, and these may significantly change the policy, the correlations present in the sequence of observations, and the correlations between the action-values and the target values [18]. We use a biologically inspired mechanism termed "experience replay" and adjust the action values toward the target values through periodic updates. To train the network, we process the data $D_t \subseteq \{d_{0t}, d_{1t} \cdots d_{it} \cdots d_{12t}\}(i = 0, 1, 2, \ldots \ldots 12)$ measured by the active sonar through regularization and noise processing to obtain the input state of the network [33]. Each neuron in the output layer corresponds to the Q-value.

The meaning of each action h is defined in Figure 4; each action corresponds to a heading change angle, with positive values representing θ increasing, negative values representing θ decreasing, 0 representing no change, and neuron 7 representing the agents causing the algorithm to search blindly in the environment for a long time. The advantage of using neuron 7 is that the DQN algorithm has the ability to find a target point in some way; thus, we can reduce the time spent in an invalid search environment and increase the probability of reaching the target point.

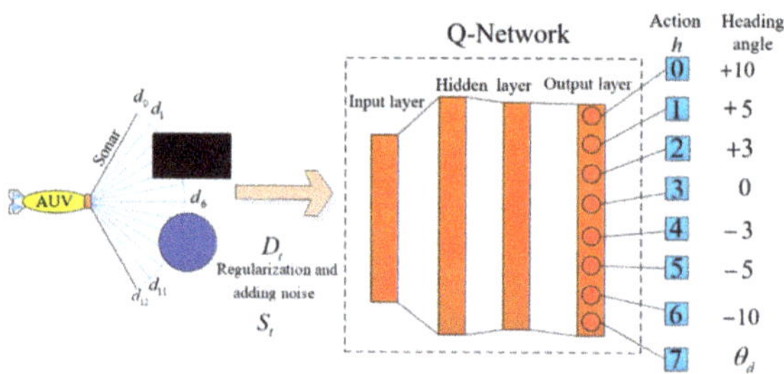

Figure 4. DQN-based AUV collision avoidance architecture.

AUV collision avoidance planning is essentially a matter of making decisions based on the state of sequential observations [34]. LSTM has proven to provide state-of-the-art performance in many sequence prediction problems. In the process of learning, the overestimation that occurs when a nonlinear network is used to approximate the Q-value seriously affects the performance of the algorithm. The D-DQN approach can be used to reduce overestimation, making the algorithm more stable and enabling reliable learning. Inspired by the literature [18], we use D-DQN, and the Q-network uses LSTM to build a new algorithm called double DQN-LSTM(D-DQN-LSTM). The structure of the Q-network in the D-DQN-LSTM algorithm is shown in Figure 5, where the output layer neurons have the same meaning as in Figure 4.

The target value y_t^{DQ} used by D-DQN-LSTM is then:

$$y_t^{DQ} = E_{D_{t+1} \sim \xi}[r_{t+1} + \gamma Q(D_{t+1}, \text{argmax}_h \hat{Q}(D_{t+1}, h; w_t); w_t')] \tag{15}$$

Similarly to Expression (13), the update amount of the weight w_t is updated according to:

$$\nabla_{w_t} L_t(w_t) = E[(r_{t+1} + \gamma Q(D_{t+1}, \text{argmax}_h \hat{Q}(D_{t+1}, h; w_t); w_t') - \hat{Q}(D_t, h_t; w_t))\nabla_{w_t}\hat{Q}(D_t, h_t; w_t)] \tag{16}$$

The concept of using reward signals to form goals is one of the most distinctive features of RL. The AUV's goal is to maximize the total amount of reward it receives while avoiding obstacles and reaching the target point in the process of motion. The reward function is:

$$r_t = \begin{cases} -3 & d_i \leq 1 \\ -0.5 & 1 < d_i < 5 \\ 5 & \text{reach target point} \\ 0 & \text{other else} \end{cases} \tag{17}$$

where d_i indicates the distance between the AUV and the obstacle. $d_i \leq 1$ indicates that the AUV has encountered an obstacle, and $1 < d_i < 5$ indicates that the AUV is too close to the obstacle, which means that the AUV does not have enough time to adjust the heading to avoid obstacles. If the AUV reaches the target point, it receives a positive reward value. In the other situations, the reward value is 0.

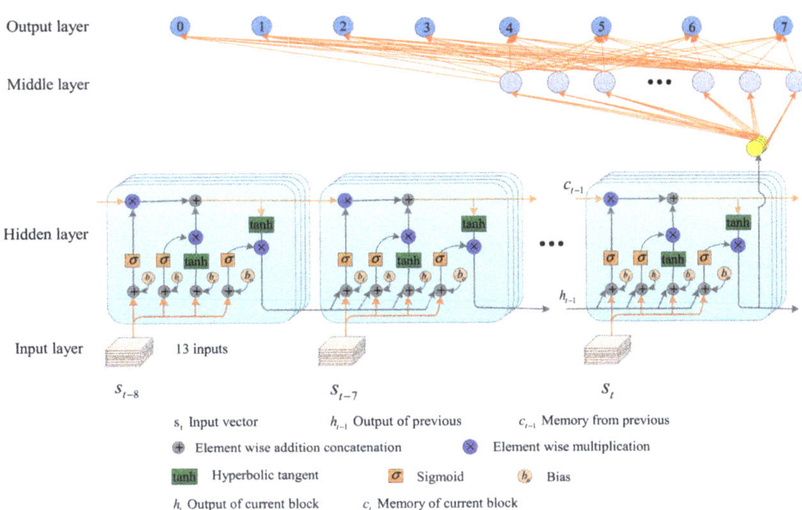

Figure 5. Using LSTM to implement the Q-network in Figure 4.

The pseudo-code of the algorithm is as follows:

Algorithm DQN-LSTM algorithm with experience replay for the AUV

Input: the linear velocities $\omega = [u, v, r]$, the position vector $\eta = [x, y, \theta]$, and the sonar information D.

    Initialize the environment E.
    Initialize replay memory M to capacity N.
    Initialize the Q-networks and the target $\hat{Q}$-networks with random weights.
    for episode = 1, K **do**
        Initialize sequence D_1 and preprocess sequencing $S_1 = \psi(D_1)$.
        for $t = 1$, T **do**
            With probability ε select a random action h_t
            otherwise select $h_t = \max_h Q^*(\psi(D_t), h; w)$.
        Execute action h_t in emulator: put ω, η, h_t into the AUV model.
        update the environment E and get the state D_{t+1}, r_{t+1}.
        Store transition $(\psi(D_t, h_t, r_t, D_{t+1}))$ in M.
        Store transition $(\psi(D_t, h_t, r_t, D_{t+1}))$ in M.
        Sample random minibatch 64 of transitions $(\psi(D_j, h_j, r_j, D_{j+1}))$ from M.
        Set $y_j = \begin{cases} r_j & for \text{ terminal state} \\ r_j + \gamma\max_{h'} Q(\psi(D_{t+1}, h'; w)) & for \text{ non} - \text{ternimal state} \end{cases}$
        Perform a gradient descent step on $(y_j - Q(\psi(D_j), h_j; w))^2$ according to Equation (13).
        Reset the target $\hat{Q}$-Networks $\hat{Q} = Q$ every 50 steps.
        end for
    end for

---

4. Simulation Experiment and Discussion

Extensive simulation studies were conducted to demonstrate the effectiveness of the newly proposed method. We used PyGame to draw the interface, AUV, and obstacles, and other entity attributes were drawn using the Pymunk library, while the network structure of the algorithm was the calling Keras library, and the programming language was Python. The results produced by the different scenarios were used to evaluate the performance of the methods under different conditions. We evaluated various network structures to determine the optimal collision avoidance planning algorithm.

4.1. Experimental Detail

The DQN algorithms were trained on a computer with an Intel i5 processor using four CPU cores and a GTX 960 M. The parameters of the Q-network, using a fully connected network and LSTM, were basically the same during the training process. We used the Adagrad optimizer with a learning rate of $\alpha = 10^{-4}$ to learn the network parameters. The hidden layer activation functions of the network were the leaky ReLU, tanh, and sigmoid functions. The activation function of the output layer used a tanh activation function, which defines the range of the action-value function. To improve the robustness and stability of the network model, the following tricks were used in the training process: (1) Gaussian noise was added to the input of the network, and (2) dropout (ratio of 0.2) was applied to each hidden layer. We used different numbers of hidden layer units $[\,[30,40], [164,150], [256,256], [512,512]\,]$, minibatch sizes of 64, and experience buffer sizes of 50,000 to verify the planning capability of the models.

The parameters used in the training phase of DRL: gamma = 0.9, number of frames to observe before training (observe = 1000), epsilon = 0.95, training_frames = 500,000, lstm_network_weights = $[-0.5, 0.5]$, nn_network_weights = $[-1, 1]$.

We can see from Table 2 that the DQN model has the shortest training time required for different models. As the number of neurons in the hidden layer increases, the time required to train each model also increases. The $[256,256]$ model gives the slowest convergence of the loss function converges, whereas the $[30,40]$ model gives the fastest convergence. The disadvantage of this early convergence is that the algorithm cannot effectively search the entire environment, and some optimal strategies may not be identified. The $[164,150]$ model is the smallest of all structures to achieve reasonable convergence accuracy. The converged value of the loss function convergence value using the LSTM network is much smaller than that given by NN, regardless of whether the DQN algorithm or the D-DQN algorithm is used. As D-DQN has one more target network than DQN, the weights must be updated at certain intervals, so a longer training time is required. The same phenomenon occurs for the DQN-LSTM and D-DQN-LSTM models. Considering these various factors, a network structure with 164 units in the hidden layer and 150 in the middle layer is considered optimal.

Table 2. Comparison of the results of different algorithm training.

	Number of Neurons in the Hidden Layer	Model Training Time(h)	Convergence Step	Convergence Accuracy
DQN	[30,40]	2.4	3797	0.089~0.092
	[164,150]	2.9	6492	0.085~0.091
	[256,256]	3.7	8045	0.093~0.096
	[512,512]	4.9	6263	0.092~0.097
DQN-LSTM	[30,40]	6.8	3095	$(7.15{\sim}7.75) \times 10^{-5}$
	[164,150]	9.6	5509	$(1.05{\sim}1.13) \times 10^{-5}$
	[256,256]	10.5	5908	$(3.43{\sim}3.57) \times 10^{-5}$
	[512,512]	11.2	4103	$(1.10{\sim}1.16) \times 10^{-5}$
D-DQN	[30,40]	2.6	3548	0.087~0.093
	[164,150]	3.3	4126	0.090~0.095
	[256,256]	4.2	6237	0.076~0.080
	[512,512]	5.1	3877	0.073~0.076
D-DQN-LSTM	[30,40]	9.1	3124	$(4.48{\sim}4.78) \times 10^{-5}$
	[164,150]	11.3	3504	$(3.15{\sim}3.37) \times 10^{-5}$
	[256,256]	12.4	5307	$(1.49{\sim}1.62) \times 10^{-5}$
	[512,512]	13.7	3440	$(5.88{\sim}6.11) \times 10^{-5}$

4.2. Testing the Performance of the Algorithm

To evaluate the collision avoidance planning capabilities of the various above algorithms, we compared them in different test environments, including static obstacles, mixed obstacles, and complex dynamic obstacles. The obstacles, start point, and endpoint positions were randomly generated in the test environment, and the GA algorithm results were used as the benchmark for comparison. During training, the environment contained only circular obstacles with a radius of 25. To test the performance of the algorithms, a variety of obstacles were added to the test environment, which did not appear in the training environment. In Test 1, the advantages of adding policy 7 were verified, and the stability of the 5 methods was also tested. In Test 2, the collision avoidance planning performance of the 5 algorithms was compared in a static environment. In Test 3, dynamic obstacles and a deep learning algorithm were added, and the collision avoidance planning performance of the 6 algorithms was compared. In the trajectory figure of the test, the hollow circle represents the safety distance of the AUV, the solid blue circle, the blue square, and the black rectangle represent the static obstacles, and the dark sea-green solid circles represent the dynamic obstacles.

4.2.1. Test 1

To verify the advantages of adding policy 7, the network model was retrained for the DQN-LSTM and D-DQN-LSTM algorithms without policy 7. Figure 6 shows the planned trajectories of the 4 algorithms in the test environment. Figure 6 shows that DQN-LSTM-None 7 cannot find the goal point. When D-DQN-LSTM-None 7 first arrived near the goal point, it did not move toward the goal point and then went through another round to arrive at the goal point. In contrast, the 2 algorithms of policy 7 move toward the target point from the beginning, and under the guidance of policy 7, the two algorithms reach the goal faster. Compared with the no policy algorithms, the policy 7 algorithms have the guidance of the goal point. When there are no obstacles around the AUV, the policy of moving to the target point is better than blind search, which can save considerable time in model training. In addition, the lack of a goal point to obtain the maximum reward also affects the accuracy of the model.

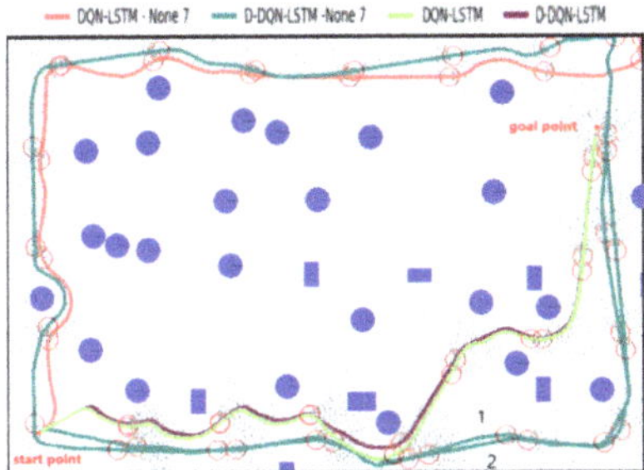

Figure 6. Comparison of DQN-LSTM and D-DQN-LSTM with or None policy 7.

Table 3 shows the results of the policy 7 algorithms and the no policy 7 algorithms. The DQN-LSTM and D-DQN-LSTM algorithms have a much higher probability of finding the goal point in the test environment than DQN-LSTM-None 7 and D-DQN-LSTM-None 7. For the former, the number of convergence steps is almost half of the latter, the training time is shorter, and the success rate of reaching the goal point is higher in the test environment. The advantage of adding policy 7 is related to shortening the training time and increasing the probability of reaching the target point.

Table 3. Comparison of algorithms performance with and without policy 7.

	Convergence Step	Training Time (h)	Success Rate (%)
DQN-LSTM-None 7	9352	13.6	37
D-DQN-LSTM-None 7	6870	18.4	46
DQN-LSTM	5509	9.6	91
D-DQN-LSTM	3504	11.3	94

Figure 7 shows that the success rates of the 4 DRL algorithms achieve higher success than the GA. The success rate of the D-DQN-LSTM algorithm has the highest success rate. DQN requires the shortest time to perform an action, whereas D-DQN-LSTM takes the longest time. The GA model is greatly affected by the complexity of the environment, which causes its running time to fluctuate more than that of other algorithms. Because the LSTM network is more complex than the NN, the time required for the DQN-LSTM and D-DQN-LSTM algorithms to perform a single action is longer than the DQN and D-DQN algorithms, and the GA algorithm takes the longest time. The environment changes randomly across 200 episodes. DRL algorithms improve the success rate of AUV reaching the target point, and the execution time of a single action is shorter than the GA algorithm.

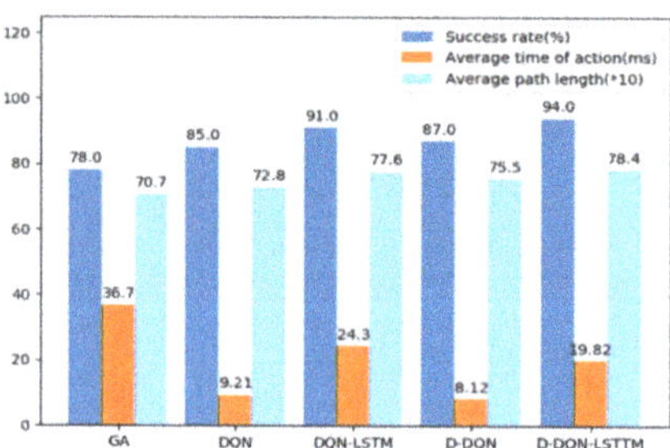

Figure 7. Comparison of the performances (success rate, single-action execution time, average path length) of different algorithms in 200 random environments.

To test the stability of the algorithms in the constant environment, each algorithm was run 100 times to obtain the trajectory, as shown in Figure 8. The path planned by the GA algorithm and the DQN algorithm is tortuous. Figures 8 and 9 show that compared to the other 3 algorithms, D-DQN and D-DQN-LSTM are more stable, and the distance of D-DQN-LSTM is shorter in the 2 optimal algorithms. The path planned by the D-DQN algorithm is very stable, but its path is not optimal. From the above 2 figures, we can conclude that the trajectory planned by the D-DQN-LSTM algorithm is optimal. The black line represents the straight-line distance from the starting point to the goal point.

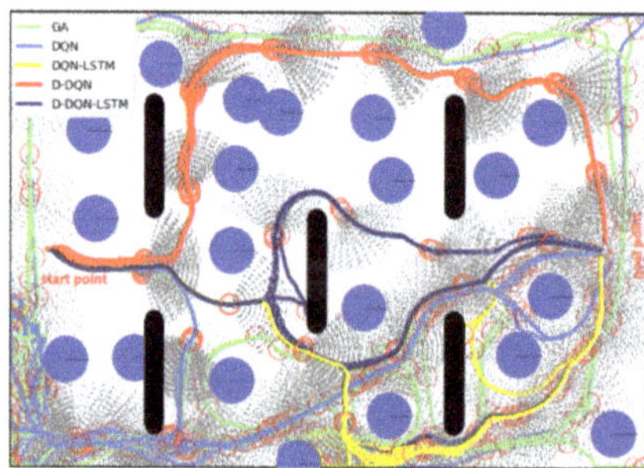

Figure 8. Trajectory graph formed by running all algorithms 100 times in the same environment.

4.2.2. Test 2

The performances of all 5 methods were tested in a static environment, as shown in Figure 10. GA's trajectory is covered by DQN. In this test, the speed of the ocean current is 0.25 m/s and the direction is $\frac{\pi}{4}$. Although the path planned by the D-DQN-LSTM algorithm is longer than that of the GA, the path planned by the D-DQN-LSTM algorithm is safe and maintains a certain distance from the obstacles, whereas the path planned by the GA is mostly nearly close to the obstacle boundaries, and there is a danger of collision.

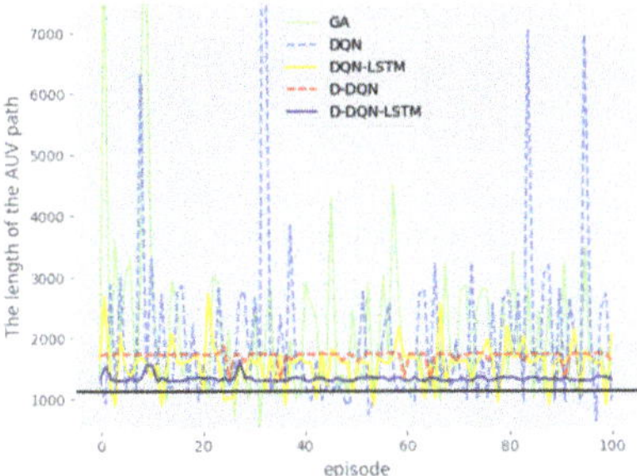

Figure 9. Comparison of the path length of 5 algorithms running 100 times in the same environment.

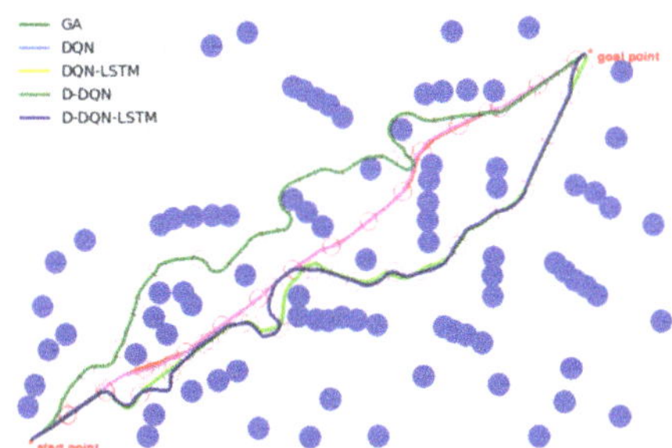

Figure 10. Path planning capabilities of five algorithms in a static environment.

Table 4 shows that the path lengths given by the GA model are optimal, and the DQN algorithm is the best of the DRL algorithms. The running times of DQN and D-DQN are faster than those of DQN-LSTM and D-DQN-LSTM. When faced with the same observation value, DQN-LSTM and D-DQN-LSTM output different actions (heading adjustment). Although the path produced by DQN-LSTM basically coincides with that of D-DQN-LSTM, the latter selects a better policy at the critical positions, and the planned path by D-DQN-LSTM gives a more secure path. The D-DQN-LSTM model has the largest number of actions and the longest path, but we can see from Figure 10 that this model produces the safest planned path by D-DQN-LSTM. During the entire movement, the distance between the AUV and all obstacles was more than or equal to the safe distance of 20 m, and no collision occurred. Recall that the security of the AUV is critical in real-world environments.

Table 4. Performance comparison of five algorithms in a static environment.

	Number of Actions	Running Time(s)	Path Length(m)
GA	–	35.81	1160
DQN	724	12.83	1162
DQN-LSTM	846	17.46	1360
D-DQN	830	13.48	1351
D-DQN-LSTM	892	18.37	1398

Figure 11 shows the policy distribution of the 4 DRL algorithms. The DQN policy is very simple, focusing on actions 3 and 7, which cause the AUV to continue along the current direction and move toward the target point, respectively. This affects the flexibility and robustness of this algorithm. When encountering a complex environment, the path planning ability will be greatly limited, and the target point will often be missed. From Figure 11, the D-DQN algorithm learns policies 1, 2, 4, 5, and 6 better than the DQN algorithm, ensuring the algorithm's production of a more uniform and reasonable policy distribution. The same result appears in the D-DQN-LSTM and DQN-LSTM algorithms, with the former policy being more uniform and robust than the latter. This can be verified from Figure 11, where the policy distributions of DQN-LSTM and D-DQN-LSTM are similar, and the generated trajectories are similar to those in Figure 10. The diversification of their policies has the advantage that it demonstrates that these algorithms offer enhanced adaptability and robustness in a given environment.

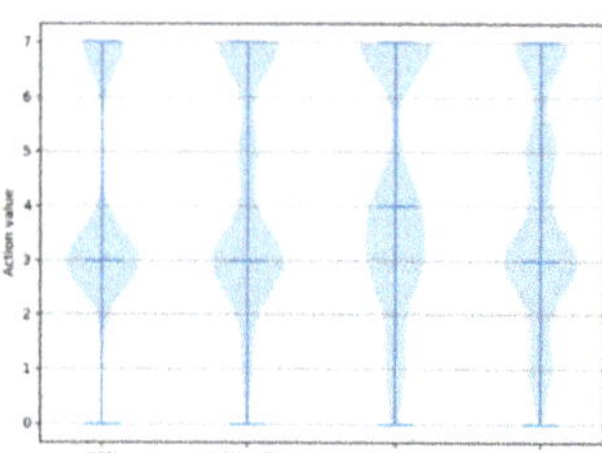

Figure 11. Policy distribution of four DRL algorithms (DQN, DQN-LSTM, D-DQN, and D-DQN-LSTM) in a static environment.

Figure 12 shows that the D-DQN and D-DQN-LSTM algorithms have a larger range of variation, which shows that these algorithms have a stronger ability to explore the environment. This is advantageous in finding the right path in an unknown and complex environment. Figure 13 shows the acceleration curve of the heading angle. It can be seen from the figure that the GA algorithm reaches the maximum acceleration value of $\pm 10^{o}$, while the D-DQN-LSTM is basically stable at $\pm 5^{o}$.

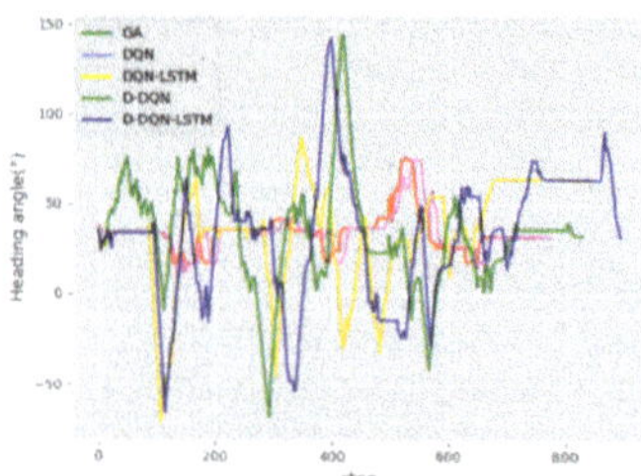

Figure 12. Heading angle curves of five algorithms in a static environment.

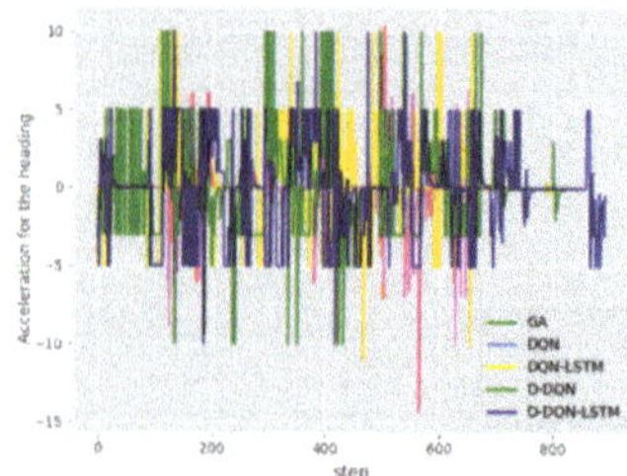

Figure 13. The curve of the acceleration for the heading of the five algorithms in a static environment.

Figures 14 and 15 show that the thrust and moment required by the 5 algorithms to complete the planning under the disturbance of ocean currents in a static environment are between 31,000 and 32,000, and the range of 31,000–32,000 corresponds to 310–320 in Figure 13. In the dynamic model simulation, there are 100 data between two time steps, which causes the horizontal axis to be enlarged by 100 times. The entire presentation will appear chaotic, so only a part of it is displayed. In Figure 14, the fluctuation range of D-DQN-LSTM is smaller than that of GA, D-DQN, D-DQN, and D-DQN-LSTM. The curve is smooth, without jagged and sharp edges. It can be seen from the two sets of partially enlarged pictures that the changing trend of the curve is relatively gentle. In Figure 15, the spikes of DQN and D-DQN-LSTM are also the smallest.

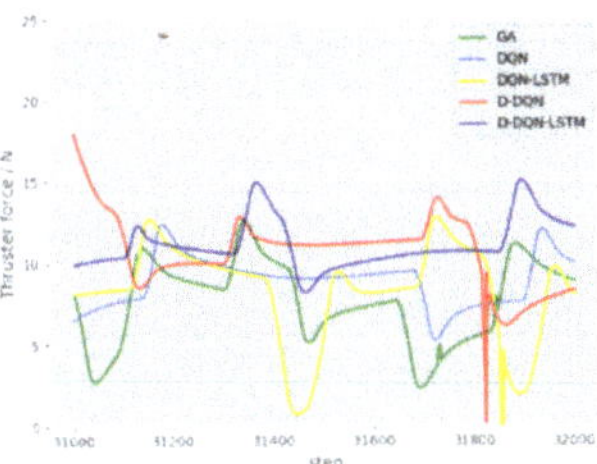

Figure 14. Thrust curve between 31,000 and 32,000 in a static environment.

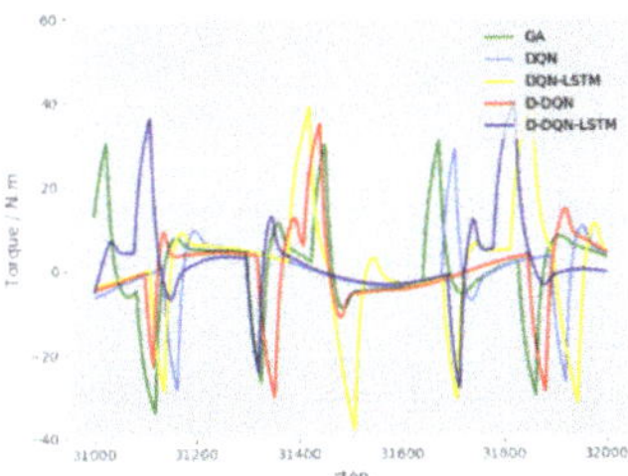

Figure 15. Torque curve between 31,000 and 32,000 in a static environment.

4.2.3. Test 3

Figure 16 illustrates the performance of the algorithms in a complex environment with moving obstacles. The rectangular box in the picture shows the range of the moving obstacles. The AUV trajectory is erased by moving obstacles in the rectangular area. Obstacle 1 moves toward 135^0 at a speed of 10, while obstacle 2 moves horizontally to the left at a speed of 20. In this test, the speed of the ocean current is 0.5 m/s and the direction is $-\frac{\pi}{6}$.

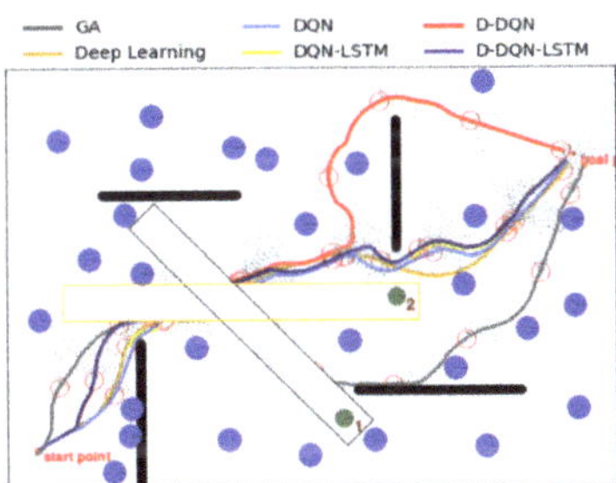

Figure 16. Planning capabilities of six algorithms in a complicated environment.

Table 5 shows the performance indicators of each algorithm. The GA has the longest running time, whereas DQN has the shortest; the longest total path length is produced by the GA algorithm, and while the D-DQN-LSTM algorithm gives the shortest. In terms of the total running time, the DQN-LSTM algorithm is longer than the DQN algorithm, and the same D-DQN-LSTM is also longer than the D-DQN algorithm. This is because the LSTM network is more time-consuming than the NN. It can also be seen from Figure 17 that the curve represents the time required to execute the single-step planning, and where the horizontal line represents the average value. The D-DQN has the shortest average execution time of a single action, while GA has the longest.

Table 5. Performance comparison of six algorithms in a complicated environment.

	Number of Actions	Running Time(s)	Path Length(m)
GA	-	45.28	1678
Deep Learning	-	18.12	1388
DQN	682	10.63	1364
DQN-LSTM	684	17.12	1368
D-DQN	806	11.62	1606
D-DQN-LSTM	695	15.78	1352

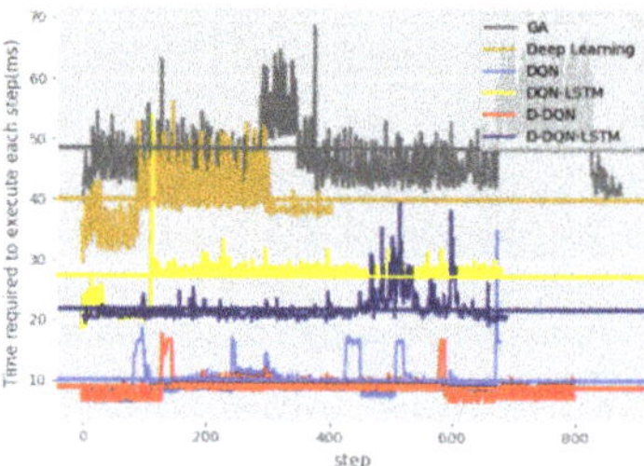

Figure 17. Comparison of the action execution times of the six algorithms in a complicated environment.

Figures 18 and 19 show the curves of the distance between the AUV and the moving obstacles over time. The horizontal red line in the picture represents a collision between the AUV and the moving obstacle. It can be seen from Figure 18 that neither D-DQN nor D-DQN-LSTM produced a collision with moving obstacle 1. Figure 19 shows that GA, D-DQN, and D-DQN-LSTM did not produce a collision with moving obstacle 2.

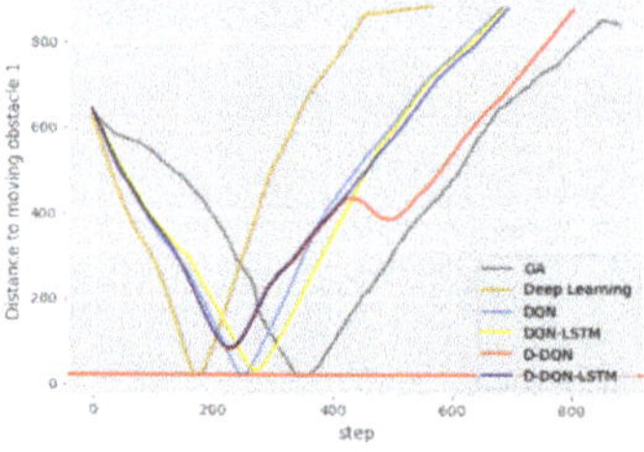

Figure 18. Change in the distance between AUV and moving obstacle 1 in a complicated environment.

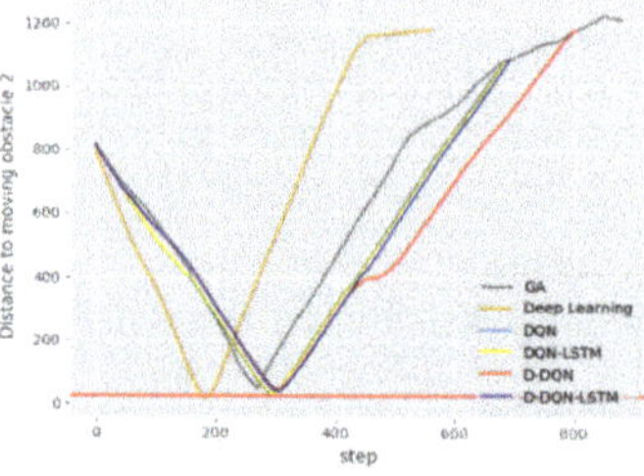

Figure 19. Change in the distance between AUV and moving obstacle 2 in a complicated environment.

Figure 20 shows the changes in the AUV heading angle. The heading angle change of the GA algorithm is the worst. It can be seen from the figure that the curve of the deep-learning algorithm gives a smoother curve and a smaller fluctuation range. However, this also results in the AUV not having sufficient flexibility to safely avoid dynamic obstacles, especially faster obstacles. D-DQN and D-DQN-LSTM produce large changes in the heading, which makes the AUV more flexible in the environment and provides it with the ability to avoid moving obstacles.

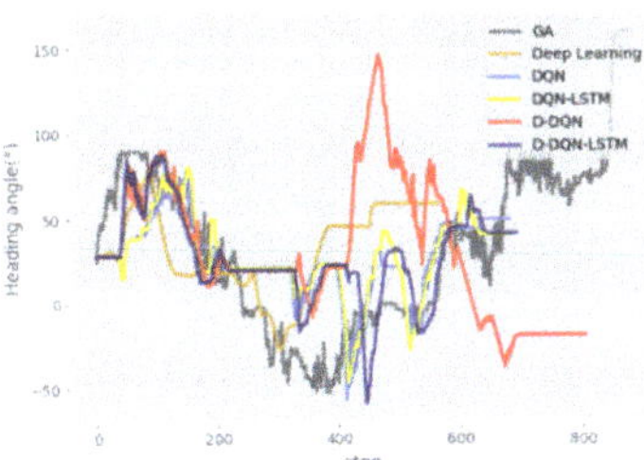

Figure 20. Heading angle curves of six algorithms in a complicated environment.

Figure 21 shows the adjustment angle to the heading at each step. The adjustment angle is more frequent in the case of the GA algorithm, and the angle adjusted change in each step is greater than $\pm 20^{0}$. This presents a significant challenge to AUV propulsion devices. In terms of heading adjustment, the performance of the four DRL algorithms is worse than that of the DL algorithm. The adjustment angle of the deep learning algorithm produces the smallest angle adjustments, but Figures 16, 18 and 19 show that this model cannot avoid static and dynamic obstacles. D-DQN-LSTM gives the smallest adjustment angles among the four methods proposed in this article, and each adjustment is less than $\pm 10^{0}$. This adjustment range is easy to achieve for AUVs.

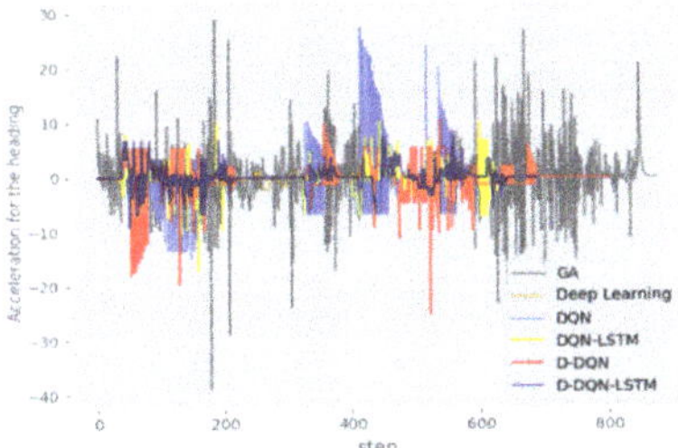

Figure 21. The curve of the acceleration for the heading of the six algorithms in a complicated environment.

Figure 22 shows the thrust curve of the thruster between 47,000 and 48,000, corresponding to steps 470–480 in Figure 21. Figure 23 shows the corresponding AUV torque curve. Other algorithms require less thrust than GA, and the curve should be smoother. Figure 22 show that the fluctuation amplitude of the force curve of D-DQN-LSTM is smaller than that of other algorithms, which also confirms the curve of heading acceleration in Figure 21.

The negative value of the thrust in Figure 22 is due to the AUV offsetting ocean currents. The function makes the AUV decelerate and makes it easier to turn. The heading of steps 470–480 in Figure 20 is basically maintained at about 60^{0}, but in Figures 22 and 23, there are fluctuations in thrust and torque, which is due to the response of the AUV to overcome the influence of ocean currents.

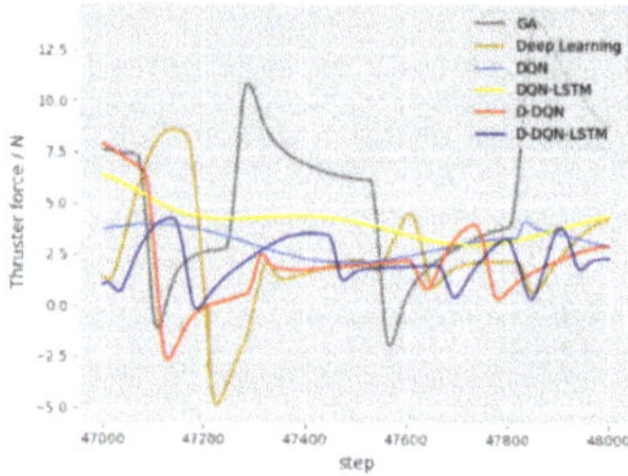

Figure 22. Thrust curve between 47,000 and 48,000 in a complicated environment.

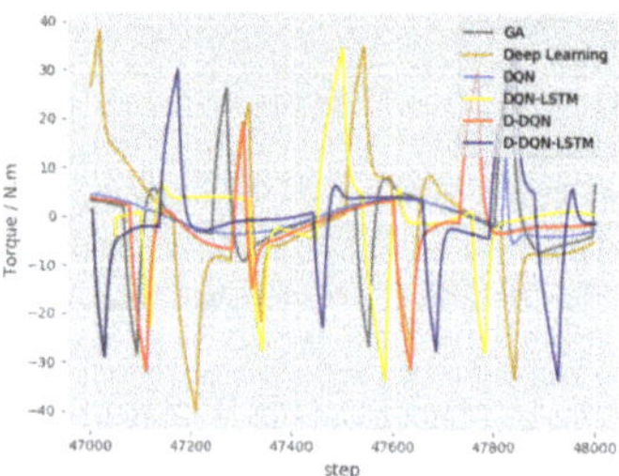

Figure 23. Torque curve between 47,000 and 48,000 in a complicated environment.

4.3. Discussion

To develop collision avoidance algorithms based on deep learning, the first step is to use GA, PSO, random-sample, or other algorithms to collect large numbers of samples, which will greatly reduce the efficiency of algorithm development. In addition, the sample quality and network training performance have a vital impact on the algorithm's collision avoidance ability. The algorithm proposed in this paper is trained in a simple random environment without the above-mentioned complicated process. After the training is completed, the AUV is placed in a complex dynamic environment and can still effectively avoid collisions. The effectiveness of the algorithm proposed in this paper was verified in two environments: random static and complex dynamic. Simulation experiments have shown that using the approach to the target point as a policy can greatly increase the probability of reaching the target point while also reducing the time required for the algorithm to blindly search the entire environment and improving the efficiency of algorithm learning. The D-DQN-LSTM algorithm is superior in stability and robustness compared to GA and other DRL algorithms. The results show that the D-DQN-LSTM algorithm has significant advantages over other algorithms in terms of success rate, collision avoidance performance, and generalization ability. Although the deep learning algorithm produces a smoother path [35], its collision avoidance ability is obviously insufficient when it encounters moving obstacles. The D-DQN-LSTM algorithm proposed in this paper is superior to GA and deep learning in terms of the running time, total path, collision avoidance performance against moving obstacles, and planning time for each step. The forces and torques required by the planning algorithms are also feasible for the actuator. According to the above analysis, we can conclude that the proposed D-DQN-LSTM method can achieve dynamic collision avoidance planning of AUVs in an unknown environment.

5. Conclusions

This paper has proposed a DRL obstacle avoidance algorithm for the AUVs in an environment with unknown dynamics. This paper has mainly studied a DRL method that realizes AUV reactive collision avoidance behavior by learning a reward function to determine the mapping between perception information and actions [36]. Several DRL-based methods were proposed to realize collision avoidance planning in an AUV. Although the algorithm proposed in this paper achieved good results, there are still many problems that have not been solved. For example, the sample utilization rate is low, the reward function is too simple [37], and it is difficult to balance the exploration and exploitation [38] so that the algorithm does not become trapped around a local minimum and instability. In addition, the heading adjustment is too frequent, and the adjustment angle is larger than the deep learning algorithm. To solve the above problems in the future, we will attempt to improve the deep deterministic strategy gradient (DDPG) [39] algorithm. The value-based approach we consider using parallel actors-critics to update the shared model would not only stabilize the learning process but also make the algorithm become on-policy, without the experience replay mechanism, while also improving the sample utilization rate, the relationship between exploration and exploitation. Discrete actions may not be sufficient to include all optimal strategies, and small changes in strategies will significantly affect the results. This motivates us to perform research on DRL algorithms with continuous state and action spaces. Continuous variants of Q-learning, DDPG, and normalized advantage function (NAF) will be introduced into the planning system of the AUV.

Author Contributions: Conceptualization, J.Y. and H.W.; methodology, J.Y. and H.W.; software, J.Y. and C.L. (Changjian Lin); investigation, D.Y. and C.L. (Chengfeng Li); data curation, D.Y. and C.L. (Chengfeng Li); writing—original draft preparation, J.Y. and C.L. (Changjian Lin); writing—review and editing, J.Y., C.L. (Changjian Lin) and H.W.; project administration, H.W., H.Z.; All authors have read and agreed to the published version of the manuscript.

Funding: This research work is supported by the National Natural Science Foundation of China (No. 61633008, No. 51609046) and the Natural Science Foundation of Heilongjiang Province under Grant F2015035.

Institutional Review Board Statement: Not applicable.

Informed Consent Statement: Not applicable.

Data Availability Statement: The data presented in this study are available on request from the corresponding author.

Acknowledgments: The authors would like to thank the anonymous reviewers and the handling editors for their constructive comments that greatly improved this article from its original form.

Conflicts of Interest: The authors declare no conflict of interest.

Nomenclature

XOY	global coordinate system
$X_R O_R Y_R$	local coordinate system
θ	angle between the X-axis of {R}and{G}
d_i	minimum distance
ϕ_i	angle corresponding to the minimum distance
r	distance from the sonar to the center of mass of the AUV
θ_g	angle between the AUV and the target point
θ_d	output as a policy
$\eta = [x, y, \theta]^T$	AUV's position and heading
$v = [u, v, r]^T$	AUV's velocity
$\mathbf{R}(\theta)$	the rotation matrix from {G} to {R}
$\tau = [\tau_u\ 0\ \tau_r]^T$	control input vector denotes the propulsion surge force and the yaw moment
$\hat{\tau}$	environmental disturbance
$\mathbf{M}$	AUV inertia matrix
$\mathbf{C}(v)$	centrifugal and Coriolis matrices
$[d_u, d_v, d_r]$	nonlinear damping coefficient
$[m_u, m_v, m_r, m_{uv}]$	coefficients in the inertia matrix $\mathbf{M}$
$[Y_r, Y_v, N_r, Y_{\lvert v\rvert v}, N_{\lvert r\rvert r}]$	viscous damping coefficient
$[X_{\dot{u}}, Y_{\dot{v}}, N_{\dot{r}}]$	added mass term

I_z	moment of inertia about the OZ axis. (OZ axis is perpendicular to XOY.)
$p(x_t, y_t)$	current position of the AUV
$p_{ac}(x_{ac}, y_{ac})$	center position of the transition arc
Δ	forward-looking vector
$\theta_d(e)$	desired error angle
R	transition arc radius
e	the cross-track error of the current AUV
D_t	observation of environment E
S_t	network input
h	network output
t	step
r	numerical reward
p	probability distribution
$Q^*(D, h)$	optimal action-value function
$L_t(w_t)$	loss function
$\rho\xi$	behavior distributions
w_t	weights
y_t	target value for iteration t
$\hat{Q}(D_t, h_t; w_t)$	estimated action value of NNs network output
$\nabla_{w_t} L_t(w_t)$	gradient of loss function to weight
α	learning rate

References

1. Persson, S.M.; Sharf, I. Sampling-based A* algorithm for robot path- planning. *Int. J. Robot. Res.* **2014**, *33*, 1683–1708. [CrossRef]
2. Kothari, M.; Postlethwaite, I. A probabilistically robust path planning algorithm for UAVs using rapidly-exploring random trees. *J. Intell. Robot. Syst.* **2013**, *2*, 231–253. [CrossRef]
3. Kuffner, J.J.; LaValle, S.M. RRT-Connect: An efficient approach to single-query path planning. In Proceedings of the 2000 ICRA Millennium Conference IEEE International Conference on Robotics and Automation Symposia Proceedings (Cat. No. 00CH37065), San Francisco, CA, USA, 24–28 April 2000; Volume 2, pp. 995–1001.
4. Yen, J.; Liao, J.C.; Lee, B.; Randolph, D. A hybrid approach to modeling metabolic systems using a genetic algorithm and simplex method. *IEEE Trans. Syst. Man Cybern.* **2012**, *28*, 173–191. [CrossRef] [PubMed]
5. Lee, J.W.; Kim, J.J.; Lee, J.J. Improved ant colony optimization algorithm by path crossover for optimal path planning. In Proceedings of the 2009 IEEE International Symposium on Industrial Electronics, Seoul, Korea, 5–8 July 2009.
6. Roberge, V.; Tarbouchi, M.; Labonté, G. Comparison of parallel genetic algorithm and particle swarm optimization for real-time UAV path planning. *IEEE Trans. Ind. Inform.* **2013**, *9*, 132–141. [CrossRef]
7. Rusu, P.; Petriu, E.M.; Whalen, T.E.; Cornell, A.; Spoelder, H.J.W. Behavior-based neuro-fuzzy controller for mobile robot navigation. *IEEE Trans. Instrum. Meas.* **2003**, *52*, 1335–1340. [CrossRef]
8. Merhy, B.A.; Payeur, P.; Petriu, E.M. Application of Segmented 2-D Probabilistic Occupancy Maps for Robot Sensing and Navigation. *IEEE Trans. Instrum. Meas.* **2008**, *57*, 2827–2837. [CrossRef]
9. Garrido, S.; Moreno, L.; Blanco, D. Exploration and mapping using the VFM motion planner. *IEEE Trans. Instrum. Meas.* **2009**, *58*, 2880–2892. [CrossRef]
10. Beom, H.R.; Cho, H.S. A sensor-based navigation for a mobile robot using fuzzy logic and reinforcement learning. *IEEE Trans. Syst. Man Cybern.* **1995**, *25*, 464–477. [CrossRef]
11. Ye, C.; Yung, N.H.C.; Wang, D. A fuzzy controller with supervised learning assisted reinforcement learning algorithm for obstacle avoidance. *IEEE Trans. Syst. Man Cybern.* **2003**, *33*, 17–27.
12. Er, M.J.; Deng, C. Obstacle avoidance of a mobile robot using hybrid learning approach. *IEEE Trans. Ind. Electroics* **2005**, *52*, 898–905. [CrossRef]
13. Fathinezhad, F.; Derhami, V.; Rezaeian, M. Supervisedfuzzy reinforcement learning for robot navigation. *Appl. Soft Comput.* **2016**, *40*, 33–41. [CrossRef]
14. Huang, B.Q.; Cao, G.Y.; Guo, M. Reinforcement learning neural network to the problem of autonomous mobile robot obstacle avoidance. In Proceedings of the 2005 International Conference on Machine Learning and Cybernetics, Guangzhou, China, 18–21 August 2005.
15. Duguleana, M.; Mogan, G. Neural networks based reinforcement learning for mobile robots obstacle avoidance. *Expert Syst. Appl.* **2016**, *62*, 104–115. [CrossRef]
16. Duguleana, M.; Barbuceanu, F.G.; Teirelbar, A. Obstacle avoidance of redundant manipulators using neural networks based reinforcement learning. *Robot. Comput.-Integr. Manuf.* **2012**, *28*, 132–146. [CrossRef]
17. Mnih, V.; Kavukcuoglu, K.; Silver, D.; Rusu, A.A.; Veness, J.; Bellemare, M.G.; Graves, A.; Riedmiller, M.; Fidjeland, A.K.; Ostrovski, G.; et al. Human-level control through deep reinforcement learning. *Nature* **2015**, *518*, 529–533. [CrossRef]

18. Mirowski, P.; Pascanu, R.; Viola, F. Learning to navigate in complex environments. *arXiv* **2016**, arXiv:1611.03673. Available online: https://arxiv.org/abs/1611.03673 (accessed on 11 November 2016).
19. Bohez, S.; Verbelen, T.; Coninck, E.D. Sensor Fusion for Robot Control through Deep Reinforcement Learning. In Proceedings of the International Conference on Intelligent Robots and Systems (IROS), Vancouver, BC, Canada, 24–28 September 2017.
20. Cheng, Y.; Zhang, W. Concise deep reinforcement learning obstacle avoidance for underactuated unmanned marine vessels. *Neurocomputing* **2018**, *272*, 63–73. [CrossRef]
21. Long, P.; Fanl, T.; Liao, X. Towards optimally decentralized multi-robot collision avoidance via deep reinforcement learning. In Proceedings of the 2018 IEEE International Conference on Robotics and Automation (ICRA), Brisbane, Australia, 21–25 May 2018.
22. Tan, Z.; Karakose, M. Comparative study for deep reinforcement learning with CNN, RNN, and LSTM in autonomous navigation. In Proceedings of the 2020 International Conference on Data Analytics for Business and Industry: Way Towards a Sustainable Economy (ICDABI), Sakheer, Bahrain, 26–27 October 2020.
23. Guo, T.; Jiang, N.; Li, B.Y. UAV navigation in high dynamic environments: A deep reinforcement learning approach. *Chin. J. Aeronaut.* **2021**, *34*, 479–489. [CrossRef]
24. Brejl, R.K.; Purwins, H.; Schoenau-Fog, H. Exploring deep recurrent Q-Learning for navigation in a 3D environment. *EAI Endorsed Trans. Creat. Technol.* **2018**, *5*. [CrossRef]
25. Zhu, Y.; Mottaghi, R.; Kolve, E. Target-driven visual navigation in indoor scenes using deep reinforcement learning. In Proceedings of the 2017 IEEE International Conference on Robotics and Automation (ICRA), Singapore, 29 May–3 June 2017.
26. Michels, J.; Saxena, A.; Ng, A.Y. High speed obstacle avoidance using monocular vision and reinforcement learning. In Proceedings of the the the 22nd International Conference on Machine Learning, Bonn, Germany, 7 August 2005.
27. Tai, L.; Paolo, G.; Liu, M. Virtual-to-real deep reinforcement learning: Continuous control of mobile robots for mapless navigation. In Proceedings of the 2017 IEEE/RSJ International Conference on Intelligent Robots and Systems (IROS), Vancouver, BC, Canada, 24–28 September 2017.
28. Lin, C.J.; Wang, H.J.; Yuan, J.Y. An improved recurrent neural network for unmanned underwater vehicle online obstacle avoidance. *Ocean. Eng.* **2019**, *189*, 106327. [CrossRef]
29. Van Hasselt, H.; Guez, A.; Silver, D. Deep Reinforcement Learning with Double Q-learning. In Proceedings of the AAAI Conference on Artificial Intelligence, Phoenix, AZ, USA, 12–17 February 2016; Volume 30. No. 1.
30. Breivik, M.; Fossen, T.I. Path following of straight lines and circles for marine surface vessels. *IFAC Proc. Vol.* **2004**, *37*, 65–70. [CrossRef]
31. Watkins, C.J.C.H.; Dayan, P. Q-learning. *Mach. Learn.* **1992**, *8*, 279–292. [CrossRef]
32. Lin, Y.; Wang, M.Y. Dynamic Spectrum Interaction of UAV Flight Formation Communication with Priority: A Deep Reinforcement Learning Approach. *IEEE Trans. Cogn. Commun. Netw.* **2020**, *6*, 892–903. [CrossRef]
33. Chen, Y.F.; Liu, M.; Everett, M. Decentralized non-communicating multiagent collision avoidance with deep reinforcement learning. In Proceedings of the 2017 IEEE International Conference on Robotics and Automation (ICRA), Singapore, 29 May–3 June 2017.
34. Wang, Z.; Schaul, T.; Hessel, M. Dueling network architectures for deep reinforcement learning. In Proceedings of the International Conference on Machine Learning, New York, NY, USA, 19–24 June 2016.
35. Wang, Z.; Yu, C.; Li, M.; Yao, B.; Lian, L. Vertical profile diving and floating motion control of the underwater glider based on fuzzy adaptive LADRC algorithm. *J. Mar. Sci. Eng.* **2021**, *9*, 698. [CrossRef]
36. Mnih, V.; Badia, A.P.; Mirza, M. Asynchronous methods for deep reinforcement learning. In Proceedings of the International Conference on Machine Learning, New York, NY, USA, 19–24 June 2016.
37. Islam, R.; Henderson, P.; Gomrokchi, M. Reproducibility of benchmarked deep reinforcement learning tasks for continuous control. In Proceedings of the International Conference on Machine Learning, Sydney, Australia, 7–11 August 2017.
38. Henderson, P.; Islam, R.; Bachman, P. Deep reinforcement learning that matters. In Proceedings of the AAAI Conference on Artificial Intelligence, New Orleans, LA, USA, 2–7 February 2018.
39. Silver, D.; Lever, G.; Heess, N. Deterministic policy gradient algorithms. In Proceedings of the International Conference on Machine Learning, Beijing, China, 21–26 June 2014.

Journal of
*Marine Science
and Engineering*

Article

Development of a Computational Model for Investigation of and Oscillating Water Column Device with a Savonius Turbine

Amanda Lopes dos Santos [1], Cristiano Fragassa [2,*], Andrei Luís Garcia Santos [3], Rodrigo Spotorno Vieira [4], Luiz Alberto Oliveira Rocha [4], José Manuel Paixão Conde [5], Liércio André Isoldi [1,2] and Elizaldo Domingues dos Santos [1,2]

[1] Ocean Engineering, School of Engineering, Federal University of Rio Grande (FURG), Italia Av., km 8, Rio Grande 96203-900, Brazil; amandalopesq@gmail.com (A.L.d.S.); liercioisoldi@furg.br (L.A.I.); elizaldosantos@furg.br (E.D.d.S.)

[2] Department of Industrial Engineering, University of Bologna, Viale Risorgimento 2, 40136 Bologna, Italy

[3] Computational Modeling, School of Engineering, Federal University of Rio Grande (FURG), Italia Av., km 8, Rio Grande 96203-900, Brazil; andrei.mec@gmail.com

[4] Department of Mechanical Engineering, Federal University of Rio Grande do Sul (UFRGS), Sarmento Leite St., 425, Porto Alegre 90040-001, Brazil; spotorno.furg@gmail.com (R.S.V.); luizrocha@mecanica.ufrgs.br (L.A.O.R.)

[5] UDEMI, NOVA School of Science and Technology, Campus de Caparica, Universidade Nova de Lisboa, 2829-516 Caparica, Portugal; jpc@fct.unl.pt

* Correspondence: cristiano.fragassa@unibo.it; Tel.: +39-347-697-4046

Citation: dos Santos, A.L.; Fragassa, C.; Santos, A.L.G.; Vieira, R.S.; Rocha, L.A.O.; Conde, J.M.P.; Isoldi, L.A.; dos Santos, E.D. Development of a Computational Model for Investigation of and Oscillating Water Column Device with a Savonius Turbine. *J. Mar. Sci. Eng.* **2022**, *10*, 79. https://doi.org/10.3390/jmse 10010079

Academic Editors: Davide Tumino and Antonio Mancuso

Received: 29 November 2021
Accepted: 5 January 2022
Published: 7 January 2022

Publisher's Note: MDPI stays neutral with regard to jurisdictional claims in published maps and institutional affiliations.

Abstract: The present work aims to develop a computational model investigating turbulent flows in a problem that simulates an oscillating water column device (OWC) considering a Savonius turbine in the air duct region. Incompressible, two-dimensional, unsteady, and turbulent flows were considered for three different configurations: (1) free turbine inserted in a long and large channel for verification/validation of the model, (2) an enclosure domain that mimics an OWC device with a constant velocity at its inlet, and (3) the same domain as that in *Case 2* with sinusoidal velocity imposed at the inlet. A dynamic rotational mesh in the turbine region was imposed. Time-averaged equations of the conservation of mass and balance of momentum with the k–ω Shear Stress Transport (SST) model for turbulence closure were solved with the finite volume method. The developed model led to promising results, predicting similar time–spatial averaged power coefficients ($\overline{C_P}$) as those obtained in the literature for different magnitudes of the tip speed ratio ($0.75 \leq \lambda \leq 2.00$). The simulation of the enclosure domain increased $\overline{C_P}$ for all studied values of λ in comparison with a free turbine (*Case 1*). The imposition of sinusoidal velocity (*Case 3*) led to a similar performance as that obtained for constant velocity (*Case 2*).

Keywords: computational model; oscillating water column; wave energy converter; turbulent flows; Savonius turbine

---

1. Introduction

The energy demand will increase by more than 1.0% per year up to 2040, increasing gases emission rates [1]. Moreover, the costs of commodities for energy generation from fossil fuels have increased significantly, leading to economic difficulties, risks associated with energy security, and geopolitical conflicts around the world [2]. Considering this scenario, there is a growing search for a better comprehension of the development of technologies and the use of devices and economic impacts of different renewable sources of energy such as wind, solar, geothermal, and ocean energy [2–11].

One of the important sources of renewable energy with high potential, but not frequently explored worldwide, is the conversion of ocean energy into electricity [12–15], i.e., wave energy conversion. Despite several signs of progress in technological development, there is no dominant main operational principle. Several devices have been proposed

---

and investigated based on various ways to convert wave energy such as point absorbers, attenuators, oscillating surge converters, overtopping, submerged plates, and oscillating water columns (OWCs) [12–14]. The efficiency and survivability of the wave energy converters (WECs) are important issues to make them more competitive and viable [16]. In this context, the OWC device has advantages as its simplicity and maintenance, e.g., the moving parts are located outside of the water, increasing the lifetime material of the power take-off (PTO) system, and the structures of buildings are robust [12–14]. Therefore, several studies and prototypes using the OWC as the main operational principle have been developed around the world: Sakata–Japan (60 kW), Mutriku–Spain (296 kW), Pico–Portugal (400 kW), Tofteshallen–Norway (500 kW), Islay island–Scotland (500 kW), and Lewis island–Scotland (4.0 MW) [17–22].

Important experimental works have sought to improve the comprehension of the fluid dynamic behavior of water/air flow in the OWC device and investigate the influence of some parameters over its performance. For instance, experiments in the laboratory and large-scale domains analyzed the influence of the inclination of the frontal wall, entrance areas of the OWC chamber, and water depth on the device efficiency, reflection, and loading of an OWC for different wave conditions [23–25]. Recently, the experimental progress extended to understand the hydrodynamic of the fluid flow into dual chambers OWC [26,27].

The numerical simulation of OWC devices has also been worth investigating. Several works have been performed since the development of computational models to represent the main operating principle of the device and the investigation of several parameters regarding the performance. For the former purpose, the representation of fluid flow in a laboratory and large-scale devices has been done without considering the effect of a turbine, as in Maciel et al. [28]. Other studies have considered it using orifice plates, plate-baffle, obstacles, or actuator disks to simulate the head loss caused by the turbine over the airflow in the hydropneumatic chamber and air duct of the device [29–33]. Recently, an interesting approach employed the numerical simulation of water and air in the chamber and considered the effects of Wells and impulse turbines by means of analytical thermodynamic models [34,35]. All models mentioned above have been used to obtain recommendations about parameters such as the depth and inclination of the frontal wall, height and length of the chamber, the diameter of the turbine, ramp placed in the seabed below the OWC chamber, and, recently, the design of multiple coupled chambers regarding device performance for different scales and wave conditions [24,32,35–40]. It is also worth mentioning the efforts made to represent the sea state in a more trustworthy form. In this field, some studies modeled the irregular waves using a wave spectrum such as JONSWAP, and others obtained the sea state from geophysical models such as TOMAWAC and used this as an input for the modeling of a channel with the device to be investigated [41–43].

Some significant advancements regarding the PTO of OWC have also been reported in the literature. For instance, Britto-Melo et al. [44] numerically investigated the influence of the aerodynamic parameters of the Wells turbine and the influence of guide vanes and the bypass valve on the pressure drop, torque, and the overall performance. In this work, the conversion of pneumatic energy into electrical energy was estimated with a computational model based on the results extrapolated from aerodynamic tests on a scale model and empirical approximations for the generator losses. Recently, Rodríguez et al. [45] proposed a computational study to understand the behavior of OutFlow Radial (OFR) turbines in both direct and reverse modes, simulating an axisymmetric domain with the flow between a blade-to-blade arrangement of the turbine. The authors observed that the outer blade angle had poor performance in reverse mode despite the improvement of the global performance due to the rotor efficiency gain in direct mode. However, there are few studies related to the numerical simulation of the OWC device turbine considering the rotational domain with the intruded turbine rotor. Prasad et al. [46] performed similar work in this direction and developed a numerical model with a Savonius turbine immersed in a channel under regular wave flow, simulating a hydrokinetic turbine. The authors also investigated the influence of some parameters such as the submergence and the rotational speed for different blade entry angles regarding the rotor power.

Despite the several above-mentioned contributions, to the authors' knowledge, the development of computational models for the simulation of OWC devices considering the rotational turbine in the air duct is an approach still little explored in the literature. Recently, however, Liu et al. [47] presented the validation of an integrated three-dimensional numerical model considering an axial-flow impulse turbine coupled with an OWC inserted in a numerical wave tank (NWT). The present work aims to perform the first step in this direction. Initially, turbulent air flow over a free turbine inserted in a long and large channel (commonly used to represent the numerical modeling of wind turbines) was simulated to verify/validate the present model (*Case 1*). The effect of the tip speed ratio (λ) on the time and spatial averaged power coefficient ($\overline{C_P}$) of a Savonius turbine was compared with numerical and experimental results available in the literature [48,49], investigating the tip speed ratios in the range $0.75 \leq \lambda \leq 2.00$. An enclosure domain that mimics an OWC device was simulated considering a constant velocity imposed at the inlet of the domain (*Case 2*) and sinusoidal velocity that simulated the alternate flow in an OWC device (*Case 3*). For all cases, the influence of λ on the turbine performance and aerodynamic coefficients was investigated.

Moreover, incompressible, two-dimensional, unsteady, and turbulent flows with $Re_D = 867{,}000$ were considered. URANS (Unsteady Reynolds-Averaged Navier Stokes) modeling was applied to all cases. Time-averaged equations of the conservation of mass, the balance of momentum, and transport equations of the k–ω SST model (used in the closure of turbulence) were solved with the finite volume method (FVM) [50–55], using the commercial code Ansys FLUENT 14.5 [56].

2. Mathematical Modeling

Incompressible, two-dimensional, unsteady, turbulent flows with constant thermo-physical properties were considered. It is worth mentioning that one of the characteristics of turbulence is the three-dimensional structure of the flow [57]. Despite this fact, the present simulations represent most of the characteristics of turbulent flows, and the modeling properly predicts the transient behavior of parameters such as drag, lift, moment, and power in the Savonius turbine.

2.1. Description of the Studied Cases

Figure 1 illustrates the computational domain of *Case 1* used for verification/validation of the present model. The domain consists of a long and large channel with an inserted free Savonius turbine. The dimensions are similar to those investigated in the work of Akwa et al. [49]. In this case, the fluid flow of air is caused by the imposition of a constant velocity (V_∞) at the inlet (left side surface). On this surface, there is also an imposed turbulent intensity of $IT = \sqrt{\overline{u'^2}}/\overline{u} = \sigma_u/\overline{u} = 1.0\%$, where u' is the fluctuation of the velocity field, $\overline{u}$ is the time-averaged velocity field, and σ_u represents the variance of u'. At the exit of the channel (right side surface), there is an imposed null gauge pressure ($p_g = 0$ atm). At the upper and lower surfaces, there is a free slip and impermeability boundary condition (also called symmetry). In the turbine region, there is an imposed constant angular velocity (n) in the gray region named the "Rotational Domain" simulating the effect of wind action over the turbine. In the turbine walls, there is an imposed no-slip and impermeability boundary condition ($u = v = 0$ m/s) related to the rotational domain. Figure 1 illustrates the details of the turbine region with the geometric variables used to design it. Table 1 presents the parameters of the fluid flow, thermo-physical properties, dimensions of the computational domain, and turbine variables used here for the four different tip speed ratios of the rotor ($\lambda = nD/2V_\infty$) of $\lambda = 0.75$, 1.00, 1.25, and 2.00; n is the angular velocity of the turbine and D is the turbine diameter. For the unsteady analysis, there was a time interval of $t_f = 3.5$ s, with the last 1.75 s being analyzed for computation of the drag, lift, momentum, and power coefficients.

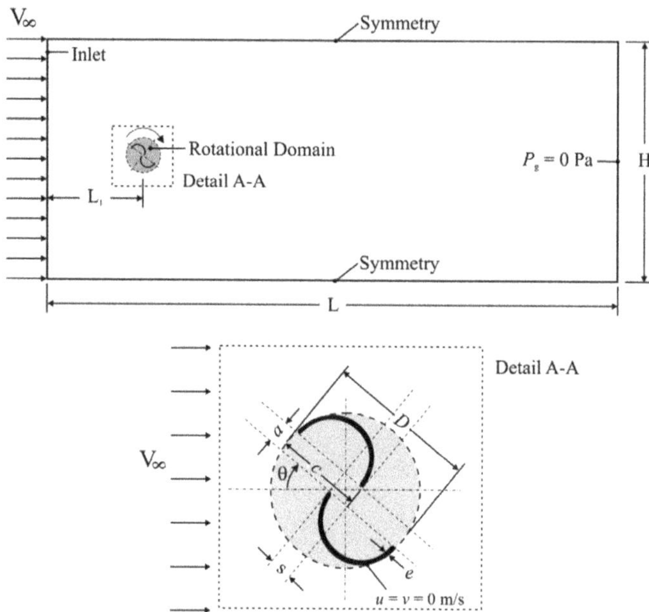

Figure 1. Schematic representation of the computational domain of turbulent air flow over a free Savonius turbine used for verification/validation of the present numerical model (*Case 1*).

Figure 2 illustrates the computational domain used in *Case 2* and *Case 3*. These cases were defined based on *Case 1*. Therefore, the thermo–physical properties and dimensions of the turbine were the same, presented in Table 1 (*Case 1*). The main differences here are the domain dimensions ($H = h = 6.0$ m, $L = 10.0$ m, and $l = 2.0$ m) and the insertion of the Savonius turbine in an enclosure domain, leading to the imposition of a no-slip and impermeability boundary condition ($u = v = 0$ m/s) in the device walls. At the lower surface, there was an imposed constant velocity of $V(t) = 1.4$ m/s and $IT = 1.0\%$ for *Case 2*. The magnitude of the imposed velocity was defined in such a way to have a value of 7.0 m/s at the inlet of the air duct (in the contraction from the chamber to the air duct), leading to a Reynolds number in the turbine similar to that reached for *Case 1*. For *Case 3*, the sole difference in comparison with *Case 2* was the imposition of a sinusoidal function that mimicked the oscillating behavior in the OWC chamber:

$$V(t) = \frac{H_w \pi}{T_w} \cos\left(\frac{2\pi t}{T_w}\right) \tag{1}$$

where $H_w = 0.4$ m, $T_w = 0.875$ s, allowing the reproduction of the piston-type movement generated by the regular waves incidence over the OWC. The range of magnitudes was limited by (-1.4 m/s $\leq V(t) \leq 1.4$ m/s). The use of real configurations of a sea wave, for example, leads to long periods of simulation, requiring a high computational effort. As the purpose of this case is to compare the imposition of the sinusoidal velocity profile with the case with constant velocity to investigate the effect of oscillating flow over the device parameters and performance, the idealized imposed velocity variation was adequate for the desired investigation. For *Case 2* and *Case 3*, the same four magnitudes of the tip speed ratio studied in the verification/validation case ($\lambda = nD/2V_1$) of $\lambda = 0.75$, 1.00, 1.25, and 2.00 were investigated. It is worth mentioning that a mean velocity was measured in the air duct and before the turbine (V_1) for the calculation of λ.

Table 1. Parameters used in the verification/validation case (*Case 1*).

Parameter	Magnitude			
Air density: ρ (kg/m^3)	1.1845			
Dynamic viscosity: μ (kg/ms)	1.7894×10^{-5}			
H (m)	21.6 (12D)			
L (m)	46.8 (26D)			
L_1 (m)	10.8 (6D)			
D (m)	1.8			
V_∞ (m/s)	7.0			
Turbulence intensity: IT (%)	1.0			
$Re_D = \rho V_\infty D/\mu$	867,000			
c (m)	0.972			
a (m)	0.0			
e (m)	7.2×10^{-3}			
s (m)	0.144			
Tip speed ratio: λ	0.75	1.00	1.25	2.00
Angular velocity of turbine: n (rad/s)	5.83	7.77	9.72	15.54
Time of simulation and statistics analysis (s)	3.50 s		1.75 s	

In the present work, results the influence of λ over the drag, lift, moment, and power coefficients (C_d, C_l, C_T, and C_P) are based on [57]:

$$C_d = \frac{F_d}{1/2\rho A_r V^2} \tag{2}$$

$$C_l = \frac{F_l}{1/2\rho A_r V^2} \tag{3}$$

$$C_T = \frac{T}{1/2\rho A_r V^2 r} \tag{4}$$

$$C_P = \frac{P}{P_{available}} = \frac{T}{1/2\rho A_r V^2 r}\frac{rn}{V} = C_T \lambda \tag{5}$$

where F_d is the drag force (N), F_l is the lift force (N), T is the rotor moment (N·m), A_r is the projected area of the Savonius rotor ($A_r = D \cdot W$), W is the depth of the domain in the z-direction (m), V is the upstream turbine velocity ($V = V_\infty$ for the *Case 1* and $V - V_1$ for *Case 2* and *Case 3*) (m/s), r is the radius of the rotor (m), P is the turbine power (W), and $P_{available}$ is the available power of the wind upstream of the rotor (W).

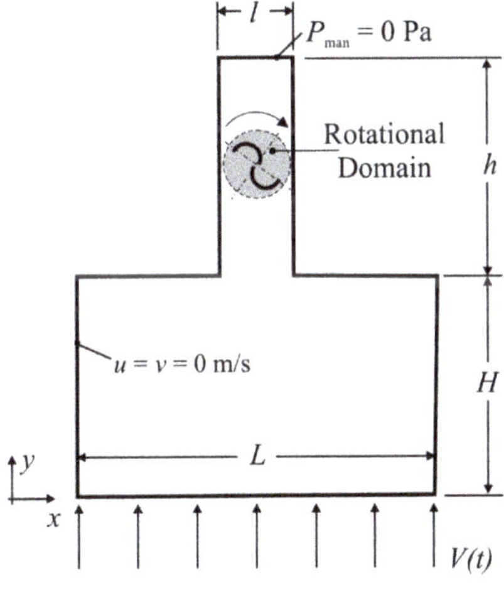

Figure 2. Schematic representation of *Case 2* and *Case 3* that simulate the OWC with a Savonius turbine.

It is worth mentioning that the use of Equation (5) is valid only for the prediction of power coefficients in turbines subjected to open flow conditions (seen in tidal or wind power devices). In OWC devices, the air flow is driven by the pneumatic power, i.e., the pressure drop in the device must also be taken into account. Therefore, for power coefficients predicted for the enclosure domain of *Case 2* and *Case 3*, the C_P is calculated as:

$$C_P = \frac{P}{P_{pneumatic}} = \frac{P}{\left(\Delta p + \frac{\rho V^2}{2}\right)Q} \tag{6}$$

where Δp is the pressure drop between the OWC chamber and the exit of the chimney (Pa), V is the air velocity at the OWC turbine-duct (m/s) ($V = V_1$), and Q is the volumetric flow rate of the air (m^3/s).

The time-averaged magnitudes of the drag, lift, moment, and power coefficients are obtained as follows:

$$\overline{C_{d,l,T,P}} = \frac{1}{t_f}\int_0^{t_f} C_{d,l,T,P}dt \tag{7}$$

2.2. Governing Equations of Turbulent Flows

For all simulations, the modeling of incompressible, two-dimensional, unsteady, and turbulent flows is given by the time-averaged conservation equation of mass and balance of momentum in the x and y directions and can be written as [50,51]:

$$\frac{\partial \overline{u}}{\partial x} + \frac{\partial \overline{v}}{\partial y} = 0 \tag{8}$$

$$\rho\left[\frac{\partial \overline{u}}{\partial t} + \overline{u}\frac{\partial \overline{u}}{\partial x} + \overline{v}\frac{\partial \overline{u}}{\partial y}\right] = -\frac{\partial \overline{p}}{\partial x} + (\mu + \mu_t)\left(\frac{\partial^2 \overline{u}}{\partial x^2} + \frac{\partial^2 \overline{u}}{\partial y^2}\right) \tag{9}$$

$$\rho\left[\frac{\partial \overline{v}}{\partial t} + \overline{u}\frac{\partial \overline{v}}{\partial x} + \overline{v}\frac{\partial \overline{v}}{\partial y}\right] = -\frac{\partial \overline{p}}{\partial y} + (\mu + \mu_t)\left(\frac{\partial^2 \overline{v}}{\partial x^2} + \frac{\partial^2 \overline{v}}{\partial y^2}\right) \tag{10}$$

where x and y are the spatial coordinates (m), u and v are the velocity components in the x and y directions, respectively (m/s), p is the pressure (N/m^2), μ is the dynamic viscosity (kg/m·s), μ_t is the turbulent viscosity (kg/m·s), and the overbar represents the time-averaged operator.

For the k–ω SST closure model, the turbulent viscosity (μ_t) is [52,53]:

$$\mu_t = \frac{\overline{\rho}\alpha_1 k}{\max(\alpha_1\omega, SF_2)} \tag{11}$$

The transport equations of turbulent kinetic energy (k) and its specific dissipation rate (ω) are as follows:

$$\frac{\partial k}{\partial t} + \frac{\partial(\overline{u}_i k)}{\partial x_i} = \widetilde{P}_k - \frac{k^{\frac{3}{2}}}{L_T} + \frac{\partial}{\partial x_i}\left[(\mu + \sigma_k\mu_t)\frac{\partial k}{\partial x_i}\right] \tag{12}$$

$$\frac{\partial \omega}{\partial t} + \frac{\partial(\overline{u}_i\omega)}{\partial x_i} = \frac{\alpha}{\mu_t}\widetilde{P}_k - \beta\omega^2 + \frac{\partial}{\partial x_i}\left[(\mu + \sigma_\omega\mu_t)\frac{\partial \omega}{\partial x_i}\right] + 2(1 - F_1)\frac{\sigma_{\omega2}}{\omega}\frac{\partial k}{\partial x_i}\frac{\partial \omega}{\partial x_i} \tag{13}$$

where $\widetilde{P}_k$ is a function that prevents the turbulence generation in stagnation regions, i represents the direction of fluid flow ($i = 1$ represents the x direction and $i = 2$ represents the y direction), $\beta = 0.09$, $\alpha_1 = 5/9$, $\beta_1 = 3/40$, $\sigma_k = 0.85$, $\sigma_w = 0.5$, $\sigma_2 = 0.44$, $\beta_2 = 0.0828$, $\sigma_{k2} = 1$,

$\sigma_{w2} = 0.856$ are ad hoc constants used in [52]. F_1 and F_2 are blending functions defined as follows:

$$F_1 = \tanh\left\{\left\{\min\left[\max\left(\frac{k^{1/2}}{\beta^* \omega y}, \frac{500v}{y^2\omega}\right), \frac{4\rho\sigma_{w2}k}{CD_{k\omega}y^2}\right]\right\}^4\right\} \tag{14}$$

$$F_2 = \tanh\left\{\left[\max\left(\frac{2k^{1/2}}{\beta^* \omega y}, \frac{500v}{y^2\omega}\right)\right]^2\right\} \tag{15}$$

In Equation (13), the term $CD_{k\omega}$ is calculated as follows:

$$CD_{k\omega} = \max\left(2\rho\sigma_{w2}\frac{1}{\omega}\frac{\partial k}{\partial x_i}\frac{\partial \omega}{\partial x_i}, 10^{-10}\right) \tag{16}$$

3. Numerical Modeling

The solution of the governing equations was performed with the FVM using the commercial package Ansys Fluent^{TM} [54–56]. To tackle the advective terms of the balance of momentum and transport of k and ω, the second-order upwind interpolation function was employed. The algorithm SIMPLE (Semi-Implicit Method for Pressure Linked Equations) was used for pressure–velocity coupling. The simulations were considered to have converged when the residuals for continuity, the balance of momentum, k and ω transport equations were less than 10^{-5}. Moreover, the maximum number of iterations per time step was 200. Concerning the time advancement, there was an implicit time advancement scheme and a fixed time step of $\Delta t = 1.75 \times 10^{-3}$ s. All simulations were performed using desktops with six core Intel$^{\circledR}$ Core ^{TM} i7 5820K @ 3.30 GHz processors and 16 Gb of RAM memory. The processing time for the simulation of $t = 3.5$ s of physical time was nearly 20×10^3 s.

Concerning spatial discretization, hybrid triangular and rectangular finite volumes were used with the domain in simulations of *Case 1*, *Case 2*, and *Case 3*. Figure 3a–c illustrates the mesh generated with software GMSH [58] for the free Savonius turbine (*Case 1*), the configuration similar to the OWC device (*Case 2* and *Case 3*), and a detail of the mesh in the blades of Savonius turbine, respectively. In detail, it is possible to observe a region around the blades with the refined rectangular mesh. The dimensions of the rectangular volumes were defined as a function of a grid independence study and the parameter for representation of the boundary layer (y^+), which must be $y^+ \leq 1.0$ in the walls; y^+ was defined as follows [51–53]:

$$y^+ = \frac{y\sqrt{\tau_w/\rho}}{v} = \frac{yu_\tau}{v} \tag{17}$$

where y is the normal distance to the wall (m), τ_w is the surface tension in the wall (N/m^2), v is the kinematic viscosity (m^2/s), and u_τ is the friction velocity (m/s).

For the grid independence study, four different meshes were simulated, and the results for the time-averaged power coefficient for the free Savonius rotor (*Case 1*) with $Re_D = 867{,}000$, $\lambda = 1.25$, and bucket overlap ratio of $R_S = s/c = 0.15$ are presented in Table 2. The mesh was considered independent when the relative difference between the results of $\overline{C_P}$ for two successive grids met the criterion given by:

$$dev\,(\%) = 100 \times \left|\frac{\overline{C_p}^j - \overline{C_p}^{j+1}}{\overline{C_p}^j}\right| \leq 5.0 \times 10^{-1} \tag{18}$$

where j represents the result obtained with the coarser mesh, and $j + 1$ represents the result obtained with the next successive refined mesh. Based on the results of Table 2, the mesh with 369,653 volumes was used in the independent grid. The same parameters used in this mesh were applied to the spatial discretization of the domain of *Case 2* and *Case 3*.

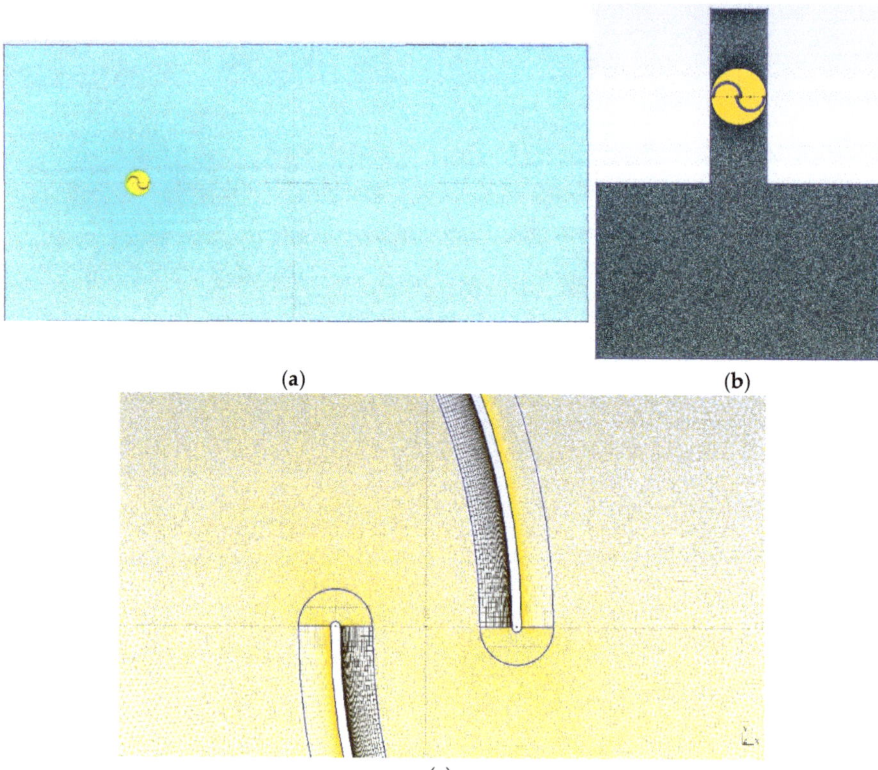

Figure 3. Employed mesh for the studied cases: (**a**) turbine in the open channel (*Case 1*), (**b**) turbine in an OWC chamber domain (*Case 2* and *Case 3*), (**c**) detail of grid refinement the surfaces of the blades.

Table 2. Grid independence test for a free turbine with $Re_D = 867,000$, $\lambda = 1.25$, and $R_S = 0.15$.

Number of Volumes	$\overline{C_P}$	*Dev* (%)
163,141	0.20726	——
168,659	0.21219	2.37×10^0
220,937	0.21030	8.91×10^{-1}
369,653	0.20963	3.18×10^{-1}

4. Results and Discussion

This section is divided into two parts, the verification/validation of the developed computational model (*Case 1*) and the investigation of the influence of the enclosure model and imposition of the sinusoidal velocity inlet over the aerodynamic and performance coefficients (*Case 2* and *Case 3*).

4.1. Verification/Validation of the Computational Model (Case 1)

Figure 4 illustrates the instantaneous drag, lift, and moment coefficients as a function of time for *Case 1* with $Re_D = 867,000$, $\lambda = 2.0$, and $R_S = 0.15$. For the first instants of time, mainly for $t \leq 0.5$ s, the coefficients had a strong variation due to the incidence of the fluid flow and the imposition of angular velocity in the rotational domain region. Therefore, the present model must be used to predict power in the turbine when the flow is stabilized, and the first cycles of rotation of the turbine were disregarded for the analysis of coefficients and power. Here, for the computation of time-averaged parameters, only the results in the range of time 1.75 s $\leq t \leq$ 3.5 s were used. Despite the complexity of the fluid flow, the results also demonstrated a regular oscillation in the magnitudes of C_d, C_l, and C_T, which had similar behavior to that previously obtained in Akwa et al. [49]. Moreover, the crest and cave magnitudes were also similar, showing the generation of regular wakes of vortices behind the turbine.

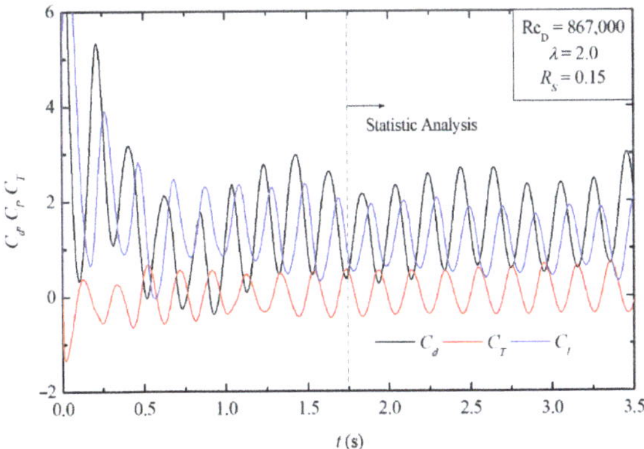

Figure 4. Transient coefficients of the lift (C_l), drag (C_d), and torque (C_T) for the free turbine case with Re_D = 867,000, Rs = 0.15, and λ = 2.0.

Figure 5 shows the comparison of power coefficients (C_P) as a function of the tip speed ratio (λ) obtained with the present computational model and the numerical predictions of Akwa et al. [49] obtained with other commercial code (Star-CCM+) also based on the FVM and the experimental results of Blackwell et al. [48]. The results predicted with the present method are in close agreement with those previously obtained in the literature, verifying and validating the method used here. Even for λ = 1.00 and 2.00, where the highest differences between the present results and the experimental ones were obtained, the deviations were lower than the uncertainty of the experiment. The results indicated that the highest magnitudes of $\overline{C_P}$ were reached at the lowest values of λ. With the increase in λ, the magnitude of $\overline{C_P}$ had a slight decrease in the range $1.00 \leq \lambda \leq 1.25$ and a step decrease for $\lambda \geq 1.25$. For λ = 2.00, negative magnitudes of $\overline{C_P}$ were obtained, indicating that the device supplies energy for the fluid flow and not the contrary.

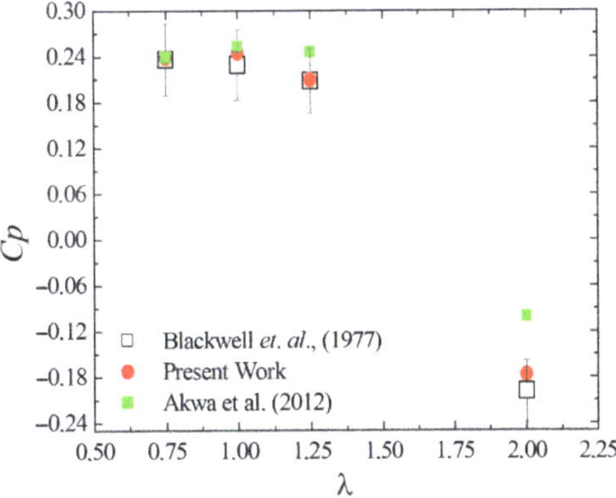

Figure 5. Comparison between the power coefficient (C_P) as a function of λ obtained in the present work and that from the literature [48,49].

Figures 6 and 7 illustrate the velocity and pressure fields, respectively, for *Case 1* with Re_D = 867,000 and λ = 1.25 for five different instants of time: (a) t = 0.0 s, (b) t = 0.53 s, (c) t = 1.05 s, (d) t = 1.58 s, and (e) t = 2.10 s, which represent different angular positions of the Savonius rotor. For the initial time step, in Figures 6a and 7a, the velocity and pressure fields noticed were generated by the imposition of angular velocity in the rotational domain by the computational model. This behavior does not represent the real condition of fluid flow over the turbine since the turbine should not be in motion before the incidence of the fluid flow. Therefore, the present computational method must be used when the flow is stabilized, which happens a few cycles later at the beginning of the fluid flow (as shown in Figure 4 for monitoring the coefficients). As the time advances, mainly for t > 1.0 s, Figures 6c–e and 7c–e show an increase in the pressure field magnitude on the concave side of the advancement blade and a pressure drop as a consequence, being the main reason for the pressure drag force that drives the turbine. The results also show the fluid flowing in the region between the two blades. The results also show the generation of wakes behind the rotor and vortices generated in the tip region of the blade. It is also worth mentioning that the behavior found here is similar to that described in previous literature, e.g., in Prasad et al. [45] and Blackwell et al. [49].

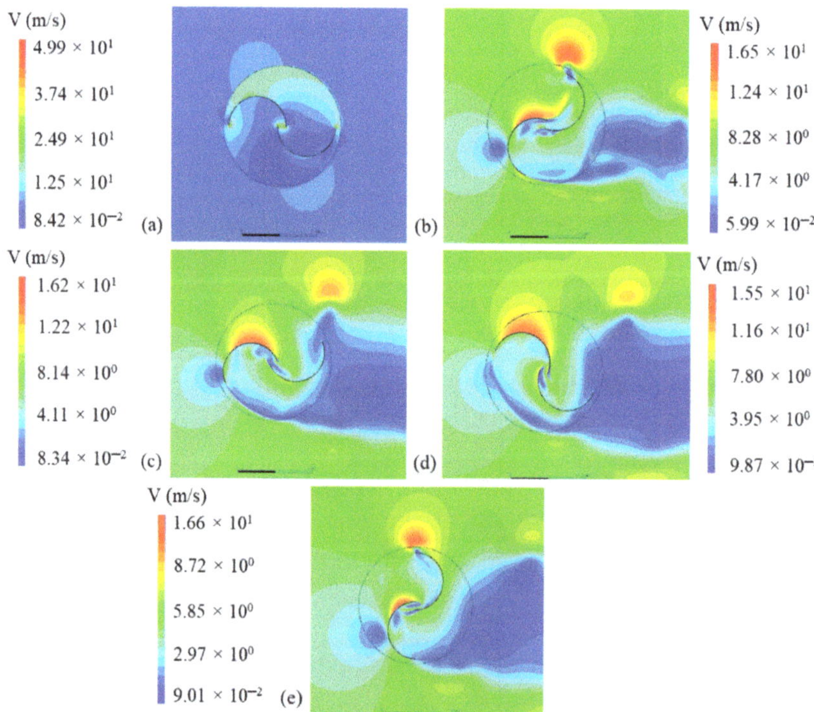

Figure 6. Velocity fields around the free turbine (*Case 1*) with Re_D = 867,000 and λ = 1.25: (**a**) t = 0.0 s, (**b**) t = 0.53 s, (**c**) t = 1.05 s, (**d**) t = 1.58 s, and (**e**) t = 2.10 s.

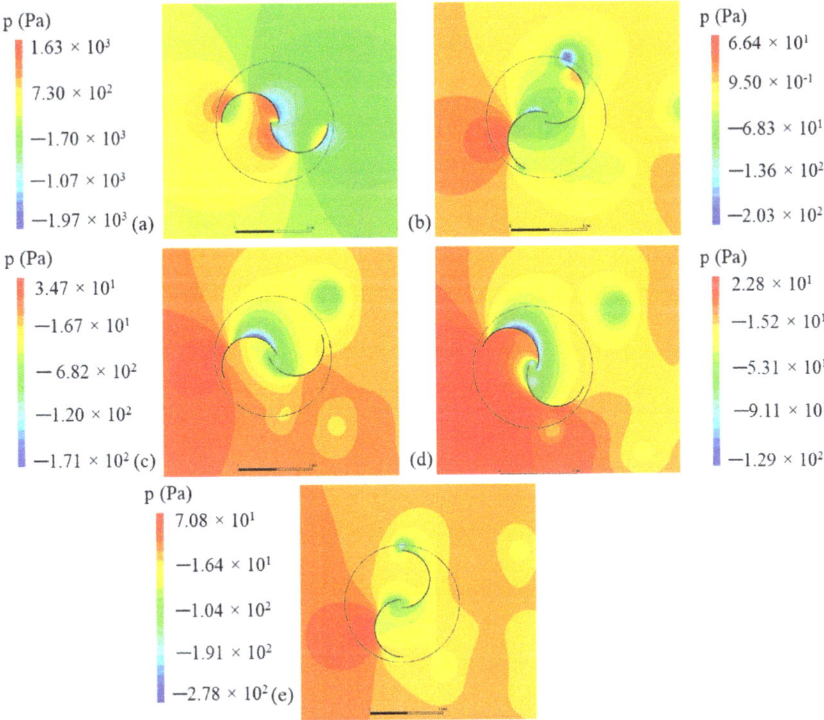

Figure 7. Pressure fields around the free turbine (*Case 1*) with Re$_D$ = 867,000 and λ = 1.25: (**a**) t = 0.0 s, (**b**) t = 0.53 s, (**c**) t = 1.05 s, (**d**) t = 1.58 s, and (**e**) t = 2.10 s.

4.2. The Results of the Savonius Turbine Inserted in an OWC Domain (Case 2 and Case 3)

For the turbulent air flow in the enclosure domain, the instantaneous aerodynamic coefficients (C_d, C_l, and C_T) were obtained as a function of time to better understand the behavior of the turbine in an OWC domain. Figure 8 depicts the coefficients for the flow with Re$_D$ = 867,000, λ = 2.0, and R_S = 0.15 for a constant imposed velocity at the inlet (*Case 2*). In general, the results show an average increase in C_d (black line) for all instants of time compared to the free turbine simulations (*Case 1*) due to the insertion of the turbine in the enclosure domain. The increase in the drag coefficient (C_d) also led to an augmentation of the moment coefficient (C_T) represented in red color in a similar form reached for the C_d, while the lift coefficient (C_l) did not suffer important modifications in its mean magnitude. The results also indicate that the transient behavior of the coefficients was strongly modified compared to *Case 1*, showing C_d and C_l with sharp peaks and smooth troughs, i.e., each cycle did not behave in a sinusoidal form as noticed in Figure 4 for the free turbine configuration. It is worth mentioning that, for other magnitudes of λ, similar behavior for the instantaneous coefficients was obtained. Therefore, for the sake of brevity, the instantaneous coefficients for other magnitudes of λ are not presented.

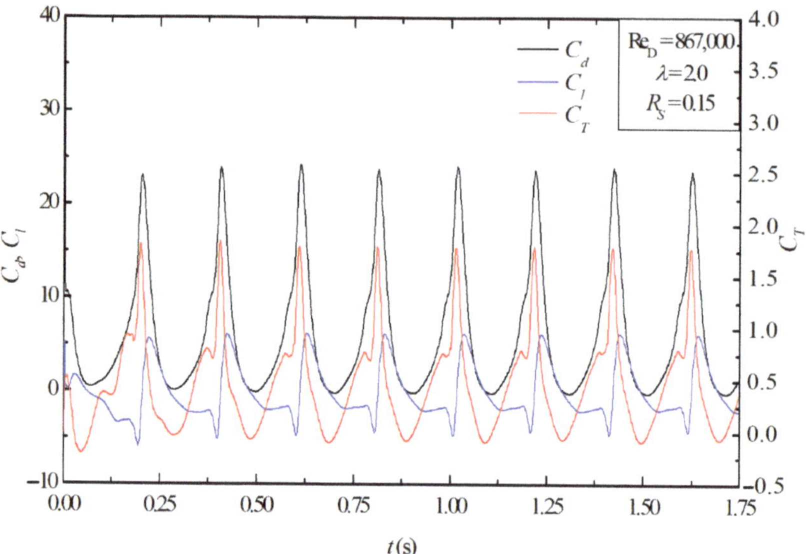

Figure 8. Instantaneous coefficients of the drag (C_d), lift (C_l), and moment (C_T) for the case of OWC with constant imposed velocity at the inlet (*Case 2*) with $Re_D = 867,000$, $\lambda = 2.0$, and $R_S = 0.15$.

As previously mentioned, the air flow in the OWC chamber is subjected to the piston-type oscillatory motion of the water column (hydropneumatic chamber). In order to simulate this effect, the results of instantaneous power coefficient obtained for *Case 3* and *Case 2*, considering the same conditions ($Re_D = 867,000$, $\lambda = 2.0$, and $R_S = 0.15$), are presented in Figure 9. It is important to reinforce that, for prediction of C_P for *Case 2* and *Case 3*, Equation (6) was used instead of Equation (5), and the time-averaged magnitudes were predicted with Equation (7). The results reveal a strong similarity between the instantaneous C_P for both cases, indicating that the imposition of sinusoidal velocity at the inlet of the domain did not have a significant influence over the pattern of the C_P investigated here, which is not intuitively expected since the mean imposed momentum decreased for *Case 3* in comparison with *Case 2*. The results also demonstrate a slight increase in the power coefficients when the sinusoidal velocity was imposed, with a difference of nearly 9.0% on average. Possible explanations for the behavior found here include the synchronization of velocity augmentation with the rotation of the turbine and the imposition of the rotational domain, as well as the acceleration of the fluid caused by the variation of the imposed velocity at the inlet for *Case 3*, which did not happen for *Case 2*. The magnitude of the imposed mean velocity at the inlet of *Case 2* and *Case 3* and acceleration at the inlet for *Case 3* are illustrated in Figure 10 to make this visualization easy. It is worth mentioning that, despite different imposed inlet velocities, the transient fields of velocity and pressure had only slight differences. Future investigations should be performed with other magnitudes of amplitude and periods of imposed velocity to corroborate this hypothesis and the development of other models where the inertia moment of the turbine is taken into account.

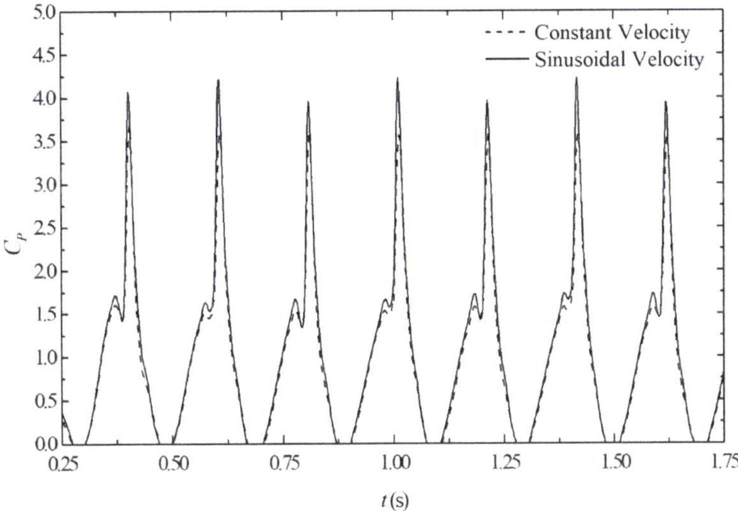

Figure 9. Effect of imposed velocity at the inlet of the OWC domain over the instantaneous power coefficient in the Savonius turbine, considering $Re_D = 867{,}000$, $\lambda = 2.0$, and $R_S = 0.15$.

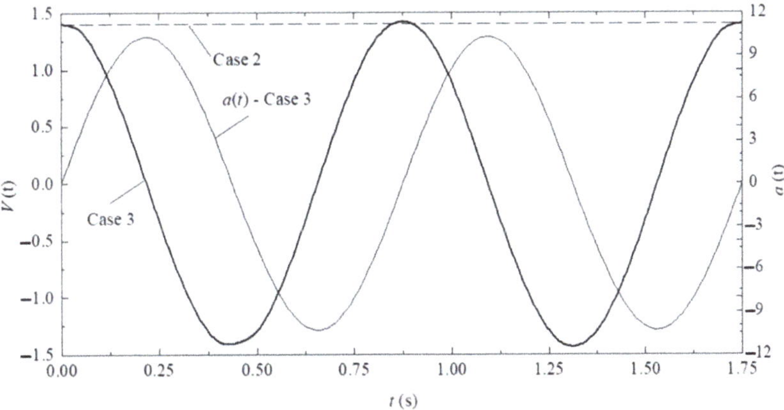

Figure 10. Magnitude of the imposed velocity at the inlet of the OWC domain for *Case 2* and *Case 3* and acceleration of the fluid at the inlet of the domain for *Case 3*.

Figures 11 and 12 illustrate the velocity and pressure fields, respectively, for the simulation of *Case 3* with $Re_D = 867{,}000$, $\lambda = 2.0$, and $R_S = 0.15$ for six different time steps: (a) $t = 0.0$ s, (b) $t = 0.26$ s, (c) $t = 0.53$ s, (d) $t = 0.79$ s, (e) $t = 1.05$ s, and (f) $t = 1.31$ s. As the behaviors of *Case 3* and *Case 2* were almost the same, only the fields for one of the cases are illustrated. As previously mentioned, for the free turbine case (see Figures 6 and 7), the initial time steps should be disregarded due to the artificial imposition of the angular velocity in the turbine region. In spite of that, when the flow was stabilized, the model properly represented the physical problem. In turn, one can note in Figures 11 and 12 an increase in the pressure difference between the concave and convex sides of the advancement blade for *Case 2* and *Case 3* compared to *Case 1*, which explains the augmentation of the pressure drag in comparison with *Case 1*.

Moreover, the pressure difference between the upstream and downstream regions of the turbine was also augmented. This influence over the water oscillation was not investigated, but the results pointed out that this aspect is worth investigation when the turbine is taken into account in the problem in order to avoid the restriction of the water column, mainly when coupling with a wave channel is performed. The results also indicated the variation of the pressure magnitude in the chamber for different instants of time, explaining the non-symmetric differences between the peaks and troughs with respect to the mean magnitudes of the coefficients. The momentum of the fluid flow intensified due to the insertion of the turbine in the enclosure domain. Moreover, the wakes generated in the turbine and secondary vortices in the tip of the blades could not be spread in the spanwise direction of the main flow due to the limitation imposed by the air duct walls. Therefore, as expected, the insertion of the turbine in the enclosure domain affected the fluid dynamic behavior of the flow considerably.

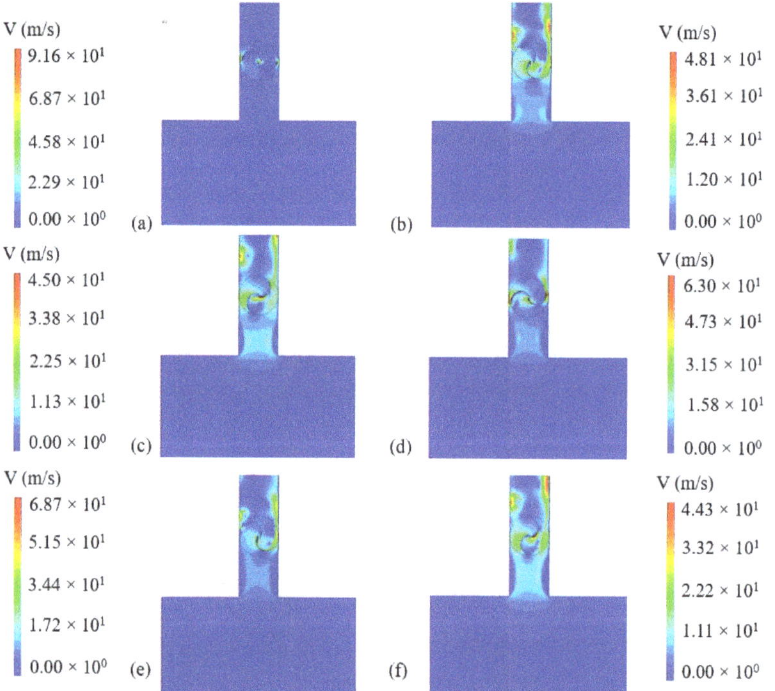

Figure 11. Velocity fields for the OWC case with sinusoidal velocity at the inlet for different instants of time for the case with $Re_D = 867{,}000$, $\lambda = 2.0$, and $R_S = 0.15$: (**a**) $t = 0.0$ s, (**b**) $t = 0.26$ s, (**c**) $t = 0.53$ s, (**d**) $t = 0.79$ s, (**e**) $t = 1.05$ s, and (**f**) $t = 1.31$ s.

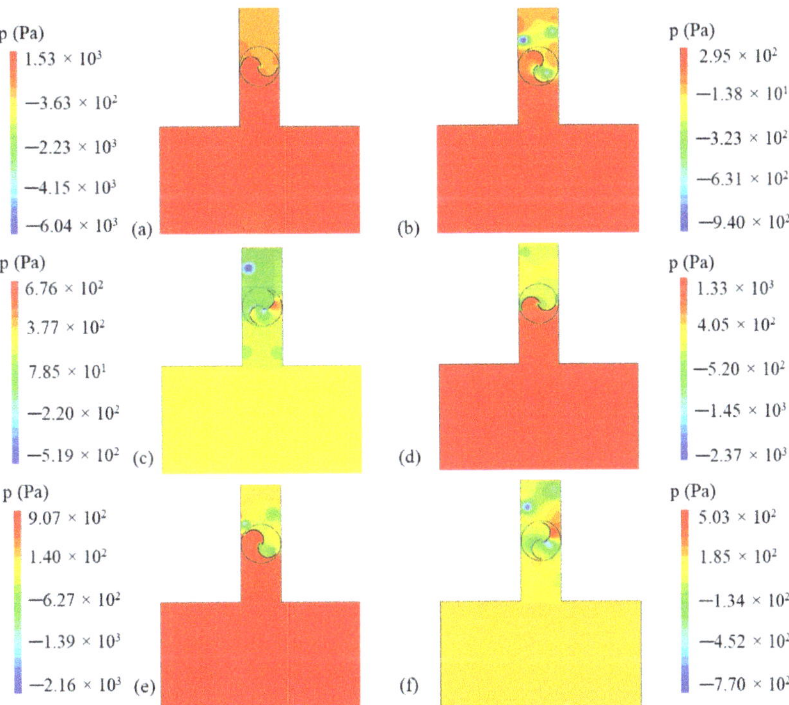

Figure 12. Pressure fields for the OWC case with sinusoidal velocity at the inlet for different instants of time for the case with $Re_D = 867,000$, $\lambda = 2.0$, and $R_S = 0.15$: (**a**) $t = 0.0$ s, (**b**) $t = 0.26$ s, (**c**) $t = 0.53$ s, (**d**) $t = 0.79$ s, (**e**) $t = 1.05$ s, and (**f**) $t = 1.31$ s.

To summarize, Figure 13 shows the effect of λ on the time-averaged magnitudes of $\overline{C_d}$, $\overline{C_l}$, $\overline{C_T}$ and $\overline{C_P}$ for *Case 2* and *Case 3*, and Figure 14 illustrates the same effect considering only $\overline{C_P}$ for *Case 1*, *Case 2*, and *Case 3*. The results indicate that the differences between the values of the aerodynamic and power coefficients obtained for *Case 2* and *Case 3* were not significant, with the highest difference being lower than 10.0%. The effect of λ on the coefficients was also similar for both cases. Concerning the magnitude of $\overline{C_d}$, it decreased in the range $0.75 \leq \lambda \leq 1.00$, and after this point, the magnitude was almost constant. The $\overline{C_l}$ and $\overline{C_T}$ magnitudes decreased with the increase in λ. However, for $\overline{C_T}$ the decrease was more significant only in the range $1.00 \leq \lambda \leq 1.25$, being almost constant in other intervals of λ. For the $\overline{C_P}$, there was an increase with the augmentation of λ. For *Case 3*, for example, the magnitude increased from $\overline{C_P} = 0.2922$ when $\lambda = 0.75$ to $\overline{C_P} = 0.5054$ when $\lambda = 2.0$. Figure 14 indicates that in the region $0.75 \leq \lambda \leq 1.25$, a similar trend of $\overline{C_P}$ was noticed when *Case 1* and the enclosure cases (*Case 2* and *Case 3*) were compared. However, for $\lambda = 2.0$, contrary to the behavior noticed for *Case 1*, *Case 2* and *Case 3* did not have a reduction of $\overline{C_P}$ for the highest magnitude of λ investigated. Therefore, the results indicated that changes in the domain where the turbine is placed are important to define its application range and the effect of λ over the problem performance. Further studies should be performed to define the range of application of λ for the enclosure domain.

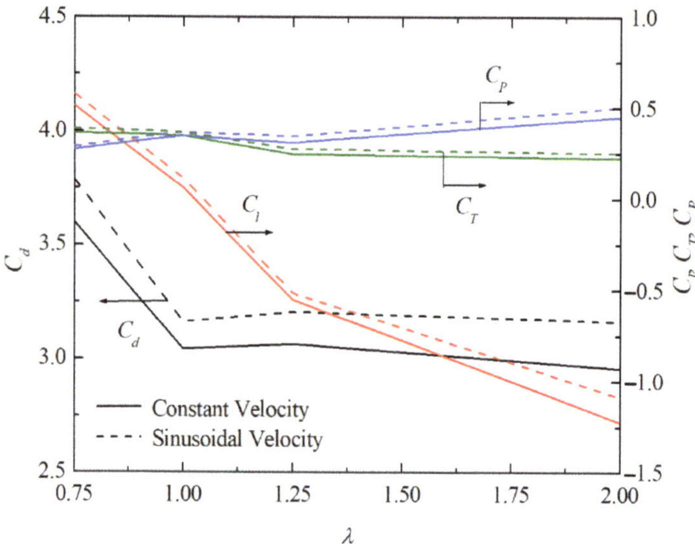

Figure 13. Averaged coefficients as a function of λ for the cases of OWC with constant- and sinusoidal-imposed velocity.

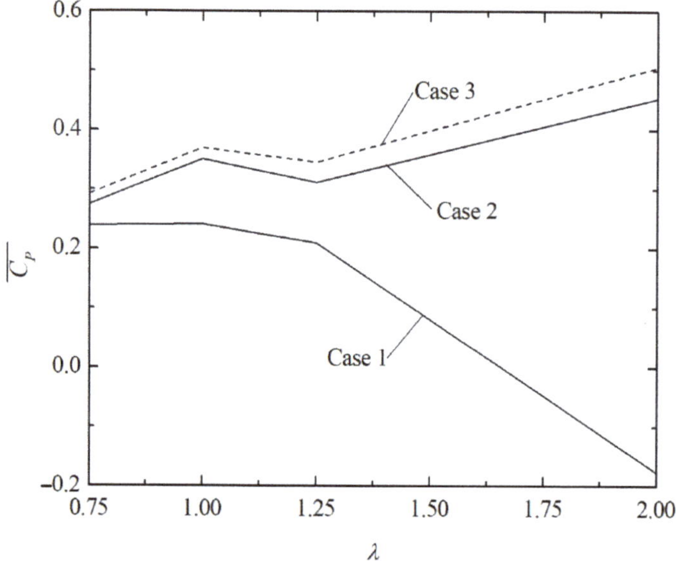

Figure 14. Power coefficients as a function of the tip speed ratio for the three studied cases (*Case 1*, *Case 2*, and *Case 3*).

5. Conclusions

The present work developed a computational model to investigate turbulent flows in enclosure domains with an inserted Savonius turbine simulating the air flow in an OWC/WEC device. Initially, a free turbine configuration (*Case 1*) was investigated in order to perform the verification/validation of the present method. Then, two different study cases were investigated (*Case 2* and *Case 3*) with constant and sinusoidal velocity imposed at the domain inlet. For all cases, Re_D = 867,000, R_S = 0.15, and four different magnitudes of λ were studied (λ = 0.75, 1.00, 1.25, and 2.00). For the simulation of incompressible, two-dimensional, transient, and turbulent flows, the time-averaged equations of mass conservation, the balance of momentum, and transport equations of the k–ω SST model were solved with the FVM.

The proposed computational model was verified and validated using a comparison of C_P with the numerical and experimental results of the literature [48,49], reproducing the effect of λ on $\overline{C_P}$ for the turbulent flow over a free Savonius turbine. Moreover, the velocity and pressure fields demonstrated that the driven force of the Savonius turbine was dominated by the pressure difference between the concave and convex sides of the advancement blade, which the present model adequately predicted.

After the verification/validation of the computational model, new recommendations were reached for the turbulent flows over the Savonius turbine inserted in the enclosure domain representing an OWC. The results demonstrated that the new configuration had a strong influence over the behavior and magnitudes of the instantaneous coefficients such as C_d and C_T, also affecting the instantaneous values of C_P. For the investigated values of λ, the results indicated that the insertion of a turbine in the enclosure domain led to an overall augmentation of $\overline{C_P}$ for all values of λ. In addition, the effect of λ on $\overline{C_d}$, $\overline{C_l}$, $\overline{C_T}$, and $\overline{C_P}$ was strongly affected by the domain change, indicating that the geometric configuration can be important to define the range of applicability of the turbine and the design of the PTO in the OWC device. The comparison between the imposition of sinusoidal velocity (*Case 3*) and constant velocity (*Case 2*) at the domain inlet led to similar performance and aerodynamic coefficients. It is important to highlight that due to the two-dimensional approach, some characteristics of the air flow phenomenology through the Savonious turbine cannot be completely revealed from the obtained results, being a limitation of the proposed computational model.

Future investigations should be performed considering a model where the inertia moment is taken into account to verify if this behavior continues to be observed. It is recommended to investigate other parameters, such as the effect of the overlap and spacing between the blades (s and a) and other tip speed ratios (λ) on the aerodynamic and performance of the OWC and the imposition of irregular velocity variation to represent the real sea state movement. It is also recommended to investigate the coupling between the present model and the wave channel for adequate simulation of the interaction between the oncoming waves, the structure of the device, and air flow over the turbine inserted in the OWC air duct.

Author Contributions: Conceptualization, A.L.d.S., R.S.V., J.M.P.C. and E.D.d.S.; methodology, A.L.d.S., A.L.G.S. and R.S.V.; software, A.L.d.S., A.L.G.S. and R.S.V.; validation, A.L.d.S. and A.L.G.S.; formal analysis, A.L.d.S., J.M.P.C. and E.D.d.S.; investigation, A.L.d.S. and A.L.G.S.; resources, C.F., L.A.O.R., L.A.I. and E.D.d.S.; data curation, J.M.P.C., R.S.V., L.A.I. and E.D.d.S.; writing—original draft preparation, A.L.d.S., A.L.G.S. and E.D.d.S.; writing—review and editing, C.F., L.A.O.R. and L.A.I.; visualization, C.F. and J.M.P.C.; supervision, R.S.V., J.M.P.C. and E.D.d.S.; project administration, L.A.O.R., L.A.I. and E.D.d.S.; funding acquisition, C.F., L.A.O.R., L.A.I. and E.D.d.S. All authors have read and agreed to the published version of the manuscript.

Funding: This research was funded by the Brazilian National Council for Scientific and Technological Development—CNPq (Processes: 306012/2017-0, 307791/2019-0, 306024/2017-9, 131487/2020 and 440010/2019-5) and the Research Support Foundation of the State of Rio Grande do Sul—FAPERGS (Public Call FAPERGS 07/2021—Programa Pesquisador Gaúcho—PqG).

Institutional Review Board Statement: Not applicable.

Informed Consent Statement: Not applicable.

Data Availability Statement: The data presented in this study are available on request from the corresponding author. The data are not publicly available due to privacy reasons.

Acknowledgments: The author A.L.d.S. thanks CNPq for the Master of Science scholarship (Process: 131487/2020). The author A.L.G.S. thanks CNPq for the doctorate scholarship (Process: 440010/2019-5). The authors L.A.I., L.A.O.R., and E.D.d.S. thank CNPq for the research grant (Processes: 306012/2017-0, 307791/2019-0 and 306024/2017-9). The authors A.L.G.S. and E.D.d.S. thank CNPq for the financial support in the CNPq/Equinor Energia Ltd., Call N° 38/2018 (Process: 440010/2019-5). L.A.I. thanks FAPERGS (Public Call FAPERGS 07/2021—PqG). J.M.P.C. thanks the Portuguese Foundation for Science and Technology funding through UNIDEMI (Process: UID/EMS/00667/2020) and the Sabbatical Leave Fellowship (Process: SFRH/BSAB/150449/2019).

References

1. International Energy Agency (IEA). *World Energy Outlook 2019*; International Energy Agency: Paris, France, 2019.
2. Jenniches, S. Assessing the regional economic impacts of renewable energy sources—A literature review. *Renew. Sustain. Energy Rev.* **2018**, *93*, 35–51. [CrossRef]
3. Cheng, M.; Zhu, Y. The state of the art of wind energy conversion systems and technologies: A review. *Energy Convers. Manag.* **2014**, *88*, 332–347. [CrossRef]
4. Olabi, A.G.; Wilberforce, T.; Elsaid, K.; Salameh, T.; Sayed, E.T.; Husain, K.S.; Abdelkareem, M.A. Selection Guidelines for Wind Energy Technologies. *Energies* **2021**, *14*, 3244. [CrossRef]
5. Kumar, K.R.; Chaitanya, N.V.V.K.; Kumar, N.S. Solar thermal energy technologies and its applications for process heating and power generation—A review. *J. Clean. Prod.* **2021**, *282*, 125296. [CrossRef]
6. Seme, S.; Štumberger, B.; Hadžiselimović, M.; Sredenšek, K. Solar Photovoltaic Tracking Systems for Electricity Generation: A Review. *Energies* **2020**, *13*, 4224. [CrossRef]
7. Nunes, B.R.; Rodrigues, M.K.; Rocha, L.A.O.; Labat, M.; Lorente, S.; Dos Santos, E.D.; Isoldi, L.A.; Biserni, C. Numerical-analytical study of earth-air heat exchangers with complex geometries guided by constructal design. *Int. J. Energy Res.* **2021**, *45*, 20970–20987. [CrossRef]
8. Khan, N.; Kalair, A.; Abas, N.; Haider, A. Review of ocean tidal, wave and thermal energy technologies. *Renew. Sustain. Energy Rev.* **2017**, *72*, 590–604. [CrossRef]
9. Melikoglu, M. Current status and future of ocean energy sources: A global review. *Ocean Eng.* **2018**, *148*, 563–573. [CrossRef]
10. Curto, D.; Franzitta, V.; Guercio, A. Sea Wave Energy. A Review of the Current Technologies and Perspectives. *Energies* **2021**, *14*, 6604. [CrossRef]
11. Chen, L.; Li, W.; Li, J.; Fu, Q.; Wang, T. Evolution Trend Research of Global Ocean Power Generation Based on a 45-Year Scientometric Analysis. *J. Mar. Sci. Eng.* **2021**, *9*, 218. [CrossRef]
12. Falcão, A.F.D.O. Wave Energy Utilization: A Review of the Technologies. *Renew. Sustain. Energy Rev.* **2010**, *14*, 899–918. [CrossRef]
13. Zabihian, F.; Fung, A.S. Review of Marine Renewable Energies: Case Study of Iran. *Renew. Sustain. Energy Rev.* **2011**, *15*, 2461–2474. [CrossRef]
14. López, I.; Andreu, J.; Ceballos, S.; De Alegría, I.M.; Kortabarria, I. Review of wave energy technologies and the necessary power-equipment. *Renew. Sustain. Energy Rev.* **2013**, *27*, 413–434. [CrossRef]
15. Liu, W.; Xu, X.; Chen, F.; Liu, Y.; Li, S.; Liu, L.; Chen, Y. A Review of Research on the Closed Thermodynamic Cycles of Ocean Thermal Energy Conversion. *Renew. Sustain. Energy Rev.* **2020**, *119*, 109581. [CrossRef]
16. Temiz, I.; Ekweoba, C.; Thomas, S.; Kramer, M.; Savin, A. Wave Absorber Ballast Optimization based on the Analytical Model for a Pitching Wave Energy Converter. *Ocean Eng.* **2021**, *240*, 109906. [CrossRef]
17. Müller, G.; Whittaker, T.J.T. Field measurements of breaking wave loads on a shoreline wave power station. *Proc. ICE Water Marit. Energy* **1995**, *112*, 187–197. [CrossRef]
18. Müller, G.; Whittaker, T.J.T. An evaluation of design wave impact pressures. *J. Waterw. Port Coast. Ocean Eng.* **1996**, *122*, 55–58. [CrossRef]
19. Patterson, C.; Dunsire, R.; Hillier, S. Development of wave energy breakwater at Siadar, Isle of Lewis. In Proceedings of the ICE Conference Coasts, Marine Structures & Breakwaters, Edinburgh, UK, 16–18 September 2009; pp. 738–749. [CrossRef]
20. Bryden, I.G. Progress towards a viable UK marine renewable energy. In Proceedings of the ICE Conference Coasts, Marine Structures & Breakwaters, Edinburgh, UK; 2009; pp. 2–13. [CrossRef]
21. Torre-Enciso, Y.; Ortubia, I.; López de Aguileta, L.I.; Marqués. Mutriku Wave Power Plant: From the thinking out to the reality. In Proceedings of the 8th European Wave and Tidal Energy Conference, Uppsala, Sweden, 7–10 September 2009; pp. 319–329.
22. Torre-Enciso, Y.; Marqués, J.; López de Aguileta, L.I. Mutriku. Lessons learnt. In Proceedings of the 3rd International Conference on Ocean Energy, Bilbao, Spain, 6–8 October 2010; pp. 1–6.
23. Dizadji, N.; Sajadian, S.E. Modeling and Optimization of the Chamber of OWC System. *Energy* **2011**, *36*, 2360–2366. [CrossRef]
24. Mahnamfar, F.; Altunkaynak, A. OWC-Type Wave Chamber Optimization under Series of Regular Waves. *Arab. J. Sci. Eng.* **2016**, *41*, 1543–1549. [CrossRef]

25. Viviano, A.; Naty, S.; Foti, E.; Bruce, T.; Allsop, W.; Vicinanza, D. Large-Scale Experiments on the Behaviour of Generalised Oscillating Water Column under Random Waves. *Renew. Energy* **2016**, *99*, 875–887. [CrossRef]

26. Gadelho, J.F.M.; Rezanejad, K.; Xu, S.; Hinostroza, M.; Soares, C.G. Experimental study on the motions of a dual chamber floating oscillating water column device. *Renew. Energy* **2021**, *170*, 1257–1274. [CrossRef]

27. Rezanejad, K.; Gadelho, J.F.M.; Xu, S.; Soares, C.G. Experimental investigation on the hydrodynamic performance of a new type floating Oscillating Water Column device with dual-chambers. *Ocean Eng.* **2021**, *234*, 109307. [CrossRef]

28. Maciel, R.P.; Fragassa, C.; Machado, B.N.; Rocha, L.A.O.; Dos Santos, E.D.; Gomes, M.N.; Isoldi, L.A. Verification and Validation of a Methodology to Numerically Generate Waves Using Transient Discrete Data as Prescribed Velocity Boundary Condition. *J. Mar. Sci. Eng.* **2021**, *9*, 896. [CrossRef]

29. Conde, J.M.; Gato, L.M.C. Numerical study of the air-flow in an oscillating water column wave energy converter. *Renew. Energy* **2008**, *33*, 2637–2644. [CrossRef]

30. Liu, Z.; Hyun, B.; Hong, K. Numerical study of air chamber for oscillating water column wave energy convertor. *China Ocean Eng.* **2011**, *25*, 169–178. [CrossRef]

31. Gomes, M.d.N.; Dos Santos, E.D.; Isoldi, L.A.; Rocha, L.A.O. Numerical Analysis including Pressure Drop in Oscillating Water Column Device. *Open Eng.* **2015**, *5*, 229–237. [CrossRef]

32. Elhanafi, A.; Macfarlane, G.; Fleming, A.; Leong, Z. Experimental and numerical investigations on the hydrodynamic performance of a floating–moored oscillating water column wave energy converter. *Appl. Energy* **2017**, *205*, 369–390. [CrossRef]

33. Moñino, A.; Medina-López, E.; Clavero, M.; Benslimane, S. Numerical Simulation of a Simple OWC Problem for Turbine Performance. *Int. J. Mar. Energy* **2017**, *20*, 17–32. [CrossRef]

34. Mahnamfar, F.; Altunkaynak, A. Comparison of Numerical and Experimental Analyses for Optimizing the Geometry of OWC Systems. *Ocean Eng.* **2017**, *130*, 10–24. [CrossRef]

35. Torres, F.R.; Teixeira, P.R.F.; Didier, E. A Methodology to Determine the Optimal Size of a Wells Turbine in an Oscillating Water Column Device by Using Coupled Hydro-Aerodynamic Models. *Renew. Energy* **2018**, *121*, 9–18. [CrossRef]

36. Gonçalves, R.A.A.C.; Teixeira, P.R.F.; Didier, E.; Torres, F.R. Numerical analysis of the influence of air compressibility effects on an oscillating water column wave energy converter chamber. *Renew. Energy* **2020**, *153*, 1183–1193. [CrossRef]

37. Gomes, M.N.; Lorenzini, G.; Rocha, L.A.O.; Dos Santos, E.D.; Isoldi, L.A. Constructal Design Applied to the Geometric Evaluation of an Oscillating Water Column Wave Energy Converter Considering Different Real Scale Wave Periods. *J. Eng. Thermophys.* **2018**, *27*, 173–190. [CrossRef]

38. Kharati-Koopaee, M.; Fathi-Kelestani, A. Assessment of oscillating water column performance: Influence of wave steepness at various chamber lengths and bottom slopes. *Renew. Energy* **2020**, *147*, 1595–1608. [CrossRef]

39. Letzow, M.; Lorenzini, G.; Barbosa, D.V.E.; Hübner, R.G.; Rocha, L.A.O.; Gomes, M.N.; Isoldi, L.A.; Dos Santos, E.D. Numerical Analysis of the Influence of Geometry on a Large Scale Onshore Oscillating Water Column Device with Associated Seabed Ramp. *Int. J. Des. Nat. Ecodyn.* **2020**, *15*, 873–884. [CrossRef]

40. Lima, Y.T.B.; Gomes, M.N.; Isoldi, L.A.; Dos Santos, E.D.; Lorenzini, G.; Rocha, L.A.O. Geometric Analysis through the Constructal Design of a Sea Wave Energy Converter with Several Coupled Hydropneumatic Chambers Considering the Oscillating Water Column Operating Principle. *Appl. Sci.* **2021**, *11*, 8630. [CrossRef]

41. Gomes, M.D.N.; Salvador, H.; Magno, F.; Rodrigues, A.A.; Santos, E.D.; Isoldi, L.A.; Rocha, L.A.O. Constructal Design Applied to Geometric Shapes Analysis of Wave Energy Converters. *Defect Diffus. Forum* **2021**, *407*, 147–160. [CrossRef]

42. Lisboa, R.C.; Teixeira, P.R.F.; Torres, F.R.; Didier, E. Numerical Evaluation of the Power Output of an Oscillating Water Column Wave Energy Converter Installed in the Southern Brazilian Coast. *Energy* **2018**, *162*, 1115–1124. [CrossRef]

43. Machado, B.N.; Oleinik, P.H.; Kirinus, E.P.; Dos Santos, E.D.; Rocha, L.A.O.; Gomes, M.N.; Conde, J.M.P.; Isoldi, L.A. WaveMIMO Methodology: Numerical Wave Generation of a Realistic Sea State. *J. Appl. Comput. Mech.* **2021**, *7*, 2129–2148. [CrossRef]

44. Brito-Melo, A.; Gato, L.M.C.; Sarmento, A.J.N.A. Analysis of Wells Turbine Design Parameters by Numerical Simulation of the OWC Performance. *Ocean Eng.* **2002**, *29*, 1463–1477. [CrossRef]

45. Rodríguez, L.; Pereiras, B.; García-Diaz, M.; Fernandez-Oro, J.; Castro, F. Flow pattern analysis of an outflow radial turbine for twin-turbines-OWC wave energy converters. *Energy* **2020**, *211*, 118584. [CrossRef]

46. Prasad, D.D.; Ahmed, M.R.; Lee, Y.-H. Studies on the Performance of Savonius Rotors in a Numerical Wave Tank. *Ocean Eng.* **2018**, *158*, 29–37. [CrossRef]

47. Liu, Z.; Xu, C.; Kim, K.; Choi, J.; Hyun, B. An integrated numerical model for the chamber-turbine system of an oscillating water column wave energy converter. *Renew. Sust. Energy Rev.* **2021**, *149*, 111350. [CrossRef]

48. Blackwell, B.F.; Sheldahl, R.E.; Feltz, L.V. *Wind Tunnel Performance Data for Two-and Three-Buckets Savonius Rotors*; Final Report SAND76-0131; Sandia Laboratories: Albuquerque, NM, USA, 1977.

49. Akwa, J.V.; Silva Júnior, G.A.; Petry, A.P. Discussion on the Verification of the Overlap Ratio Influence on Performance Coefficients of a Savonius Wind Rotor using Computational Fluid Dynamics. *Renew. Energy* **2021**, *38*, 141–149. [CrossRef]

50. Schlichting, H.; Gersten, K. *Boundary-Layer Theory*, 8th ed.; Springer: Berlin/Heidelberg, Germany, 2000.

51. Wilcox, D.C. *Turbulence Modeling for CFD*, 3rd ed.; DWC Industries: La Cañada Flintridge, CA, USA, 2006.

52. Menter, F.R. Zonal Two Equation j–x Turbulence Models for Aerodynamic Flows. In Proceedings of the AIAA 24th Fluid Dynamics Conference, Orlando, FL, USA, 6–9 July 1993; AIAA 93-2906.

53. Menter, F.R.; Kuntz, M.; Langtry, R. Ten Years of Industrial Experience with the SST Turbulence Model. In *Turbulence, Heat and Mass Transfer*; Hanjalić, K., Nagano, Y., Tummers, M., Eds.; Begell House: Danbury, CT, USA, 2003; Volume 4, pp. 625–632.
54. Versteeg, H.K.; Malalasekera, W. *An Introduction to Computational Fluid Dynamics—The Finite Volume Method*, 2nd ed.; Longman: London, UK, 2007.
55. Patankar, S.V. *Numerical Heat Transfer and Fluid Flow*; McGraw-Hill: New York, NY, USA, 1980.
56. ANSYS Inc. *Ansys Fluent Theory Guide*; ANSYS Inc.: Cannonsburg, PA, USA, 2013.
57. Khaligh, A.; Onar, O.C. *Energy Harvesting—Solar, Wind and Ocean Energy Conversion Systems*; CRC Press: Boca Raton, FL, USA; Taylor and Francis: Abingdon, UK, 2010.
58. Geuzaine, C.; Remacle, J.-F. Gmsh: A 3-D Finite Element Mesh Generator with Built-in Pre and Post-Processing Facilities. *Int. J. Numer. Meth. Eng.* **2009**, *11*, 1309–1331. [CrossRef]

MDPI

St. Alban-Anlage 66

4052 Basel

Switzerland

Tel. +41 61 683 77 34

Fax +41 61 302 89 18

www.mdpi.com

Journal of Marine Science and Engineering Editorial Office

E-mail: jmse@mdpi.com

www.mdpi.com/journal/jmse